Introduction to Biostatistics Using R

Introduction to Biostatistics Using R

Mohsen Nady

ARCLER
PRESS

www.arclerpress.com

Introduction to Biostatistics Using R

Mohsen Nady

Arcler Press
224 Shoreacres Road
Burlington, ON L7L 2H2
Canada
www.arclerpress.com
Email: orders@arclereducation.com

ISBN: 978-1-77469-040-6 (Hardcover)

ABOUT THE AUTHOR

Mohsen Nady is a pharmacist with a M.D. in Microbiology and a diploma in Industrial Pharmacy. In addition, Mohsen has more than 4 years experience using R programming language. Mohsen has applied his skills in R programming to different projects related to Genomics, Microbiology, Biostatistics, Six Sigma, Data Analytics, Data Visualization, Building Apps, Geography, Market Analysis, Business Analysis,.....etc. Mohsen also published his thesis in high impact journal that attracted many citations, where all the statistical analysis were performed by him in addition to the methodological part. Furthermore, Mohsen has earned additional certificates, from top universities (Harvard, Johns Hopkins, Denmark,...etc) in R programming, Python, Excel, and Minitab that highlight his outstanding programming skills.

TABLE OF CONTENTS

LIST OF ABBREVIATIONS

ANOVA	Analysis of Variance
ART	Antiretroviral Therapy
AUC	Area Under the Curve
CI	Confidence Interval
CLT	Central Limit Theorem
CVD	Cardiovascular Disease
HTN	Hypertension
IQR	Interquartile Range
IR	Incidence Rate
IRR	Incidence Rate Ratio
OR	Odds Ratio
PBC	Primary Biliary Cirrhosis
PH	Proportional Hazards
RMSE	Root Mean Squared Error
ROC Curve	Receiver Operator Characteristic Curve
RR	Relative Risk
SE	Standard Error

PREFACE

This book covers some introductory steps in biostatistics using R programming language. Biostatistics is the branch of statistics that applies statistical methods to medical and biological problems. Biostatistics has become more important recently for studying the great amount of data that is produced from census data, genome sequencing, gene expression data, medical bioinformatics, and medical imaging data. With the help of R programming, statistical analysis, data cleaning, data visualization, and machine learning has become a relatively easy tasks for these huge datasets. R is now considered the centerpiece language for doing all these data science skills because it has many useful packages that not only can perform all these tasks, but also, has additional packages that were specifically designed for several statistical tasks related to biology and medical data. In addition, many scientific journals require the data analysis R scripts to ensure reproducibility of the submitted results.

The first chapter of this book introduces many statistical concepts used in scientific research like study designs, sample, and population, and data types. Chapters 2, 4, and 5 cover the three main data types which are continuous data, categorical data, and time to event data. Chapter 3 discusses the popular continuous distribution that is the normal distribution along with its application to sample data. Chapter 6 is about the sampling distribution of different sample estimates along with a discussion of the famous central limit theorem (CLT). Chapters 7 and 8 are involved in confidence interval (CI) calculations, and Chapters 9–11 discuss several types of statistical tests like t-test, ANOVA, Chi-square, log-rank, etc.

Finally, Chapters 12–14 deal with different regression types; linear regression for continuous outcomes, logistic regression for binary outcomes, and Cox regression for time to event outcomes.

In all these chapters, many examples from many scientific journal articles or built in data sets along with different codes and outputs are given to help your understanding of these numerous statistical concepts. I hope this book will be a great addition to your future biostatistical projects.

CHAPTER 1

INTRODUCTION TO STATISTICS

CONTENTS

1.1. THE ROLE OF STATISTICS IN BIOLOGY

Statistics is the science whereby inferences are made about a specific population on the basis of relatively limited sample material. Biostatistics is the branch of statistics that applies statistical methods to medical and biological problems.

Biostatistics has become more important recently for studying the great amount of data that is produced from genome sequencing, gene expression, medical bioinformatics, and medical imaging data.

Biostatistics allows us to collect good data (related to our question of interest); then, analyze and summarize it to give useful information. On the other hand, bad data can be collected; and then, analyzed and summarized it to give harmful or non-useful information.

1.2. RESEARCH PROJECT STEPS

Biostatistics can play a role in each of the following steps and not only the data analysis part. However, biostatistics are sometimes called upon for the data analysis part when there is no chance to correct for the bad data that are coming in.

1.2.1. Design of the Study

Biostatistics can help us in:

1. Formatting our primary question of interest: for example, to quantify the information about a single group or comparing two or more groups.
2. Sample size calculation.
3. Selecting the study participants: randomly selected from a complete list of the population or selected from a pool of interested persons.
4. Assigning people to different treatments by randomization or they are self-selected by their characteristics as comparing smokers to non-smokers.

1.2.2. Data collection

Where the data are collected according to the recommended biostatistical design.

1.2.3. Data Analysis

Biostatistics can help us in:

1. Descriptive statistics: where we summarize the information from raw data using numerical summaries and graphical displays.
2. Differentiate between real data patterns from chance variation because important patterns in data can be obscured by sample variability.
3. Inferential statistics: where using the information from a single sample combined with sampling variability to make conclusions about the larger population from which the sample is collected.

1.2.4. Presentation

Using the summary measures that best address the question of interest. In addition, we also present the uncertainty in our estimates.

1.2.5. Interpretation

What the results mean in terms of practical application

1.3. SAMPLE AND POPULATION

The sample is a subset of a larger group (population) from which information is collected to learn about that larger group or population.

The population is the entire group we want to study. This may not be possible in many cases due to the great resources it needs to collect data from the whole population. An example of population data is the census data.

In the last section of this chapter, we will explore a data set that contains some information about the USA states in 1977. This dataset may be considered as a sample or as a population. It is a sample if we want to study the facts about the US states in the 1970s or from 1970–1980. It is a population for the facts about USA states in 1977 and from which we can take a random sample (some USA states) to make inference about the whole states if this is a huge data set and we do not want to examine all the rows. In fact, it is a small dataset with 50 rows so we can examine it.

It is optimal if the sample is representative of the population under study, but this is not always possible.

1.3.1. Representative Sampling Strategies (Simple Random Sampling)

In simple random sampling, every unit in our population is equally likely to be selected. Therefore, the sample characteristics should mimic those of the population. In this scenario,

In random selection, each member of the population is assigned a number. Then using a computer program, a random subset of any size can be selected.

In random assignment to placebo or drug treatment, each participant is assigned a number and we select a random half subset of these numbers

One issue in obtaining random samples on the computer is whether the samples are obtained with or without replacement. The default option is sampling without replacement, whereby the same data point from the population cannot be selected more than once in a specific sample. In sampling with replacement (sometimes called bootstrap sampling), repetitions are permissible within a particular sample.

The sample() function in R can help us in random selection and random assignment. For reproducibility, you can use the set.seed() function with any numbers of your choice to ensure that the same subset of random numbers will be produced.

1.3.2. Example of Random Selection

Suppose we want to study how effective a hypertension treatment program is in controlling the blood pressure of its participants. We have a list of all 1000 participants in the program, but because of limited resources, only 20 can be surveyed. We would like the 20 people chosen to be a random sample from the population of all participants in the program.

Create a vector of numbers from 1 to 1000, then select a random 20 numbers from it

```
# vector of 1000 numbers

x<-1:1000

sample(x,20)

## [1] 61 547 372 11 538 722 666 523 463 708 352 100 737 246 322 972
898 58 619
```

```
## [20] 225

# another random sample

sample(x,20)

## [1] 133 782 652 289 219 947 202 191 695 921 722 657 660 535 564 568
25 974 340

## [20] 610

## reproducibility

set.seed(123)

sample(x,20)

## [1] 415 463 179 526 195 938 818 118 299 229 244 14 374 665 602 603
768 709 91

## [20] 953

# another sample

set.seed(123)

sample(x,20)

## [1] 415 463 179 526 195 938 818 118 299 229 244 14 374 665 602 603
768 709 91

## [20] 953
```

The sample() function has produced different subset in each time you execute it unless you use the set.seed() function before it.

1.3.3. Example of Random Assignment

Suppose we want to conduct a clinical trial to compare the effectiveness of an oral hypoglycemic agent for diabetes with standard insulin therapy. A small study of this type will be conducted on 20 patients: 10 patients will be randomly assigned to the oral agent and 10 to insulin therapy

Create a vector of numbers from 1 to 20, then select a random 10 numbers from it for the oral treatment and the other numbers will be for the standard treatment

```
x<-1:20

# patient numbers for oral treatment

set.seed(123)

sample(x,10)

## [1] 15 19 14 3 10 2 6 11 5 4
```

patient numbers; 15, 19, 14, 3, 10, 2, 6, 11, 5, 4, will be selected for the new oral treatment while other numbers;1,7,8,9,12, 13, 16, 17, 18, 20, will be selected for the standard treatment.

1.3.4. Non-Representative Sampling Strategies

Bias or systematic difference between the sample and population can be introduced when non-random sampling is used as in the following cases.

1. Random digit dialing for voters sampling, as not every voter has a phone.
2. Convenience sampling as by setting a camp at a certain location (for homeless people, IV drug users,…) and enroll people who are willing to participate in the study or survey. The people enrolled in the study may be systematically different from our population of interest. This is called "selection bias."
3. Respondent-driven sampling where for the example above, you asked people participating to bring more people who they know. The bias introduced will be the same as above in addition to bias introduced as they will invite-only people they know who may be different from the whole population of interest.
4. Random sampling from a hospital population to represent patients with certain diseases. The bias introduced in this case is due to that the hospital population may be different from the entire population with a certain disease. This sample only represents the population who have access to that hospital.

Sampling procedure needs to be considered when making inference from the sample to the population.

1.4. STUDY DESIGNS

1.4.1. Prospective Cohort Studies

A group of disease-free individuals is identified at one point in time and are classified according to their exposure variables. Then, they are followed over a period of time until some of them develop the disease. The development of disease over time is then related to the exposure variables measured at baseline.

The exposure variables can also be called: predictors or independent variables. On the other hand, the outcome (as developing disease) can also be called the dependent variable.

There are two types of prospective cohort studies:

1. Randomized cohort studies; and
2. Observational cohort studies.

Example:

Designing a study to look at factors associated with flu. The outcome of interest (disease) is flu while the exposure variables are the factors that are associated with flu.

In randomized cohort studies, we randomly assign the participants to an exposure of interest as the diet type while in observational cohort studies, people cannot be assigned to an exposure as they already have it, for example the smoking status, the race, the immunity status,…, etc. The participants are self-selected into the exposure category.

The randomized studies account for many systematic differences between the different exposed groups (different diet groups). Therefore, we are assured that the only difference between the different groups is the exposure type or the diet type.

Example of observational study, when comparing males to females for the association with breast cancer (outcome). Females may have other factors as smoking that are related to breast cancer and so it is difficult to assess if female sex is the only contributing factor. As participants are self-selected to the exposure, any biases from these observational studies are called "self-selection biases."

Example of randomized study, when comparing the association between certain diet types and cancer. By randomly assigning the diet types to participants, we negate other factors (other than diet) that may be different between the groups that contribute to cancer.

Confounders are the factors that cannot be controlled in observational studies. Confounders are characteristics that are related to an exposure and an outcome, and ignoring them may lead to overestimate or underestimate of the association between the exposure and the outcome.

One possible treatment for confounding is by adjustment for possible confounders associated with the outcome as age, sex, diet type, that are measured during the study. The problem is that we cannot measure all factors that are different between the exposure groups.

1.4.2. Case-Control Studies (Retrospective Research)

Where participants are classified according to their outcome (disease) status and then the exposure that occurred before the outcome is assessed.

Used in epidemiology for studying of rare diseases. The Cohort studies cannot be applied as they need many participants to see any outcome.

Example, for studying the factors associated with lung cancers. We have two groups, one that has lung cancer and the other that do not have lung cancer. We ask the participants for factors that occurred before that time point of the study that may be associated with lung cancer as smoking habits, diet, etc.

The disadvantage of this design is also confounding as we do not randomize the participants to the variables that may be associated with cancer (as smoking habits, diet, etc.) but they are self-selected. A possible treatment is by adjustment.

1.4.3. Cross-Sectional Study

Where a study population is ascertained at a single point in time. All participants in the study population are asked about their current disease status and their current or past exposure status.

This type of study is sometimes called a prevalence study (screening programs) because the prevalence of disease at one point in time is compared between exposed and unexposed individuals. This contrasts with a prospective study, in which one is interested in the incidence rather than the prevalence of disease.

Example, the participants of a study are asked for the current smoking status (exposure) and the current flu status (outcome).

- **Disadvantages:**
  - Confounding as no randomization was done
  - Timing issue because everything is assessed at the same time point so we get only a snapshot of the association.

1.5. DATA TYPES

1.5.1. Numerical Data

Have two types:

1. Continuous data that can take infinite values within a certain range (decimal values). *Example:* height, age, income, weight. A certain weight can be 70.5 kg but with increasing balance accuracy, we can have a value of 70.5321458 kg and so infinite values; and
2. Discrete data take only integer values within a certain range (integer values). They cannot take decimal values. *Example:* Count data as the number of fruits consumed every day, the number of patients in a hospital in a certain time period. These values cannot take decimal values, so the number of patients cannot be 300.5 patients.

1.5.2. Categorical Data

Have two types:

1. Nominal categorical data, where the categories have no inherent ordering. A special type is the Binary data that can take only two values. Example, race, country of birth, sex (binary), polio status (binary, with polio or not).
2. Ordinal categorical data, where the categories are ordered. Examples, the degree of agreement (strongly disagree, moderately disagree, neutral, moderately agree, strongly agree), the academic ranks (professors, assistant professors, lecturers, assistant lecturers).

1.5.3. Time to Event Data

They are a hybrid of continuous and binary data. Each value has two pieces of information, the event occurred or not (binary) and the time to that occurrence (continuous data).

1.6. EXAMINING DIFFERENT DATA TYPES USING R

You can download [R] https://cran.r-project.org/bin/windows/base/ or [RStudio] https://www.rstudio.com/products/rstudio/download/

For using R, RStudio is preferable as it has many functionalities that help the user. If you intend to use RStudio, you should install R first then install R studio.

The grey blocks within the text are code chunks, and the text following them is the code outputs so you can try the codes yourself. In addition, the text within the code chunks that starts with # is comments to understand the role of each code line.

We will examine the states.x77 dataset in R which contains continuous and discrete data.

```
#activating the necessary packages in R which is datasets

library(datasets)

# Showing the first rows of state.x77 dataset

head(state.x77)

##       Population Income Illiteracy Life Exp Murder HS Grad Frost  Area

## Alabama     3615  3624   2.1  69.05  15.1  41.3  20 50708

## Alaska      365  6315    1.5  69.31  11.3  66.7  152 566432

## Arizona     2212  4530   1.8  70.55  7.8  58.1  15 113417

## Arkansas    2110  3378   1.9  70.66  10.1  39.9  65 51945

## California  21198  5114   1.1  71.71  10.3  62.6  20 156361

## Colorado    2541  4884   0.7  72.06  6.8  63.9  166 103766

class(state.x77)
```

```
## [1] "matrix" "array"

# convert the matrix to dataframe

state<-data.frame(state.x77)

class(state)

## [1] "data.frame"

# Examining the dataset

str(state)

## 'data.frame':  50 obs. of 8 variables:

## $ Population: num 3615 365 2212 2110 21198 ...

## $ Income   : num 3624 6315 4530 3378 5114 ...

## $ Illiteracy: num 2.1 1.5 1.8 1.9 1.1 0.7 1.1 0.9 1.3 2 ...

## $ Life.Exp : num 69 69.3 70.5 70.7 71.7 ...

## $ Murder   : num 15.1 11.3 7.8 10.1 10.3 6.8 3.1 6.2 10.7 13.9 ...

## $ HS.Grad  : num 41.3 66.7 58.1 39.9 62.6 63.9 56 54.6 52.6 40.6 ...

## $ Frost    : num 20 152 15 65 20 166 139 103 11 60 ...

## $ Area     : num 50708 566432 113417 51945 156361 ...
```

It is a dataframe with 50 obs.(observations or rows) and 8 (variables or columns). The observations are the USA states, each one state is represented by a single row.

The columns are the data or variables obtained from each state:

1. **Population:** This estimate as of July 1, 1975.
2. **Income:** Per capita income (1974).
3. **Illiteracy:** Illiteracy (1970, percent of population).
4. **Life Exp.:** Life expectancy in years (1969–71).
5. **Murder:** Murder rate per 100,000 population (1976).
6. **HS Grad:** Percent high-school graduates (1970).

7. **Frost:** Mean number of days with minimum temperature below freezing (1931–1960) in capital or large city.
8. **Area:** Land area in square miles.

1.6.1. Numerical Data

Numerical continuous data (decimal)

```
#The $ is used to access the data in only one column of the dataframe

head(state$Illiteracy)

## [1] 2.1 1.5 1.8 1.9 1.1 0.7

head(state$Life.Exp)

## [1] 69.05 69.31 70.55 70.66 71.71 72.06

head(state$Murder)

## [1] 15.1 11.3 7.8 10.1 10.3 6.8

head(state$HS.Grad)

## [1] 41.3 66.7 58.1 39.9 62.6 63.9
```

Illiteracy, life expectancy, murder rates, and percentage of high school graduates are examples of continuous numerical data. They have a numeric class as evident from str() function

Numerical discrete data (count data, integer)

```
head(state$Population)
## [1] 3615  365 2212 2110 21198 2541
#Although the income can be continuous but the data is represented as integers

head(state$Income)

## [1] 3624 6315 4530 3378 5114 4884

head(state$Frost)

## [1] 20 152 15 65 20 166
```

```
head(state$Area)

## [1] 50708 566432 113417 51945 156361 103766
```

population, income, frost, and area are examples of discrete data (no decimals are present). They can have numeric or integer class in R.

1.6.2. Categorical Data

Example of ordinal categorical data is the month.abb and month.name vectors in R

```
# examine the month.add and month.name vectors

month.abb
## [1] "Jan" "Feb" "Mar" "Apr" "May" "Jun" "Jul" "Aug" "Sep" "Oct"
"Nov" "Dec"

str(month.abb)

## chr [1:12] "Jan" "Feb" "Mar" "Apr" "May" "Jun" "Jul" "Aug" "Sep"
"Oct" ...

month.name

## [1] "January" "February" "March" "April" "May" "June"

## [7] "July" "August" "September" "October" "November" "December"

str(month.name)

## chr [1:12] "January" "February" "March" "April" "May" "June" "July"
...
```

1.6.3. Time to Event Data

Time to event data start with a cohort of observations and follow them over time till they have a well-defined binary outcome, so each data point has two pieces of information, the event occurred or not (binary) and the time to that occurrence (continuous data).

We will use survival package in R with its accompanying dataset veteran. Veteran data is randomized trial of two treatment regimens for lung cancer. It consists of 137 rows and 8 columns which are

1. **trt:** treatment type, 1=standard 2=test
2. **celltype:** cancer cell type, 1=squamous, 2=smallcell, 3=adeno, 4=large
3. **time:** survival time, the continuous data
4. **status:** censoring status,1=death was observed, 0=censoring observation, binary outcome. Censored observation is for subjects dropping out of the study or ending the study without having the outcome or death.
5. **karno:** Karnofsky performance score (100=good)
6. **diagtime:** months from diagnosis to randomization
7. **age:** in years
8. **prior:** prior therapy, 0 = no, 10 = yes

In R, time to event data is represented by survival object which combines the time (continuous) and outcome (binary) columns. This is done using Surv() function in R from the survival package

```
library(survival)

data("veteran")

str(veteran)

## 'data.frame':  137 obs. of 8 variables:

## $ trt    : num 1 1 1 1 1 1 1 1 1 1 ...

## $ celltype: Factor w/ 4 levels "squamous,""smallcell,"..: 1 1 1 1 1 1 1 1 1 ...

## $ time   : num 72 411 228 126 118 10 82 110 314 100 ...

## $ status : num 1 1 1 1 1 1 1 1 1 0 ...

## $ karno  : num 60 70 60 60 70 20 40 80 50 70 ...

## $ diagtime: num 7 5 3 9 11 5 10 29 18 6 ...

## $ age    : num 69 64 38 63 65 49 69 68 43 70 ...

## $ prior  : num 0 10 0 10 10 0 10 0 0 0 ...

head(veteran,10)
```

```
## trt celltype time status karno diagtime age prior
## 1 1 squamous 72 1 60 7 69 0
## 2 1 squamous 411 1 70 5 64 10
## 3 1 squamous 228 1 60 3 38 0
## 4 1 squamous 126 1 60 9 63 10
## 5 1 squamous 118 1 70 11 65 10
## 6 1 squamous 10 1 20 5 49 0
## 7 1 squamous 82 1 40 10 69 10
## 8 1 squamous 110 1 80 29 68 0
## 9 1 squamous 314 1 50 18 43 0
## 10 1 squamous 100 0 70 6 70 0
#Surv function takes the time and status column to build the survival object
km<-Surv(veteran$time,veteran$status)
head(km,10)
## [1] 72 411 228 126 118 10 82 110 314 100+
class(km)
## [1] "Surv"
# km is a survival object
head(veteran[,3:4],10)
## time status
## 1 72 1
## 2 411 1
## 3 228 1
```

```
## 4  126   1
## 5  118   1
## 6  10   1
## 7  82   1
## 8  110   1
## 9  314   1
## 10 100   0
```

The values of the survival object "km" are the same values as the time column but with "+" sign to indicate censoring.

Each of these data types will be examined more deeply in the following chapters.

CHAPTER 2

NUMERICAL DATA

CONTENTS

Instead of looking at the whole sample data, we look at useful numerical summaries that can better describe that data. These sample summaries are called statistics and they are estimates of the unknown population summaries. The population values of these summaries are called parameters.

Sample summaries or statistics are measures of location (mean, median, percentiles) and measures of spread (standard deviation, variance, interquartile range (IQR)).

2.1. MEASURES OF LOCATION FOR UNIVARIATE NUMERICAL DATA

2.1.1. Mean (Average or Arithmetic Mean)

It is obtained by adding up the data values and dividing by their number. The population mean is denoted by the Greek letter, mu (μ), while the sample mean is denoted by x bar ().

For example, for the numbers (1,2,3), their mean would be=(1+2+3)/3=2.

The mean is sensitive to extreme values in small samples. For a sample of two observations, each one observation will affect the mean by 50% while for a sample of 100 observations, each one observation will affect the mean by 1%. Therefore, increasing the sample size will lead to more stable sample mean.

For example, for the values (1,2,10), their mean = (1+2+10)/3 = 4.333. The mean has doubled compared to the previous example although we still have three values. The mean has the same unit of the data, so if we have blood pressure measurements in mmHg (110,120,130) their mean will be 120 mmHg.

Some properties of the arithmetic mean:

1. If yi = xi + c, i = 1, . . . , n, then = + c
2. If yi = c * xi, i = 1, . . . , n then =
3. If , i = 1, . . . , n then =

We can use the mean() function of R to verify these properties.

```
# first property
x<-1:10
x
```

```
## [1] 1 2 3 4 5 6 7 8 9 10
mean(x)
## [1] 5.5
y<-x+5
y
## [1] 6 7 8 9 10 11 12 13 14 15
mean(y)
## [1] 10.5
# second property

x<–1:10
x
## [1] 1 2 3 4 5 6 7 8 9 10
mean(x)
## [1] 5.5
y<-x*5
y
## [1] 5 10 15 20 25 30 35 40 45 50
mean(y)
## [1] 27.5
# third property
x<–1:10
x
## [1] 1 2 3 4 5 6 7 8 9 10
mean(x)
## [1] 5.5
y<-x*5+10
y
## [1] 15 20 25 30 35 40 45 50 55 60
mean(y)
## [1] 37.5
```

In the first example, the mean(y) = mean(x)+5 because y = x+5;

In the second example, the mean(y) = mean(x)*5 because y = x*5;

In the third example, the mean(y) = mean(x)*5+10 because y = x*5+10.

Another example, The regicor dataframe, from the package compareGroups, contains data from three different cross-sectional surveys of individuals representative of the population from a north-west Spanish province (Girona), REGICOR study. Visit [www.regicor.org].

We will first install package compareGroups using install.packages() function, then activate the package in our working session using the library() function. Then load the regicor data in our working session using the data() function. As I have already installed that package in R so I put # before that line to not activate that code line again.

```
#install.packages("compareGroups")
library(compareGroups)
data("regicor")

# data structure
str(regicor)
## 'data.frame':  2294 obs. of 25 variables:
## $ id   : num 2.26e+03 1.88e+03 3.00e+09 3.00e+09 3.00e+09 ...
##  ..- attr(*, "label")= Named chr "Individual id"
##  .. ..- attr(*, "names")= chr "id"
## $ year  : Factor w/ 3 levels "1995,""2000,"..: 3 3 2 2 2 2 2 1 3 1 ...
##  ..- attr(*, "label")= Named chr "Recruitment year"
##  .. ..- attr(*, "names")= chr "year"
## $ age   : int 70 56 37 69 70 40 66 53 43 70 ...
##  ..- attr(*, "label")= Named chr "Age"
##  .. ..- attr(*, "names")= chr "age"
## $ sex   : Factor w/ 2 levels "Male,""Female": 2 2 1 2 2 2 1 2 2 1 ...
##  ..- attr(*, "label")= chr "Sex"
## $ smoker : Factor w/ 3 levels "Never smoker,"..: 1 1 2 1 NA 2 1 1 3 3 ...
##  ..- attr(*, "label")= Named chr "Smoking status"
```

```
## .. ..- attr(*, “names”)= chr “smoker”
## $ sbp   : int 138 139 132 168 NA 108 120 132 95 142 ...
## ..- attr(*, “label”)= Named chr “Systolic blood pressure”
## .. ..- attr(*, “names”)= chr “sbp”
## $ dbp   : int 75 89 82 97 NA 70 72 78 65 78 ...
## ..- attr(*, “label”)= Named chr “Diastolic blood pressure”
## .. ..- attr(*, “names”)= chr “dbp”
## $ histhtn : Factor w/ 2 levels “Yes,””No”: 2 2 2 2 2 2 1 2 2 2 ...
## ..- attr(*, “label”)= Named chr “History of hypertension”
## .. ..- attr(*, “names”)= chr “histbp”
## $ txhtn  : Factor w/ 2 levels “No,””Yes”: 1 1 1 1 1 1 2 1 1 1 ...
## ..- attr(*, “label”)= chr “Hypertension treatment”
## $ chol  : num 294 220 245 168 NA NA 298 254 194 188 ...
## ..- attr(*, “label”)= Named chr “Total cholesterol”
## .. ..- attr(*, “names”)= chr “chol”
## $ hdl   : num 57 50 59.8 53.2 NA ...
## ..- attr(*, “label”)= Named chr “HDL cholesterol”
## .. ..- attr(*, “names”)= chr “hdl”
## $ triglyc : num 93 160 89 116 NA 94 71 NA 68 137 ...
## ..- attr(*, “label”)= Named chr “Triglycerides”
## .. ..- attr(*, “names”)= chr “triglyc”
## $ ldl   : num 218.4 138 167.4 91.6 NA ...
## ..- attr(*, “label”)= Named chr “LDL cholesterol”
## .. ..- attr(*, “names”)= chr “ldl”
## $ histchol: Factor w/ 2 levels “Yes,””No”: 2 2 2 2 NA 2 1 2 2 2 ...
## ..- attr(*, “label”)= chr “History of hyperchol.”
## $ txchol : Factor w/ 2 levels “No,””Yes”: 1 1 1 1 NA 1 1 1 1 1 ...
## ..- attr(*, “label”)= Named chr “Cholesterol treatment”
## .. ..- attr(*, “names”)= chr “txchol”
## $ height : num 160 163 170 147 NA ...
## ..- attr(*, “label”)= Named chr “Height (cm)”
```

```
## .. ..- attr(*, "names")= chr "height"
## $ weight : num 64 67 70 68 NA 43.5 79.2 45.8 53 62 ...
## ..- attr(*, "label")= Named chr "Weight (Kg)"
## .. ..- attr(*, "names")= chr "weight"
## $ bmi   : num 25 25.2 24.2 31.5 NA ...
## ..- attr(*, "label")= Named chr "Body mass index"
## .. ..- attr(*, "names")= chr "bmi"
## $ phyact : num 304 160 553 522 NA ...
## ..- attr(*, "label")= Named chr "Physical activity (Kcal/week)"
## .. ..- attr(*, "names")= chr "phyact"
## $ pcs   : num 54.5 58.2 43.4 54.3 NA ...
## ..- attr(*, "label")= Named chr "Physical component"
## .. ..- attr(*, "names")= chr "pcs"
## $ mcs   : num 58.9 48 62.6 57.9 NA ...
## ..- attr(*, "label")= chr "Mental component"
## $ cv   : Factor w/ 2 levels "No,""Yes": 1 1 1 1 NA 1 1 1 1 1 ...
## ..- attr(*, "label")= chr "Cardiovascular event"
## $ tocv  : num 1025 2757 1906 1055 NA ...
## ..- attr(*, "label")= chr "Days to cardiovascular event or end of follow-up"
## $ death  : Factor w/ 2 levels "No,""Yes": 2 1 1 1 NA 1 2 1 1 1 ...
## ..- attr(*, "label")= chr "Overall death"
## $ todeath : num 1299.2 39.3 858.4 1833.1 NA ...
## ..- attr(*, "label")= chr "Days to overall death or end of follow-up"
```

```
# first 6 rows
head(regicor)
##        id year age  sex        smoker sbp dbp histhtn txhtn
## 6101    2265 2005 70 Female     Never smoker 138 75   No  No
## 5762    1882 2005 56 Female     Never smoker 139 89   No  No
## 2992 3000105616 2000 37  Male Current or former < 1y 132 82   No  No
```

```
## 2611 3000103485 2000 69 Female     Never smoker 168 97   No  No
## 2762 3000103963 2000 70 Female           <NA> NA NA   No  No
## 1516 3000100883 2000 40 Female Current or former < 1y 108 70   No
No
##    chol   hdl triglyc    ldl histchol txchol height weight   bmi
## 6101 294 57.00000    93 218.40000    No    No   160   64.0 25.00000
## 5762 220 50.00000   160 138.00000    No    No   163   67.0 25.21736
## 2992 245 59.80429    89 167.39571    No    No   170   70.0 24.22145
## 2611 168 53.17571   116 91.62429    No    No   147   68.0 31.46837
## 2762  NA       NA    NA       NA   <NA>  <NA>    NA     NA       NA
## 1516  NA 68.90000    94       NA    No    No   158   43.5 17.42509
##     phyact    pcs    mcs  cv     tocv death   todeath
## 6101 304.2000 54.455 58.918  No 1024.882  Yes 1299.16343
## 5762 160.3000 58.165 47.995  No 2756.849  No  39.32629
## 2992 552.7912 43.429 62.585  No 1905.969  No 858.42203
## 2611 522.0000 54.325 57.900  No 1055.380  No 1833.07619
## 2762       NA     NA     NA <NA>       NA <NA>        NA
## 1516 386.9505 57.315 47.869  No 3239.241  No 877.61155
# last 6 rows
tail(regicor)
##         id year age  sex          smoker sbp dbp histhtn txhtn
## 7334     3672 2005 58  Male      Former >= 1y 110 77   No  No
## 62  1000001076 1995 42  Male Current or former < 1y 112 64   No  No
## 3863 3000107763 2000 68  Male     Never smoker 120 80   Yes  No
## 3686 3000107330 2000 64  Male     Never smoker 198 110   No  No
## 4335      258 2005 49 Female     Never smoker 155 100   No  No
## 6786     3032 2005 42  Male     Never smoker 107 65   No  No
##    chol   hdl triglyc    ldl histchol txchol height weight   bmi
## 7334 235 50.00000    93 166.4000   Yes    No   165   62.0 22.77319
## 62  215 42.30000   118 149.1000   Yes    No   178   68.0 21.46194
## 3863 191 50.89000    91 121.9100   Yes    No   170   80.0 27.68166
```

```
## 3686 237 46.43286   122 166.1671   No  No  170  82.5 28.54671
## 4335 186 37.00000   155 118.0000   No  No  158  67.0 26.83865
## 6786 243 71.00000   89 154.2000   No  No  165  66.0 24.24242
##    phyact  pcs   mcs cv   tocv death  todeath
## 7334 346.7000 59.49900 39.28400 No 3617.2435  No 441.3844
## 62     NA 58.01476 35.03757 No 1451.6946  No 3487.0368
## 3863 389.6703 51.90300 55.14400 No 1003.7665  No 836.4889
## 3686 157.1209 50.36100 51.26000 No 1220.7234  Yes 1010.7434
## 4335 1093.5000 57.63100 42.67800 No 331.7726  No 3133.9241
## 6786 353.7000 50.46100 57.43000 No 698.8185  No 1899.4416
```

The data contains 2294 observations (rows) on the following 25 variables (columns):

1. **id:** Individual id
2. **year:** Recruitment year, a factor with levels 1995, 2000, 2005.
3. **age:** Patient age at recruitment date.
4. **sex:** Sex, a factor with levels male, female.
5. **smoker:** Smoking status, a factor with levels Never smoker, Current or former < 1y, Never or former >= 1y.
6. **sbp:** Systolic blood pressure.
7. **dbp:** Diastolic blood pressure.
8. **histhtn:** History of hypertension, a factor with levels Yes, No.
9. **txhtn:** Hypertension (HTN) treatment, a factor with levels No, Yes.
10. **chol:** Total cholesterol (mg/dl).
11. **hdl:** HDL cholesterol (mg/dl).
12. **triglyc:** Triglycerides (mg/dl).
13. **ldl:** LDL cholesterol (mg/dl).
14. **histchol:** History of hypercholesterolemia, a factor with levels Yes, No.
15. **txchol:** Cholesterol treatment, a factor with levels No, Yes.
16. **height:** Height (cm).
17. **weight:** Weight (Kg).
18. **bmi:** Body mass index.

19. **phyact:** Physical activity (Kcal/week).
20. **pcs:** Physical component summary.
21. **mcs:** Mental component summary.
22. **death:** Overall death, a factor with levels No, Yes.
23. **todeath:** Days to overall death or end of follow-up.
24. **cv:** Cardiovascular event, a factor with levels No, Yes.
25. **tocv:** Days to cardiovascular event or end of follow-up.

We can use the mean() function in R to calculate the mean of any numerical column of this dataset. In addition, the na.rm argument is used when you have NA (missing values) values to remove them before mean calculation.

```
# for data containing no missing values
x<-1:20
x
## [1] 1 2 3 4 5 6 7 8 9 10 11 12 13 14 15 16 17 18 19 20
mean(x)
## [1] 10.5
# for data containing missing values, for the column height
mean(regicor$height)
## [1] NA
mean(regicor$height, na.rm = TRUE)
## [1] 162.9156

# for column weight
mean(regicor$weight)
## [1] NA
mean(regicor$weight, na.rm = TRUE)
## [1] 73.43586

# for physical activity column
mean(regicor$phyact)
## [1] NA
```

```
mean(regicor$phyact, na.rm = TRUE)
## [1] 398.8314
```

The results show that the mean for x that contains numbers from 1 to 20 is 10.5. The mean for participants' height in regicor data = 162.92 cm, the mean weight = 73.44 kg, and the mean physical activity = 398.8 Kcal/week.

2.1.2. Median (50th Percentile)

It is the value the divides the data equally, so it is less than 50% of data points and greater than 50% of the other data points.

For an odd number of data points, it is the central data point after arranging them. For an even number of data points, it is the average of the two central data points.

An example of an odd number of data points (1,2,3,4,5), their median = 3. Another example of even number of data points (1,2,3,4,5,6), their median = (3+4)/2=3.5.

The median is not sensitive to extreme values as the mean. For example, for the two datasets (1,2,3) and (1,2,10), the median for both =2. The median has the same unit of the data, so if we have blood pressure measurements in mmHg,(110,120,130) their median will be 120 mmHg.

In R, the median is calculated using the median() function and use na.rm=TRUE argument if your data contain NA values

```
median(regicor$height, na.rm = TRUE)
## [1] 162.5
median(regicor$weight, na.rm = TRUE)
## [1] 73
```

The median for participants' height in regicor data = 162.5 cm, the median weight = 73 kg, and the median physical activity = 303.7 Kcal/week.

We can compare the mean and median value of numerical columns to assess their skewness. For right skewed or positive skewed data, the most frequent values are small values with rare large values. Therefore, the mean will be larger than the median because the mean is pulled up by the large values. On the other hand, left skewed or negative skewed data, the most frequent values are large values with rare small values. So, the mean is smaller than the median as the mean is pulled down by the small values. This will be more evident when we examine these columns visually.

From the results above, both the height and weight columns are not skewed as nearly the mean equals the median, while the physical activity column is right skewed as the mean is greatly larger than the median.

2.1.3. Percentiles

The pth sample percentile is that data point such that p% of data points are below that value and (100-p) % of data points are larger than that value.

25th percentile is that data point that nearly 25% of data is below that value and 75% are larger than that value. While 75th percentile is that data point that nearly 75% of data points are below that value and 25% are larger than that value. Recall that 50th percentile is the median.

To calculate the percentiles, use the quantile() function and the percentage as a proportion. The quantile() function produces quantiles corresponding to the given probabilities or proportions. The smallest observation (minimum) corresponds to a probability (percentile) of 0 and the largest (maximum) to a probability (percentile) of 1.

Example:

```
x<–1:100
x
## [1] 1 2 3 4 5 6 7 8 9 10 11 12 13 14 15 16 17 18
## [19] 19 20 21 22 23 24 25 26 27 28 29 30 31 32 33 34 35 36
## [37] 37 38 39 40 41 42 43 44 45 46 47 48 49 50 51 52 53 54
## [55] 55 56 57 58 59 60 61 62 63 64 65 66 67 68 69 70 71 72
## [73] 73 74 75 76 77 78 79 80 81 82 83 84 85 86 87 88 89 90
## [91] 91 92 93 94 95 96 97 98 99 100
quantile(x)
## 0% 25% 50% 75% 100%
## 1.00 25.75 50.50 75.25 100.00
```

The default quantile() function produces 0%, 25%, 50%, 75%, and 100%s. 0% is minimum data value and 100% is maximum data value. 25% = 25.75 means that 25% of data values in x are lower than 25.75.

25% is called the 1st quantile, 50% = 2nd quantile = median and 75% = 3rd quantile.

Any quantile or percentile can be produced. For example, the 65% and 83%s for the weight column of regicor data

```
quantile(regicor$weight, probs = c(0.65,0.83), na.rm = TRUE)
## 65% 83%
## 78 86
```

The 65% and 83%s for the weight column is 78 and 86 respectively. This means that 65% of weight values in regicor data are less than 78 kg, while 83% of weight values in regicor data are less than 86 kg.

2.2. MEASURES OF SPREAD FOR UNIVARIATE NUMERICAL DATA

2.2.1. Standard Deviation (s) and Variance (s^2)

The variance (s^2) is the average square deviations of data points about its mean. It has the squared unit of the data points.

The standard deviation (s) is the square root of variance so it has the same unit of the data points; and so, it is a more meaningful number.

s =

For example, a group of blood pressure measurements has a variance of 255 $mmHg^2$ and a standard deviation of 16 mmHg. So on average, each single data point has a blood pressure of 16 mmHg away from the sample mean.

The sample variance () is an estimate of the population variance () and the sample standard deviation (s) is an estimate of the population standard deviation ().

A standard deviation and variance of 0 can occur if all data points have the same value.

Some properties of the standard deviation and variance:

1. If yi = xi + c, i = 1, . . . , n, then = and

2. If yi = c * xi, i = 1, . . . , n and c > 0, then = and

We can use the sd() and var() functions in R to compute the standard deviation and variance respectively and verify these assumptions

```
# the first property
x<–1:10
x
## [1] 1 2 3 4 5 6 7 8 9 10
sd(x)
## [1] 3.02765
var(x)
## [1] 9.166667
y<-x + 10
y
## [1] 11 12 13 14 15 16 17 18 19 20
sd(y)
## [1] 3.02765
var(y)
## [1] 9.166667
# the second property
x<–1:10
x
## [1] 1 2 3 4 5 6 7 8 9 10
sd(x)
## [1] 3.02765
var(x)
## [1] 9.166667
y<-x * 10
y
## [1] 10 20 30 40 50 60 70 80 90 100
sd(y)
```

```
## [1] 30.2765
var(y)
## [1] 916.6667
```

In the first example, the standard deviation and variance of y is the same as x although y = x+10.

In the second example, because y = x*10 so the standard deviation of y = 10 times the standard deviation of x and the variance of y = 100 times the variance of x.

Examine the regicor data

```
sd(regicor$weight, na.rm = TRUE)
## [1] 13.67845
sd(regicor$height, na.rm = TRUE)
## [1] 9.216404
```

From the results, each participant in our data has a weight that is 13.67 kg away from the mean on the average. On the other hand, each participant in our data has a height that is 9.2 cm away from the mean on the average.

2.2.2. Interquartile Range (IQR)

It is the difference between the 25% and 75%s of our data. Standard deviation and mean are sensitive to extreme values, while the IQR and median are not sensitive to extreme values. Therefore, median and IQR are robust estimates of population parameters (not affected by sample extreme values) while the mean and standard deviation are not.

The IQR contains the middle 50% of our data and can be calculated by the IQR() function of R.

```
IQR(regicor$weight, na.rm = TRUE)
## [1] 18.5
IQR(regicor$height, na.rm = TRUE)
## [1] 13.15
```

So the IQR for regicor weights = 18.5 kg while for the regicor heights = 13.15 cm.

2.3. GRAPHICAL METHODS FOR UNIVARIATE NUMERICAL DATA

The mean, standard deviation, and percentiles do not tell the whole story of the data distribution or the distribution shape. Instead, we can use different graphical methods to quickly look at the different data distributions.

In this chapter and subsequent chapters, we will rely on ggplot2 package (a member of the tidyverse package) that allows us to quickly visualize and explore data. In ggplot2, it is recommended that everything you want to plot is included in a dataframe as a column, and the first argument to ggplot must be a dataframe. In addition, many of other packages as ggpubr that produces publication ready graphs also depends on ggplot2.

Instead of loading ggplot2 package, we will load tidyverse package as it includes several other useful packages (tidyr, dplyr) in addition to ggplot2.

2.3.1. geom_histogram()

geom_histogram() is used to plot histograms. Histograms are used to visualize the distribution of a single continuous variable by dividing the x axis into bins, counting the number of observations in each bin, and displaying the counts with bars.

geom_histogram() requires only one aesthetic (x) or the continuous variable you want to visualize. One useful argument is bins which default to 30 bins.

For example, plot a histogram of regicor weight, height, and physical activity. Add col = “red” to see bins.

```
library(tidyverse)
# weight
ggplot(data = regicor, aes(x = weight))+geom_histogram(col=”red”)
```

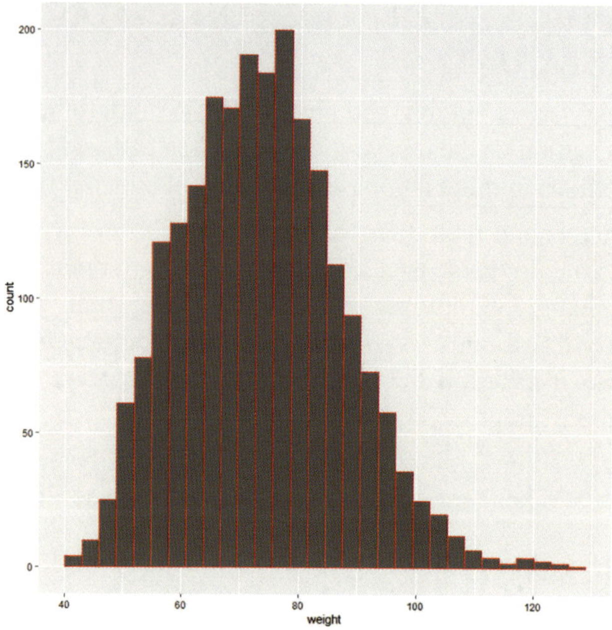

change the number of bins to 20

ggplot(data = regicor, aes(x = weight))+geom_histogram(bins = 20, col="red")

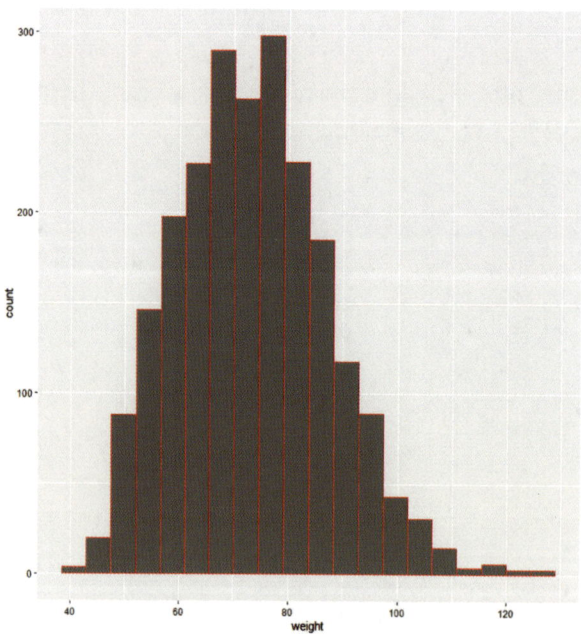

 # height

```
ggplot(data = regicor, aes(x = height))+geom_histogram(col="red")
```

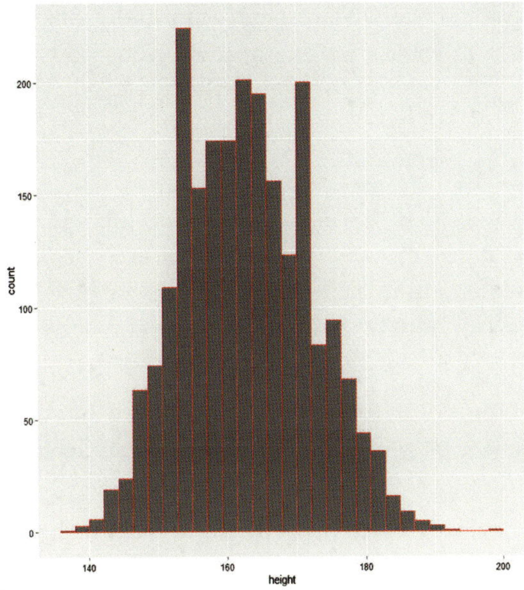

 # physical activity

```
ggplot(data = regicor, aes(x = phyact))+geom_histogram(col="red")
```

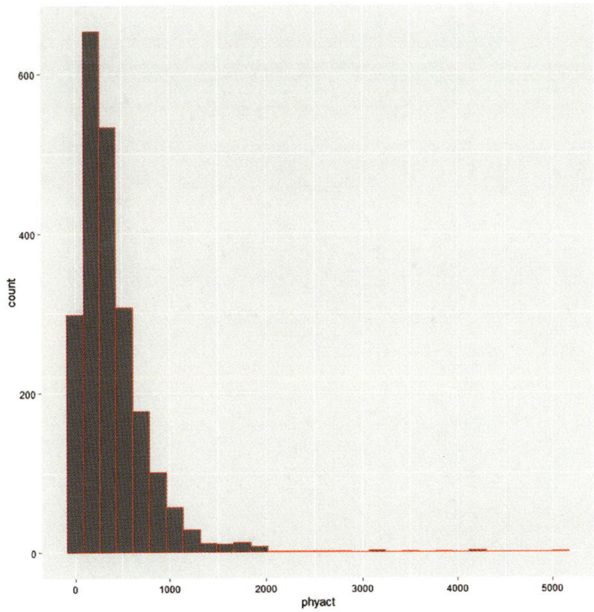

You see that the maximum frequency for weight is in the range = 70–80 kg, the maximum frequency for height is in the range = 150–160 cm, and the maximum frequency for physical activity is in the range = 0–500 Kcal/week. Both weight and height are nearly normally distributed (symmetric bell shaped) while phyact is an example of right skewed data (rare large values).

2.3.2. geom_boxplot()

geom_boxplot() is used to plot boxplots. Boxplots are used to compactly displays the distribution of a continuous variable. It visualizes the summary statistics (the median, first, and third quartile as box boundary), and all "outlying" points individually from which you can determine the minimum and maximum.

geom_boxplot() requires two aesthetics (y for the continuous variable you want to visualize on y-axis and group = 1 to indicate that all points are within a single group).

For example, plot a boxplot of regicor weight, height, and physical activity

```
# weight
ggplot(data = regicor, aes(y = weight, group=1))+geom_boxplot()
```

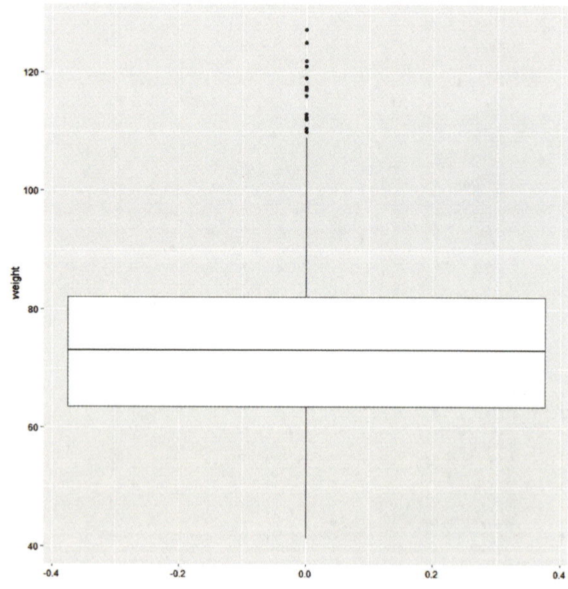

```
quantile(regicor$weight, na.rm = TRUE)
##  0%  25%  50%  75% 100%
## 41.2 63.5 73.0 82.0 127.2
```

```
# height
ggplot(data = regicor, aes(y = height, group=1))+geom_boxplot()
```

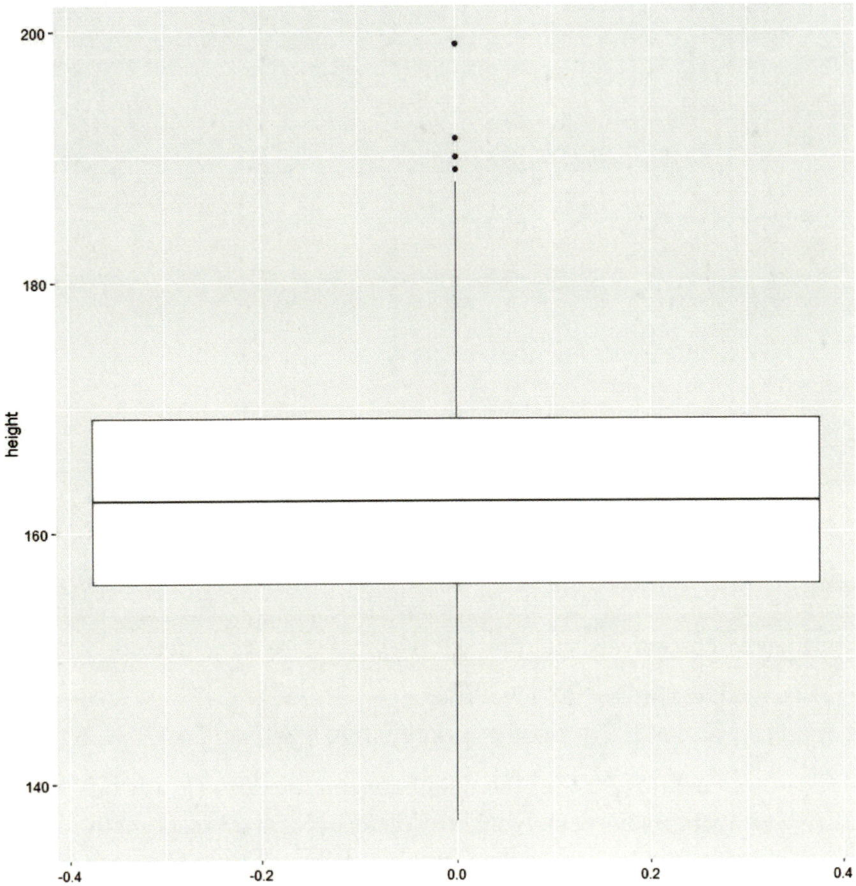


```
quantile(regicor$height, na.rm = TRUE)
##   0%  25%  50%  75%  100%
## 137.00 156.00 162.50 169.15 199.00
```

```
# physical activity
   ggplot(data = regicor, aes(y = phyact, group=1))+geom_boxplot()
```

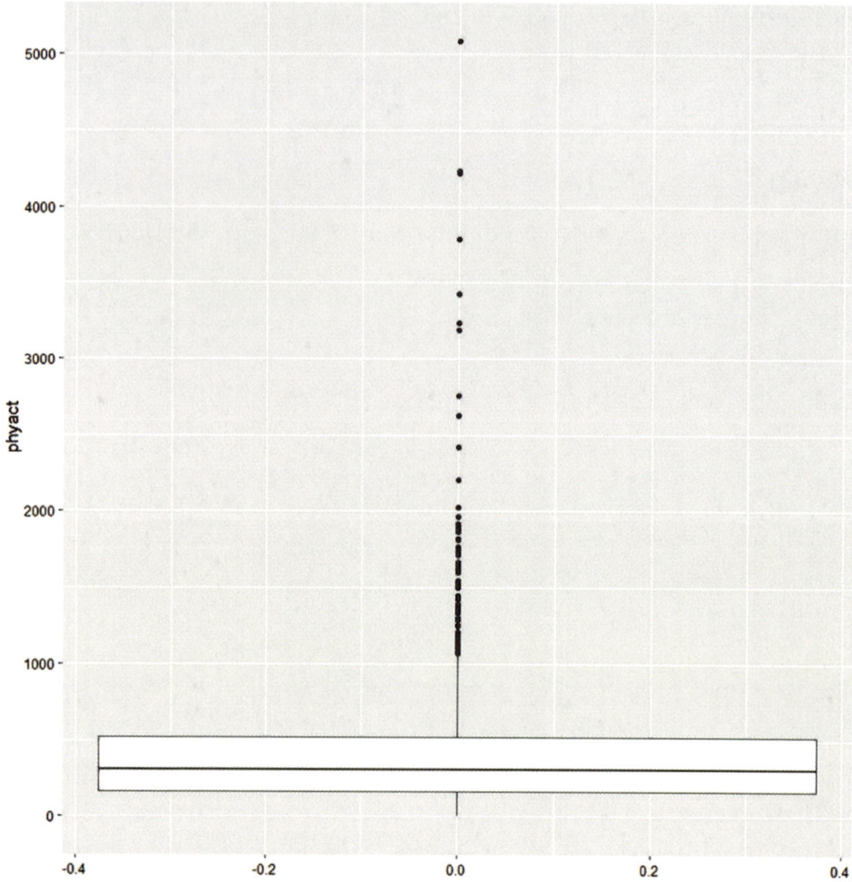


```
quantile(regicor$phyact, na.rm = TRUE)
##    0%     25%     50%     75%    100%
## 0.0000 159.4750 303.7088 521.9000 5083.1868
```

The box contains 50% of the data between the 25% (lower box border) and 75% (upper box border). Note how these borders correspond to 25% and 75%s obtained from the quantile() function. For example, the 25% quantile of weight = 63.5 kg and 75% quantile of weight = 82 kg.

The central line in the box is the median or 50%. Note how these central lines correspond to 50%s obtained from the quantile() function.

The boxplots of weight and height indicate a symmetric distribution where the median line splits the box in equal halves and the lines extending from the box have similar lengths.

The boxplot of physical activity indicates a right skewed distribution where the median line splits the box in unequal halves. The upper half is longer than lower half and the lines extending from the 75% quantile is longer than the line extending from the 25% quantile.

2.3.3. geom_density()

geom_density() is used to plot kernel density estimate. Kernel density estimate is a smoothed version of the histogram. This is a useful alternative to the histogram for continuous data that comes from an underlying smooth distribution. The default kernel used is the gaussian one.

geom_density() requires only one aesthetic (x) or the continuous variable you want to visualize.

For example, plot a density plot of regicor weight, height, and physical activity

```
# weight
ggplot(data = regicor, aes(x = weight))+geom_density()
```

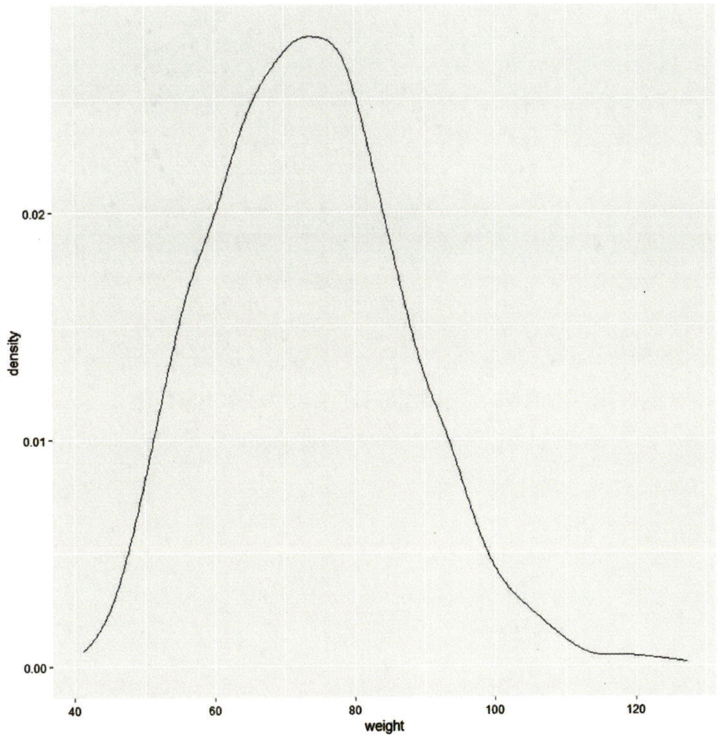


```
# height
```

```
ggplot(data = regicor, aes(x = height))+geom_density()
```

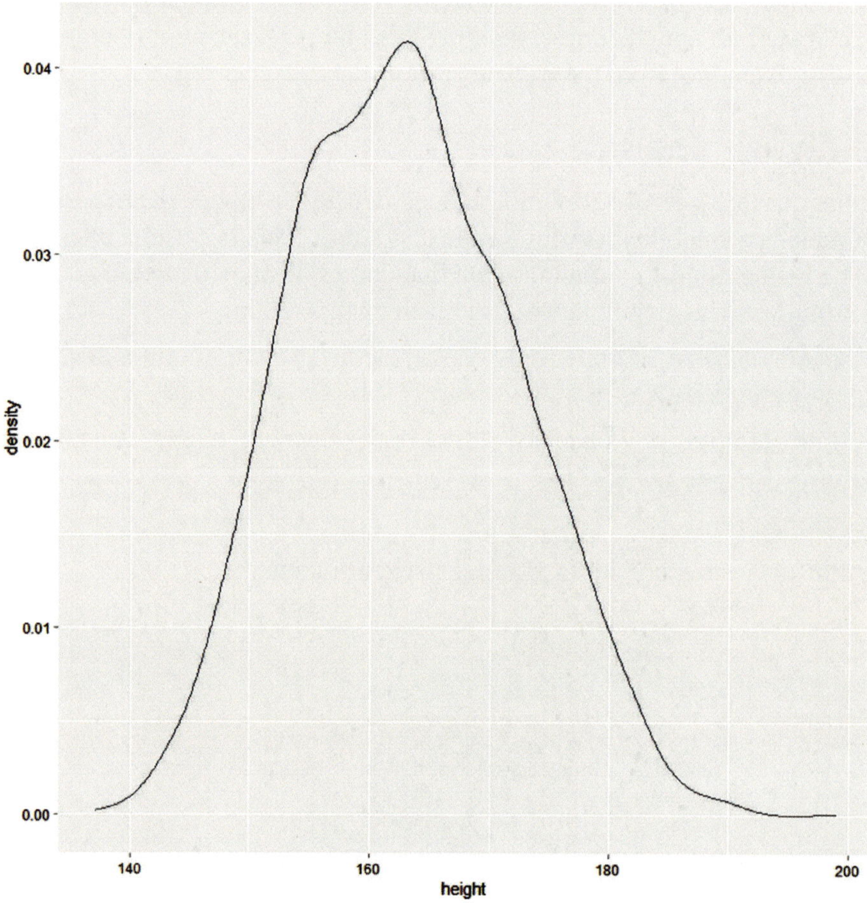


```
# physical activity
ggplot(data = regicor, aes(x = phyact))+geom_density()
```

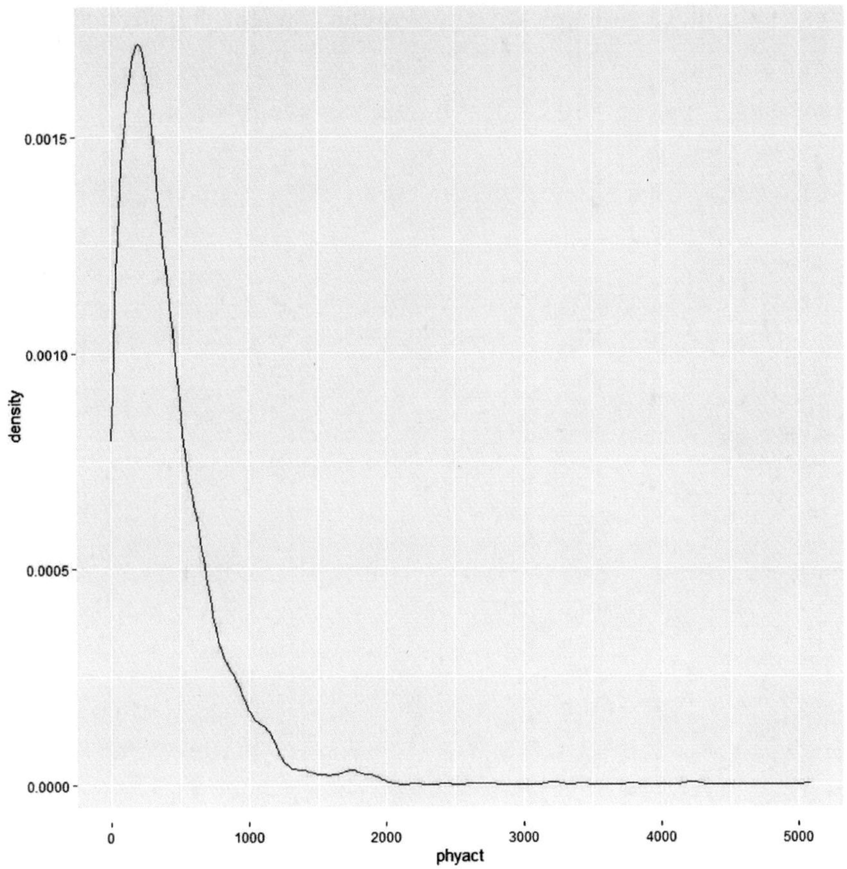

The results show the nearly symmetric bell shape distributions of weight and height. In contrast, the physical activity is right skewed with a long tail to the right. The peak of the density plot corresponds to the most frequent values.

2.3.4. geom_violin()

geom_violin() is used to plot violin plot. A violin plot is a compact display of a continuous distribution that is a mirrored density plot displayed in the same way as a boxplot.

geom_violin() requires two aesthetics (x for locating the symmetry line of the violin plot and y for the continuous variable you want to visualize on y-axis).

For example, plot a violin plot of regicor weight, height, and physical activity

```
# weight
ggplot(data = regicor, aes(x = 1, y = weight))+geom_violin()
```

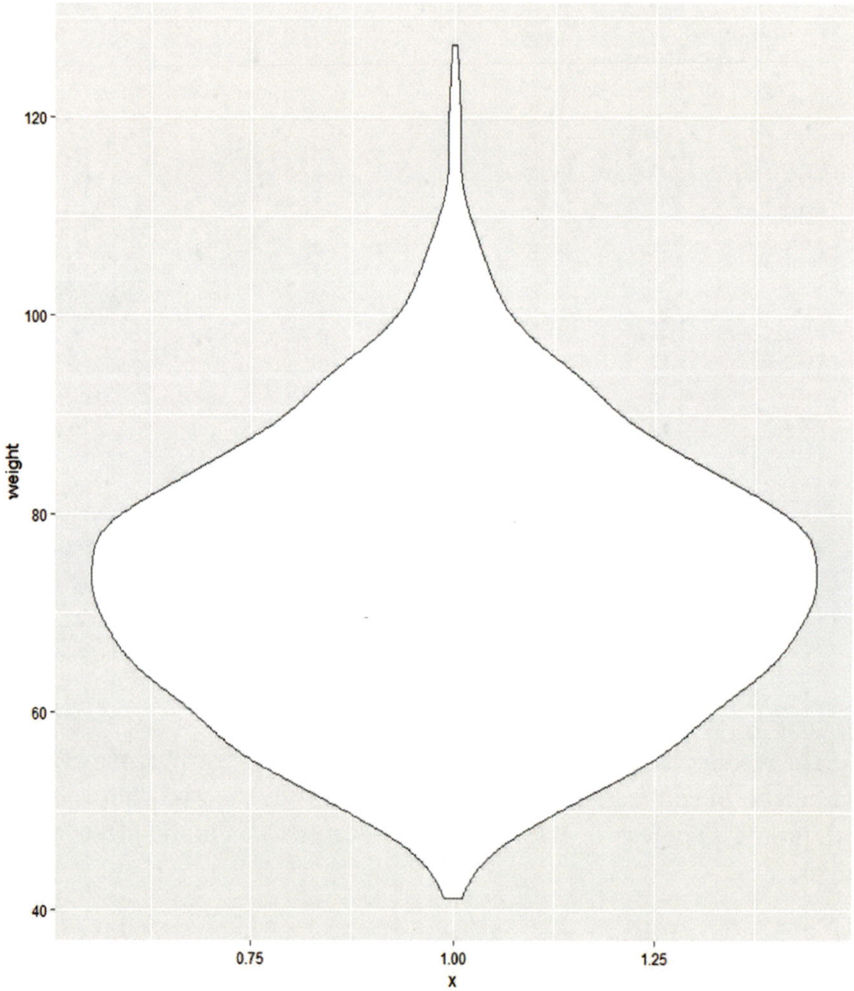


```
# to see that violin plot is a mirrored density plot
ggplot(data = regicor, aes(x = 1, y = weight))+geom_violin()+
 geom_vline(xintercept = 1) + coord_flip()
```

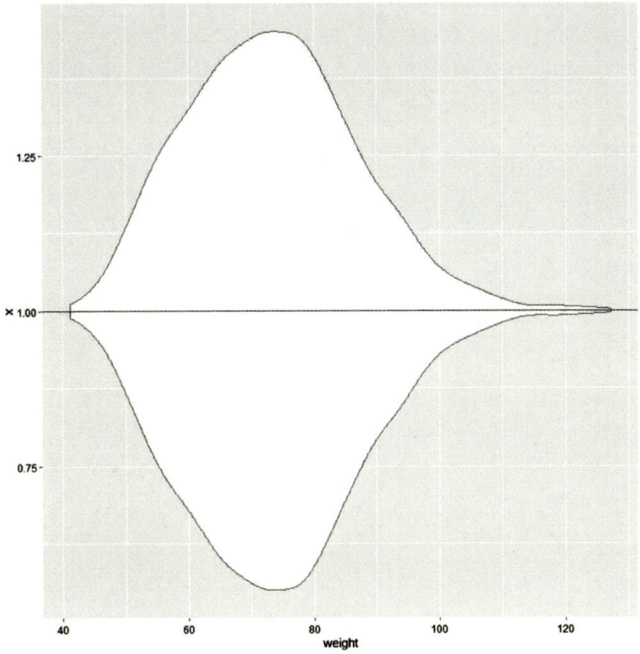

 # height

```
ggplot(data = regicor, aes(x = 1, y = height))+geom_violin()
```

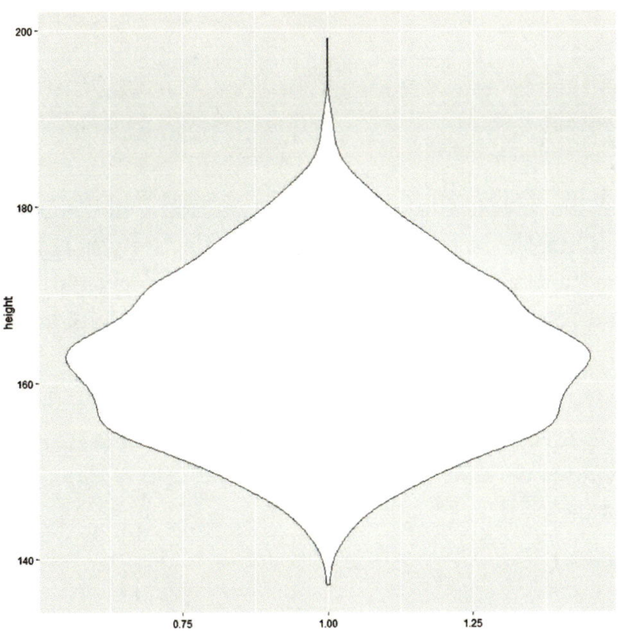

```
# physical activity
ggplot(data = regicor, aes(x = 1, y = phyact))+geom_violin()
```

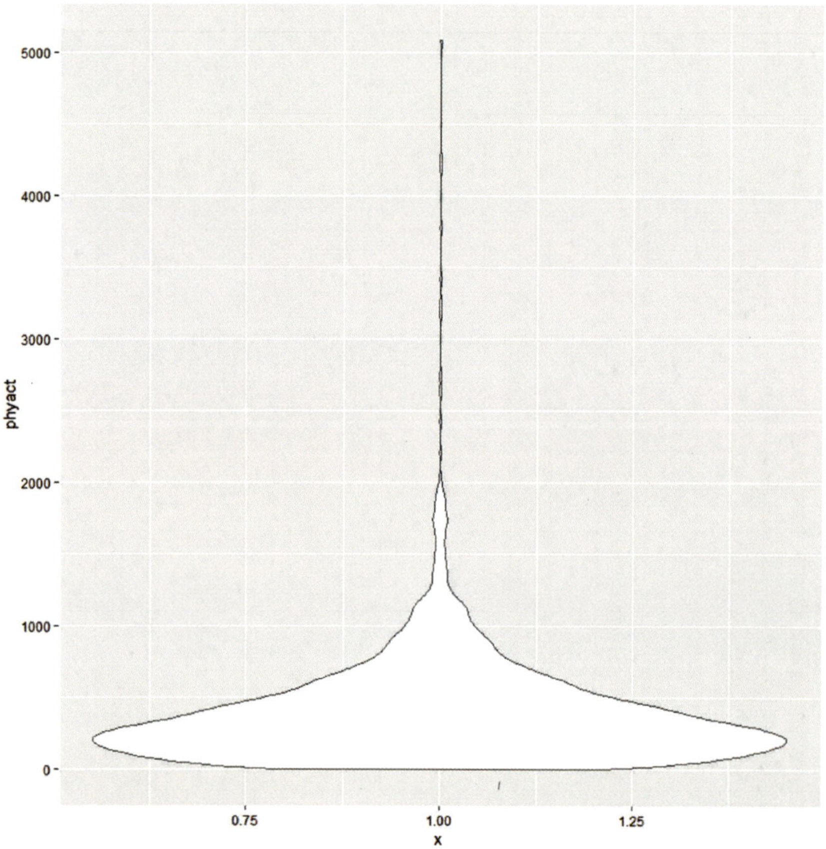

Similarly, the results show the nearly symmetric bell shape distributions of weight and height. In contrast, the physical activity is right skewed with a long tail to the right. We have add some functions (adding a line at the symmetry and flipping the chart) to show that a violin plot is a mirror image of density plot.

To complete our explanation, we will examine a left skewed data. The freeny income level in the freeny dataset is an example of left skewed data (rare small values or long tail to the left). The freeny dataset is part of the datasets package of R.

```
library(datasets)
data("freeny")
```

 # histogram

```
ggplot(data = freeny, aes(x = income.level))+geom_histogram(col="red")
```

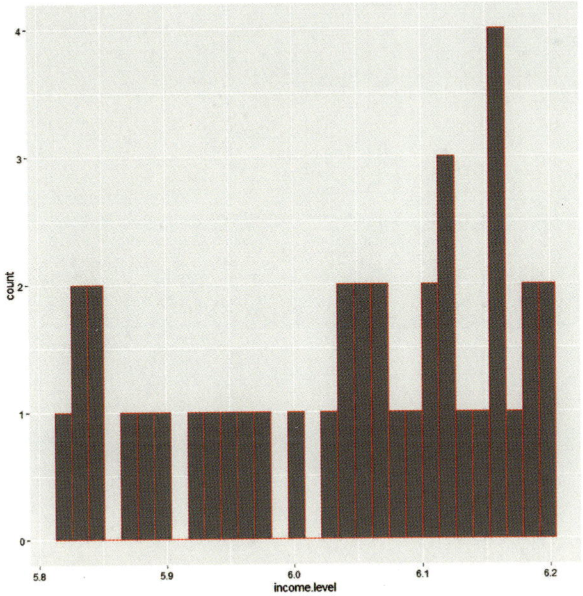

 # boxplot

```
ggplot(data = freeny, aes(y = income.level , group=1))+geom_boxplot()
```

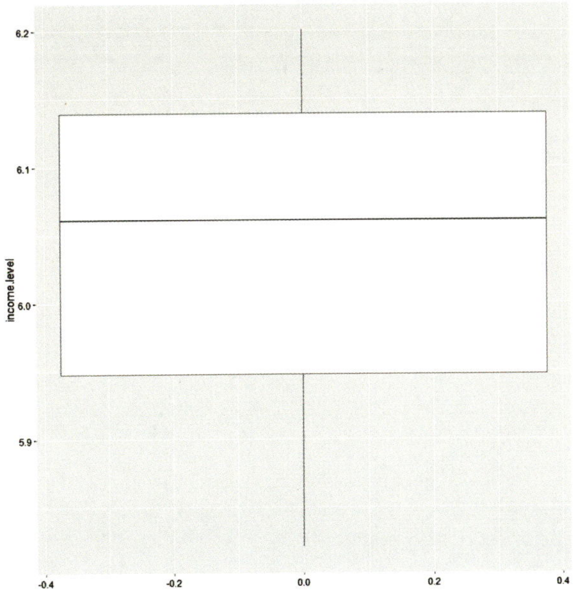

```
# density plot
ggplot(data = freeny, aes(x = income.level))+geom_density()
```

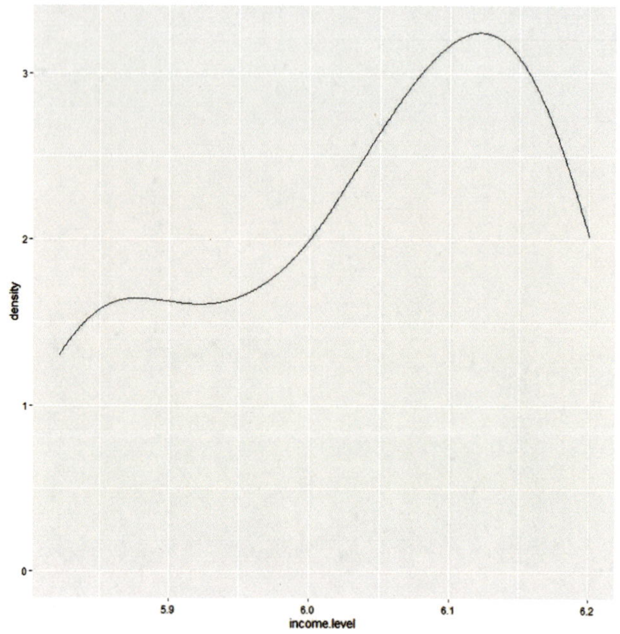


```
# violin plot
ggplot(data = freeny, aes(x = 1, y = income.level))+geom_violin()
```

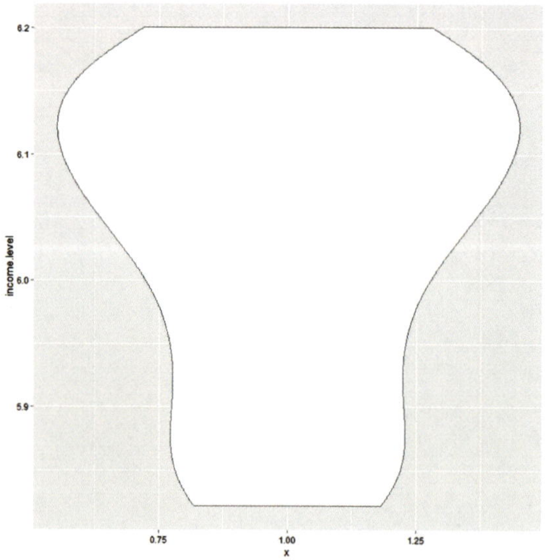

The most frequent values are large values with infrequent small values as seen from the histogram, density plot and violin plot. The density and violin plots shows the long tail to the left.

In the boxplot, the median line splits the box into unequal halves. The lower half is longer than upper half and the lines extending from the 25% quantile is longer than the line extending from the 75% quantile.

2.4. COMPARING TWO NUMERICAL VARIABLES, NUMERICAL MEASURES

The correlation coefficient is used to examine the association between two numerical variables without necessarily inferring a cause-and-effect relationship. The correlation coefficient ranges between +1 and –1.The two variables (x,y) are positively correlated if the correlation coefficient is greater than 0. As x increases, y tends to increase and as x decreases, y tends to decrease too.The two variables (x,y) are negatively correlated if the correlation coefficient is less than 0. If x increases, y tends to decrease, whereas as x decreases, y tends to increase.Two variables (x, y) are uncorrelated if the correlation coefficient is nearly 0. This occurs if there is no linear relationship between x and y.As a rule of thumb, generally correlation coefficients greater than 0.5 or smaller than –0.5 indicate strong positive correlation or strong negative correlation respectively.

2.4.1. Types of Correlation Coefficients

There are different correlation coefficients:

1. Pearson correlation (r): It can be used only when x and y have normal bell shaped distribution.
2. Kendall tau and Spearman rho which are correlation coefficients between the ranks of x and y values. They are used when x or y are not having a normal bell shaped distribution. Kendall tau is used for small samples less than 20 or 30.

cor() function is used to calculate different correlation coefficients. If some data are NA values, use the argument use =“complete.obs.” correlation between weight and height of the regicor dataset. Both have normal bell shaped curve so Pearson correlation can be calculated which is the default of cor() function

```
cor(regicor$weight, regicor$height, use = “complete.obs”)
## [1] 0.5093961
```

```
# spearman
cor(regicor$weight, regicor$height, use = "complete.obs," method = "spearman")
## [1] 0.5094698
```

Nearly there is no difference between Pearson and spearman correlation coefficients

Correlation between height and physical activity. Pearson cannot be used as physical activity is highly skewed

```
# spearman
cor(regicor$height, regicor$phyact, use = "complete.obs," method = "spearman")
## [1] –0.05207527
```

The correlation is nearly zero.

Correlation between weight and physical activity. Pearson cannot be used as physical activity is highly skewed

```
# spearman
cor(regicor$weight, regicor$phyact, use = "complete.obs," method = "spearman")
## [1] –0.07197231
```

The correlation is nearly zero.

cor() function can be used to form a table of correlation matrix

For example, state.x77 dataset. Use the spearman because we have not examined the data for normality

```
cor(state.x77, method = "spearman")
## Population Income Illiteracy Life Exp Murder HS Grad
## Population 1.0000000 0.12460984 0.3130496 –0.1040171 0.3457401 –0.3833649
## Income 0.1246098 1.00000000 –0.3145948 0.3241050 –0.2174623 0.5104809
## Illiteracy 0.3130496 –0.31459482 1.0000000 –0.5553735 0.6723592 –0.6545396
## Life Exp –0.1040171 0.32410498 –0.5553735 1.0000000 –0.7802406 0.5239410
## Murder 0.3457401 –0.21746230 0.6723592 –0.7802406 1.0000000
```

```
-0.4367330
## HS Grad -0.3833649 0.51048095 -0.6545396 0.5239410 -0.4367330
1.0000000
## Frost -0.4588526 0.19686382 -0.6831936 0.2983910 -0.5438432
0.3985351
## Area -0.1206723 0.05709484 -0.2503721 0.1275002 0.1064259 0.4389752
## Frost Area
## Population -0.4588526 -0.12067227
## Income 0.1968638 0.05709484
## Illiteracy -0.6831936 -0.25037208
## Life Exp 0.2983910 0.12750018
## Murder -0.5438432 0.10642590
## HS Grad 0.3985351 0.43897520
## Frost 1.0000000 0.11228778
## Area 0.1122878 1.00000000
```

Each column has a correlation coefficient of 1 with itself. The population has weak negative correlation with life expectancy (–0.10), while illiteracy has strong negative correlation with high school graduates (–0.65).

For regicor data, select only the numeric columns. Use the spearman because we have not examined the data for normality

```
cor(regicor[,c(3,6,7,10:13,16:21,23,25)], method = "spearman," use = "complete.obs")
## age sbp dbp chol hdl
## age 1.000000000 0.50614352 0.2237962783 0.107535252
0.01426149
## sbp 0.506143518 1.00000000 0.6731131121 0.182479671
-0.10101111
## dbp 0.223796278 0.67311311 1.0000000000 0.165095860
-0.08672871
## chol 0.107535252 0.18247967 0.1650958597 1.000000000
0.21722052
## hdl 0.014261494 -0.10101111 -0.0867287143 0.217220520
1.00000000
## triglyc 0.172645404 0.28024577 0.2349708252 0.339401841
```

```
–0.39609182
## ldl 0.089822137 0.18332277 0.1692060528 0.943092943 –0.01065800
## height –0.220821963 –0.02603402 0.0751861832 –0.072117357
–0.22774359
## weight 0.080643971 0.29641602 0.3344572144 0.032238016
–0.33781263
## bmi 0.272206403 0.37833375 0.3469122386 0.079292902 –0.23805245
## phyact 0.071877810 0.05499217 0.0009647736 0.048550671 0.03782503
## pcs –0.297536757 –0.17059335 –0.0421464761 –0.013392702
0.02662504
## mcs 0.073608178 0.12393664 0.0780139952 0.015884210 –0.07743757
## tocv 0.001221869 –0.00931904 –0.0067718293 –0.004297945
0.02171248
## todeath –0.033249289 –0.04278461 –0.0240941992 0.024777974
0.04117635
## triglyc ldl height weight bmi
## age 0.17264540 0.089822137 –2.208220e–01 0.08064397 0.27220640
## sbp 0.28024577 0.183322767 –2.603402e–02 0.29641602 0.37833375
## dbp 0.23497083 0.169206053 7.518618e–02 0.33445721 0.34691224
## chol 0.33940184 0.943092943 –7.211736e–02 0.03223802 0.07929290
## hdl –0.39609182 –0.010658001 –2.277436e–01 –0.33781263
–0.23805245

## triglyc 1.00000000 0.292388531 9.303758e–02 0.33620592 0.32948760
## ldl 0.29238853 1.000000000 –1.993212e–02 0.08832035 0.10807898
## height 0.09303758 –0.019932124 1.000000e+00 0.49223410
–0.10634864
## weight 0.33620592 0.088320352 4.922341e–01 1.00000000 0.78511887
## bmi 0.32948760 0.108078976 –1.063486e–01 0.78511887 1.00000000
## phyact –0.03401799 0.049842142 –5.444555e–02 –0.07347617
–0.02938858
## pcs –0.07369963 –0.011950008 2.227288e–01 –0.07031414 –0.24072964
## mcs 0.03632974 0.027395330 1.548218e–01 0.11139684 0.03804744
## tocv –0.03175978 –0.006400891 –3.747418e–03 –0.02465190
```

–0.02243060

todeath –0.02813590 0.015482578 9.557585e–06 –0.05032118 –0.06453251

phyact pcs mcs tocv todeath

age 0.0718778101 –0.29753676 0.073608178 0.001221869 –3.324929e–02

sbp 0.0549921650 –0.17059335 0.123936642 –0.009319040 –4.278461e–02

dbp 0.0009647736 –0.04214648 0.078013995 –0.006771829 –2.409420e–02

chol 0.0485506707 –0.01339270 0.015884210 –0.004297945 2.477797e–02

hdl 0.0378250297 0.02662504 –0.077437570 0.021712481 4.117635e–02

triglyc –0.0340179861 –0.07369963 0.036329736 –0.031759784 –2.813590e–02

ldl 0.0498421422 –0.01195001 0.027395330 –0.006400891 1.548258e–02

height –0.0544455517 0.22272876 0.154821846 –0.003747418 9.557585e–06

weight –0.0734761662 –0.07031414 0.111396835 –0.024651900 –5.032118e–02

bmi –0.0293885820 –0.24072964 0.038047439 –0.022430597 –6.453251e–02

phyact 1.0000000000 0.04952339 0.084658502 0.018253431 –1.843596e–02

pcs 0.0495233912 1.00000000 –0.094088059 –0.007172060 2.063117e–02

mcs 0.0846585019 –0.09408806 1.000000000 0.014207975 –9.732282e–03

tocv 0.0182534308 –0.00717206 0.014207975 1.000000000 6.930303e–03

todeath –0.0184359645 0.02063117 –0.009732282 0.006930303 1.000000e+00

Age has strong positive correlation with systolic blood pressure and nearly weak correlation with all other numerical variables

2.4.2. Visualization of the Correlation Matrix

A more compact way to visualize the correlation matrix is by using ggcorr() function from the GGally package

```
library(GGally)
# complete.obs and spearman method
ggcorr(state.x77, method = c("complete.obs,""spearman"))
```

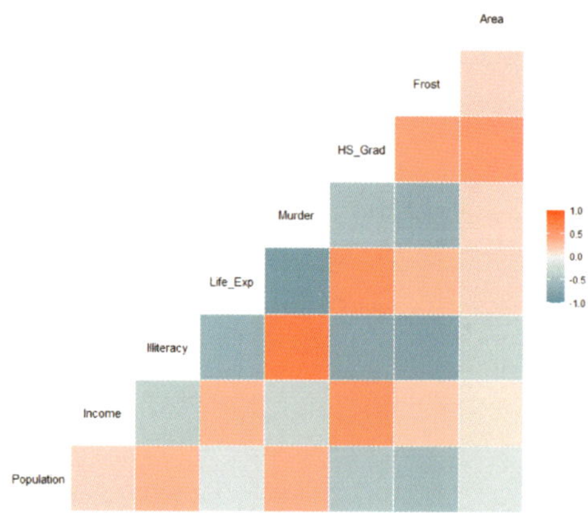


```
ggcorr(regicor[,c(3,6,7,10:13,16:21,23,25)], method = c("complete.obs,""spearman"))
```

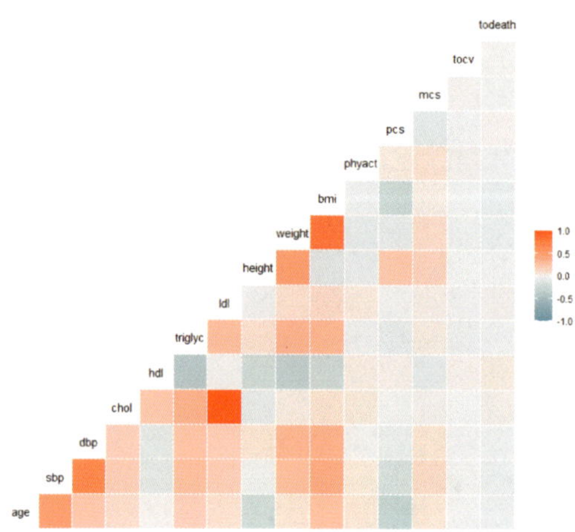

```
# add correlation coefficients within tiles
ggcorr(state.x77, method = c("complete.obs,""spearman"), label = TRUE)
```

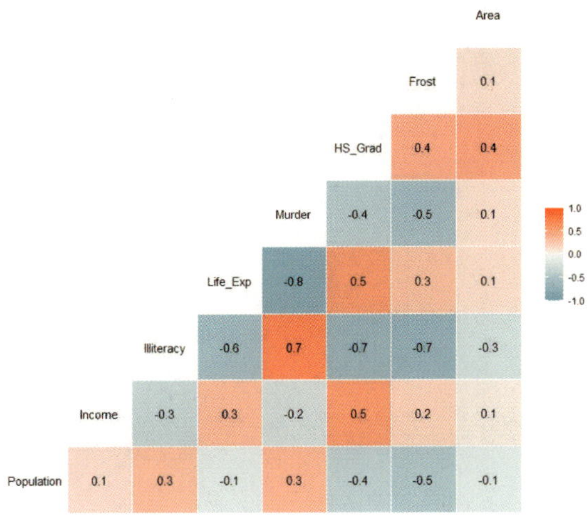


```
ggcorr(regicor[,c(3,6,7,10:13,16:21,23,25)], method = c("complete.obs,""spearman"), label = TRUE)
```

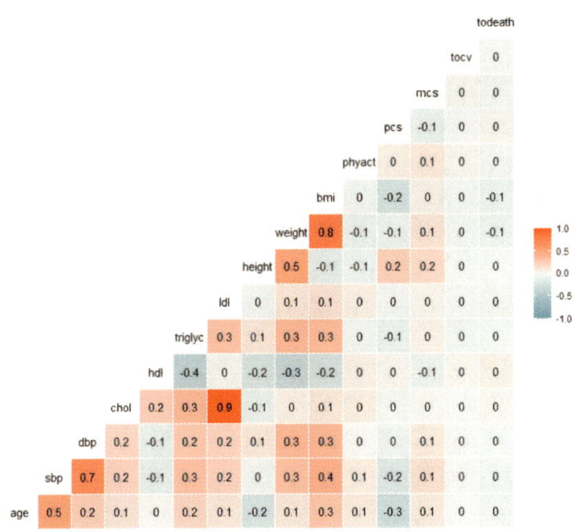


```
# add breaks for correlation coefficients
```

```
ggcorr(state.x77, method = c("complete.obs,""spearman"), label = TRUE,
nbreaks = 10)
```

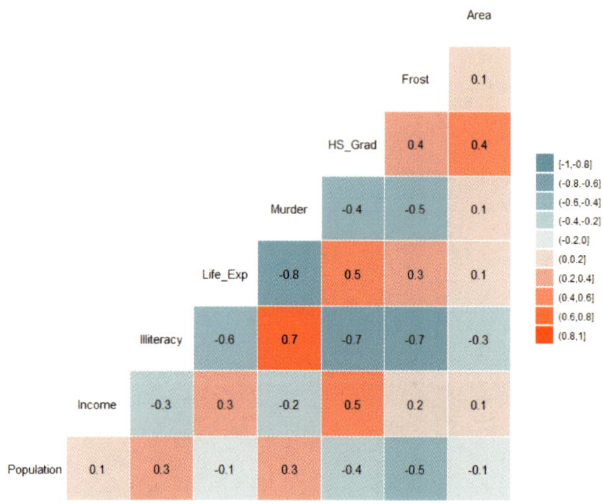


```
ggcorr(regicor[,c(3,6,7,10:13,16:21,23,25)], method = c("complete.
obs,""spearman"),
  label = TRUE, nbreaks = 10)
```

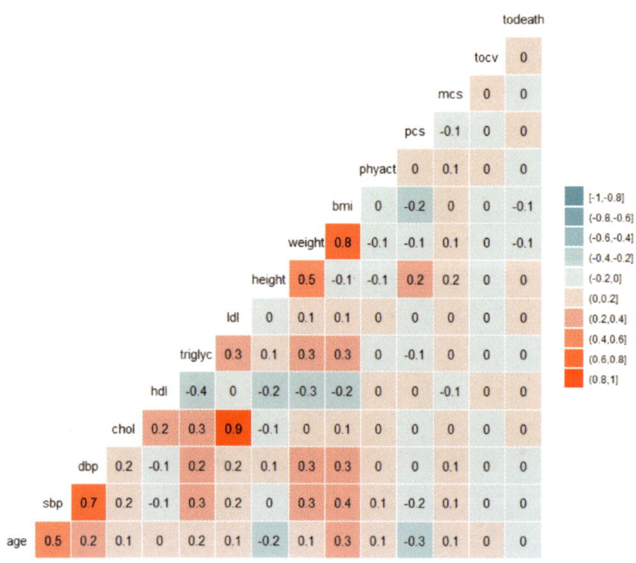

 # display correlation coefficients as two digits

```
ggcorr(state.x77, method = c("complete.obs,""spearman"),
  label = TRUE, nbreaks = 10, label_round = 2)
```

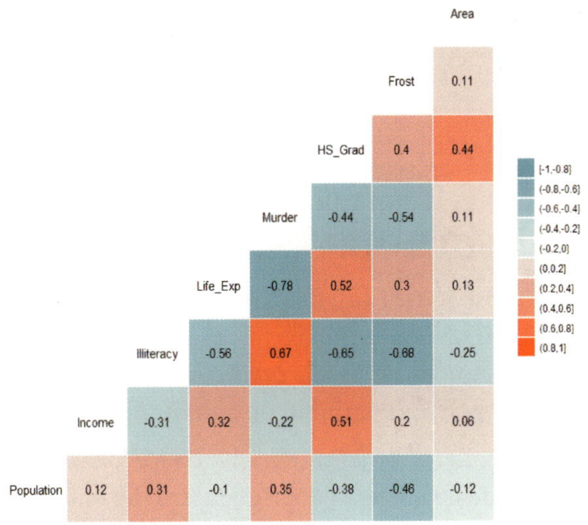


```
ggcorr(regicor[,c(3,6,7,10:13,16:21,23,25)], method = c("complete.
obs,""spearman"),
  label = TRUE, nbreaks = 10, label_round = 2)
```

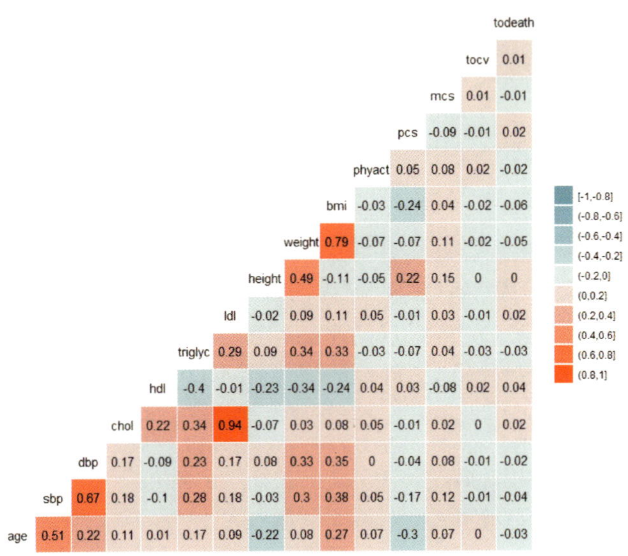

For the state data, the maximum positive correlation was between murder and illiteracy (0.7) and the maximum negative correlation was between murder and life expectancy (–0.8). As we said before, this does not mean causation

For the regicor data, the maximum positive correlation was between ldl (low-density lipoprotein) and chol (cholesterol) (0.9), while all the negative correlations were weak and never exceeds –0.3. As we said before, this does not mean causation.

2.5. COMPARING TWO NUMERICAL VARIABLES, GRAPHICAL METHODS

The scatterplot is most useful for displaying the relationship between two continuous variables. The scatterplot requires one of the continuous variables on x-axis (independent variable or predictor) and the other continuous variable on y-axis (dependent variable or outcome).

We will use ggscatter() function from ggpubr package because it also allows us to display the correlation coefficient on the scatter plot. ggscatter() requires the data to be a dataframe.

2.5.1. For example, state.x77 Data

```
library(ggpubr)
# scatterplot of illiteracy and murder
state<-data.frame(state.x77)
head(state)
##      Population Income Illiteracy Life.Exp Murder HS.Grad Frost  Area
## Alabama     3615  3624   2.1 69.05 15.1 41.3  20 50708
## Alaska       365  6315   1.5 69.31 11.3 66.7 152 566432
## Arizona     2212  4530   1.8 70.55 7.8 58.1  15 113417
## Arkansas    2110  3378   1.9 70.66 10.1 39.9  65 51945
## California 21198  5114   1.1 71.71 10.3 62.6  20 156361
## Colorado    2541  4884   0.7 72.06 6.8 63.9 166 103766
ggscatter(data = state, x = “Illiteracy,” y = “Murder,”
   cor.coef = TRUE,
   cor.coeff.args = list(method= “spearman,” cor.coef.name=”rho”))
```

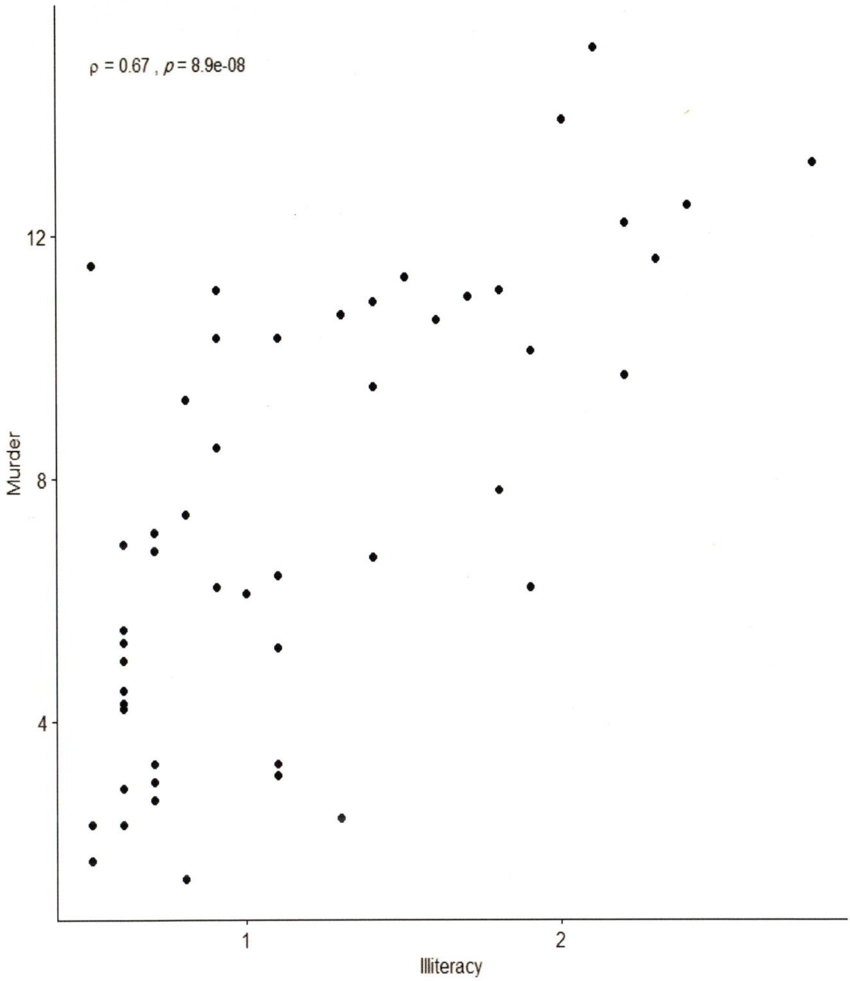


```
# add regression line
ggscatter(data = state, x = "Illiteracy," y = "Murder," add = "reg.line,"
    cor.coef = TRUE,
    cor.coeff.args = list(method= "spearman," cor.coef.name="rho"))
```

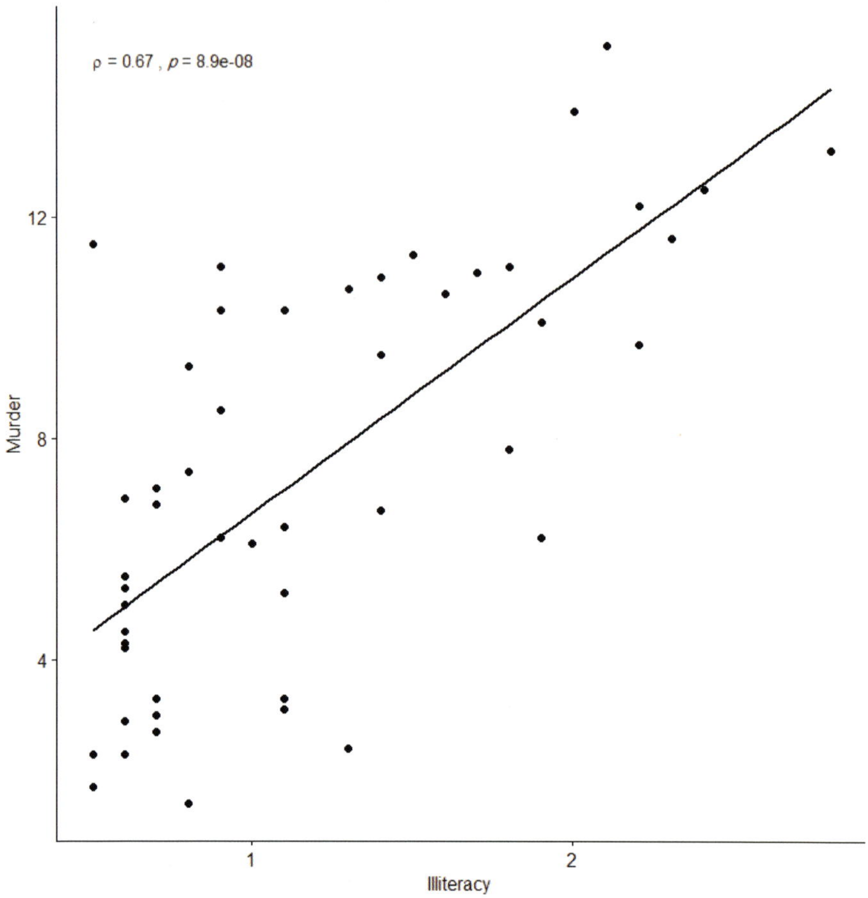


```
# scatterplot of murder and life expectancy
ggscatter(data = state, x = "Life.Exp," y = "Murder,"

    cor.coef = TRUE,

    cor.coeff.args = list(method= "spearman," cor.coef.name="rho"))
```

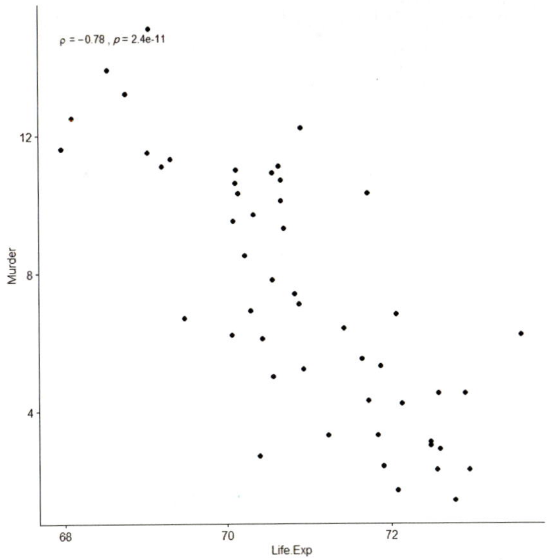


```
# add regression line
ggscatter(data = state, x = "Life.Exp," y = "Murder," add = "reg.line,"
    cor.coef = TRUE,

    cor.coeff.args = list(method= "spearman," cor.coef.name="rho"))
```

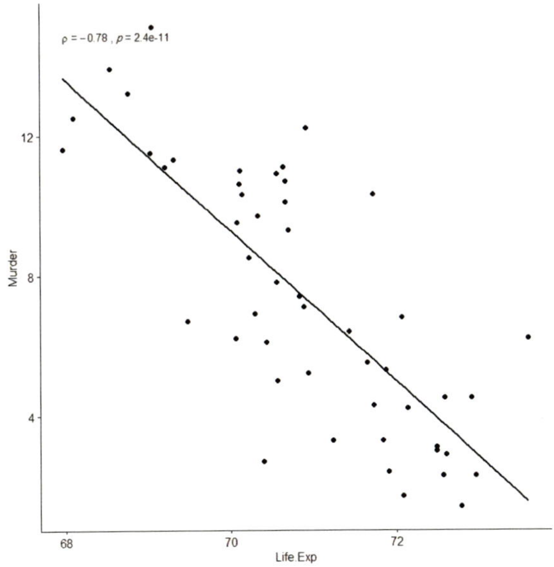

In the scatterplot between illiteracy and murder, there was positive correlation and this is also evident from the positive slope of regression line. Regression will be discussed in later chapters

In the scatterplot between life expectancy and murder, there was negative correlation and this is also evident from the negative slope of regression line.

2.5.2. For example, Regicor Data

```
# correlation between cholesterol (chol) and low density lipoprotein (ldl)
ggscatter(data = regicor, x = "ldl," y = "chol," add = "reg.line,"
    cor.coef = TRUE,
    cor.coeff.args = list(method= "spearman," cor.coef.name="rho"))
```

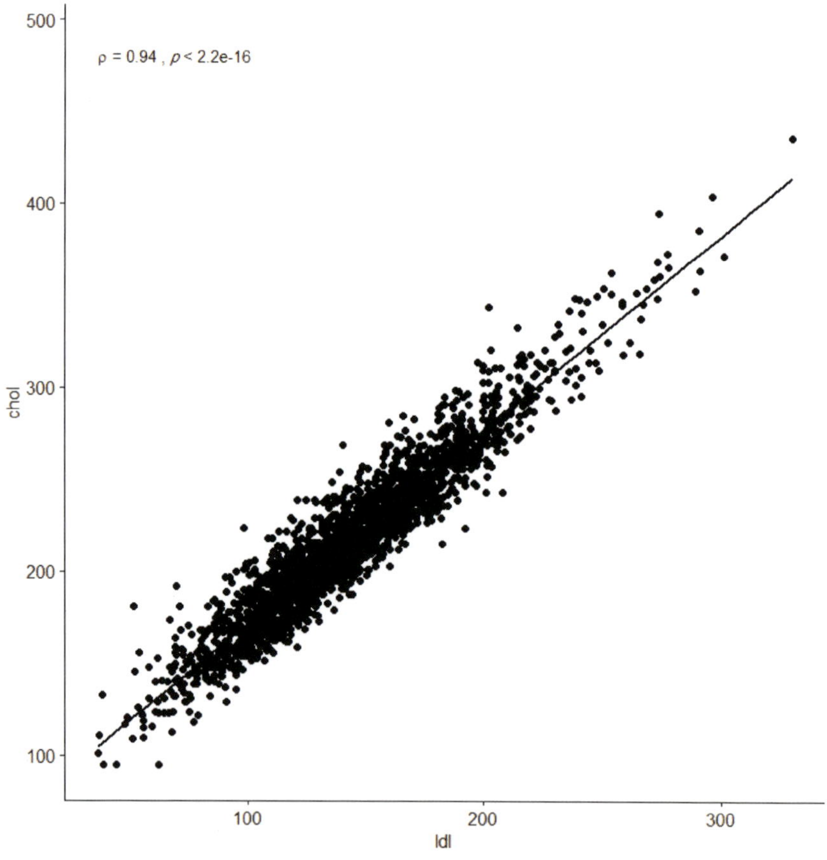


```
# correlation between weight and height
```

```
ggscatter(data = regicor, x = "weight," y = "height," add = "reg.line,"

    cor.coef = TRUE,

    cor.coeff.args = list(method= "spearman," cor.coef.name="rho"))
```

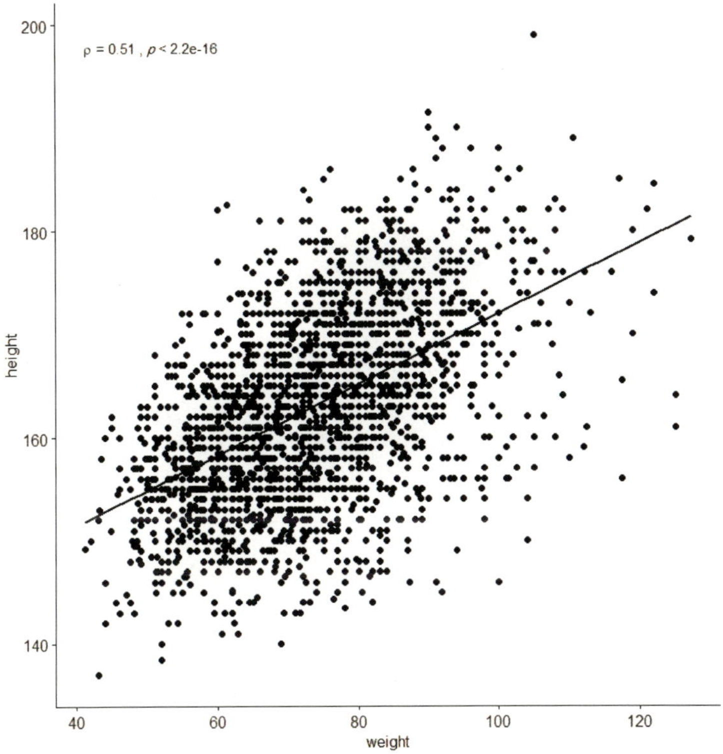


```
# correlation between weight and physical activity
ggscatter(data = regicor, x = "weight," y = "phyact," add = "reg.line,"

    cor.coef = TRUE,

    cor.coeff.args = list(method= "spearman," cor.coef.name="rho"))
```

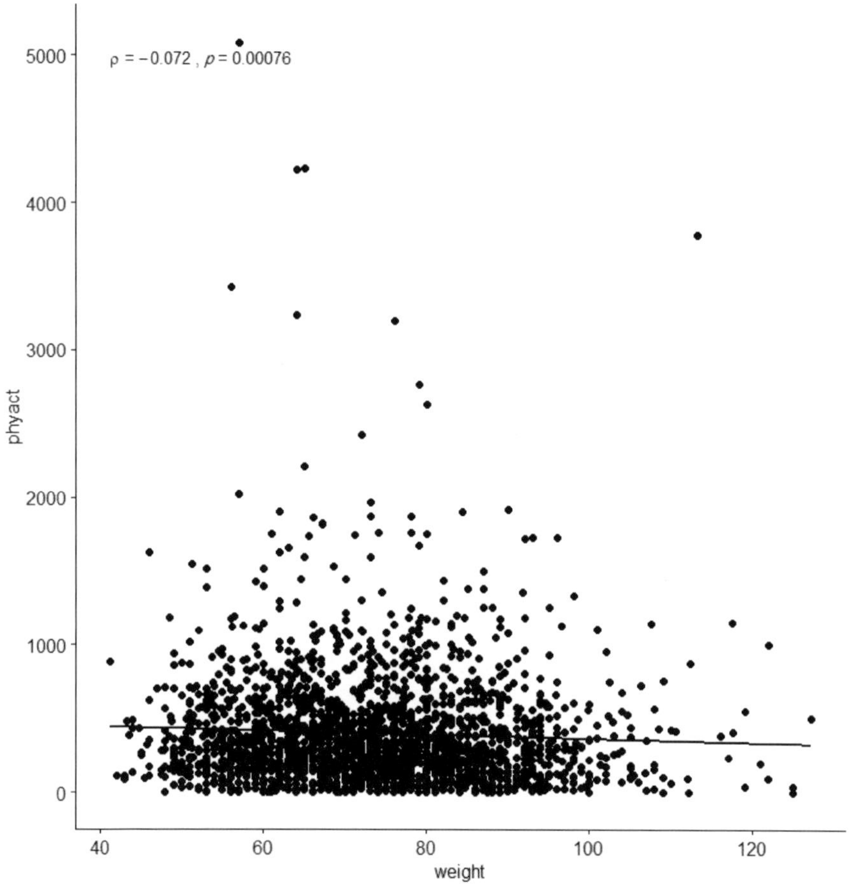

In the scatterplot between cholesterol and low density lipoprotein, there was a strong positive correlation and this is also evident from the positive slope of regression line.

In the scatterplot between weight and height, there was also a relatively strong positive correlation and this is also evident from the positive slope of regression line.

In the scatterplot between weight and physical activity, there was a very weak negative correlation (–0.072) and this is also evident from the nearly horizontal regression line. This indicates that these two variables are nearly independent although the correlation coefficient is statistically significant (p-value < 0.05) which means that this correlation is really smaller than zero. To plot all possible scatterplots with their correlation coefficients, you can use chart.Correlation() function from the PerformanceAnalytics package.

```
library(PerformanceAnalytics)
chart.Correlation(state, method = "spearman")
```

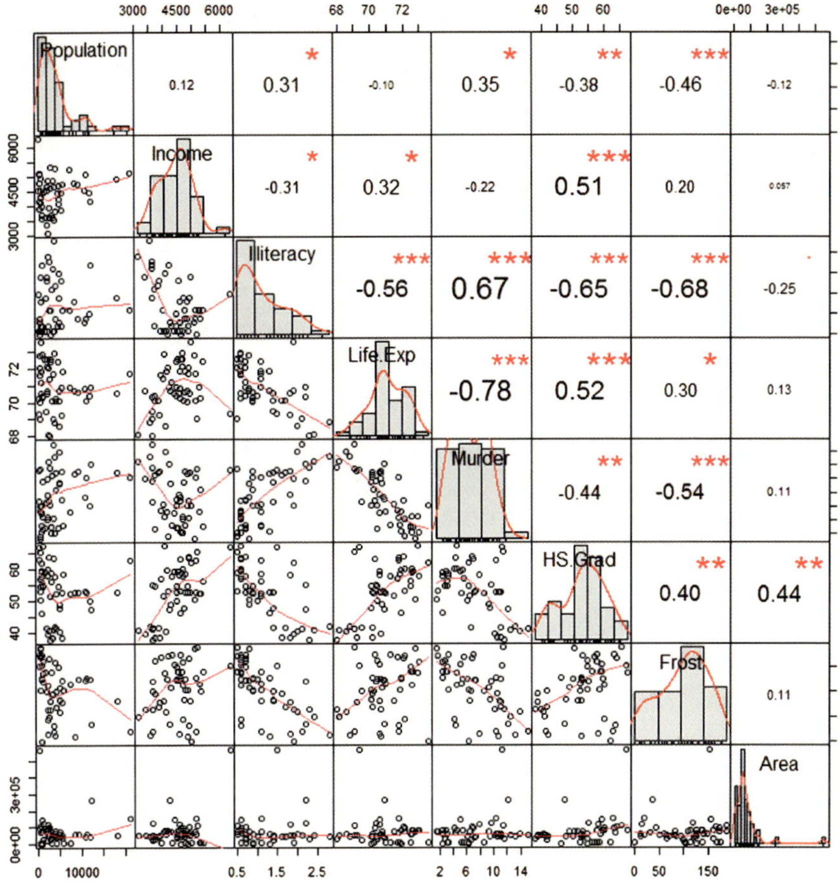

The distribution of each variable is shown on the diagonal. For example, population, illiteracy, and area are right skewed variables.

On the bottom of the diagonal, the bivariate scatter plots with a fitted lowess (locally weighted scatterplot smoothing) line are displayed. For example, the fitted line between population and income is nearly horizontal with correlation = 0.12 indicating that the population and income are independent of each other.

On the top of the diagonal, the value of the spearman correlation plus the significance level as stars. Each significance level is associated to a symbol: the p-values (0, 0.001, 0.01, 0.05, 0.1, 1) are indicated by symbols ("***", "**", "*", ".", " "), respectively

2.6. COMPARING ONE NUMERICAL AND ONE CATEGORICAL VARIABLE, NUMERICAL MEASURES

In research, we commonly compare between two or more populations (categories) by collecting samples from these populations and measuring different continuous variables on them.

The most commonly used measure is the mean difference because the mean difference quantifies how the distribution of your continuous data shifts up or down across the two samples. In addition, the statistical theory allows us to quantify the uncertainty of the estimated mean difference for the underlying mean difference in populations that we have sampled from.

For any dataframe, you can use the pipeline, group_by() for categorical variable followed by summarize to obtain the mean of continuous data for each category. If you have more than two categories (groups), designate one category as a reference category and compare each category to it

2.6.1. Regicor Data, the Mean Age Difference for Females to Males

```
dat<- regicor %>% group_by(sex) %>%
 summarise(mean_age = mean(age, na.rm = TRUE))
dat
## # A tibble: 2 x 2
##  sex  mean_age
##  <fct>  <dbl>
## 1 Male    54.8
## 2 Female   54.7
# calculate the mean difference
x<-dat$mean_age[2]-dat$mean_age[1]
x
## [1] –0.0932072
```

The mean age difference for females to males = –0.09 years. This means that females in our sample data (2294 rows) are younger than males by 0.09 years on the average. In other words, not every female is younger than every male but on the average the females are younger by 0.09 years or males are older by 0.09 years on the average.

2.6.2. Regicor Data, the Mean Weight Difference for Females to Males

```
dat<- regicor %>% group_by(sex) %>%
 summarise(mean_weight = mean(weight, na.rm = TRUE))
dat
## # A tibble: 2 x 2
##  sex  mean_weight
##  <fct>    <dbl>
## 1 Male     79.7
## 2 Female    67.6
# calculate the mean difference
x<-dat$mean_weight[2]-dat$mean_weight[1]
x
## [1] –12.13563
```

The mean weight difference for females to males = –12.1 kg. This means that females in our sample data (2294 rows) weigh less than males by 12.1 kg on the average. In other words, males weigh more than females by 12.1 kg on the average.

2.6.3. Regicor Data, the Mean Days to Cardiovascular Event for Different Smoker Groups

Assume the never smoker as our reference group.

```
dat<- regicor %>% group_by(smoker) %>%

 summarise(mean_tocv = mean(tocv, na.rm = TRUE))
dat
## # A tibble: 4 x 2
##  smoker        mean_tocv
##  <fct>          <dbl>
## 1 Never smoker        1736.
## 2 Current or former < 1y   1692.
## 3 Former >= 1y        1870.
```

```
## 4 <NA>          1899.
# as smoker contains NA values so we should exclude them
dat<- regicor %>% filter(!is.na(smoker)) %>% group_by(smoker) %>%
 summarise(mean_tocv = mean(tocv, na.rm = TRUE))
dat
## # A tibble: 3 x 2
##  smoker        mean_tocv
##  <fct>         <dbl>
## 1 Never smoker       1736.
## 2 Current or former < 1y   1692.
## 3 Former >= 1y       1870.
# calculate the mean difference
diff_never_current<-dat$mean_tocv[1]-dat$mean_tocv[2]
diff_never_current
## [1] 44.06174
diff_never_former<-dat$mean_tocv[1]-dat$mean_tocv[3]
diff_never_former
## [1] –134.1475
```

The mean days to cardiovascular event difference for never smokers to current smokers = 44.06 days. This means that never smokers in our sample data have 44.06 more days to cardiovascular event than current smokers on the average. On the other hand, never smokers have 134.15 less days to cardiovascular event than former smokers on the average.

It appears that former smokers have more days to cardiovascular event than other smoker groups; however, this may be caused by censoring or end of follow-up for never smokers. In addition, statistical tests are needed to prove this hypothesis.

2.6.4. Regicor Data, the Mean Weight for Different Smoker and Sex Groups

Assume that female never smokers are our reference group.

```
dat<- regicor %>% filter(!is.na(smoker)) %>% group_by(smoker, sex)
%>%
```

```
summarise(mean_weight = mean(weight, na.rm = TRUE))
dat
## # A tibble: 6 x 3
## # Groups:  smoker [3]
##   smoker          sex  mean_weight
##   <fct>           <fct>  <dbl>
## 1 Never smoker    Male    79.6
## 2 Never smoker    Female  68.3
## 3 Current or former < 1y Male    79.1
## 4 Current or former < 1y Female  64.1
## 5 Former >= 1y    Male    80.6
## 6 Former >= 1y    Female  67.7
# calculate the mean difference
diff_neverf_neverm<-dat$mean_weight[2]-dat$mean_weight[1]
diff_neverf_neverm
## [1] -11.36229
diff_neverf_currentm<-dat$mean_weight[2]-dat$mean_weight[3]
diff_neverf_currentm
## [1] -10.82239
diff_neverf_currentf<-dat$mean_weight[2]-dat$mean_weight[4]
diff_neverf_currentf
## [1] 4.111232
diff_neverf_formerm<-dat$mean_weight[2]-dat$mean_weight[5]
diff_neverf_formerm
## [1] -12.31051
diff_neverf_formerf<-dat$mean_weight[2]-dat$mean_weight[6]
diff_neverf_formerf
## [1] 0.5642619
```

- Never smoker females weigh less than never smoker males by 11.36 kg on the average.
- Never smoker females weigh less than current smoker males by 10.82 kg on the average.

- Never smoker females weigh more than current smoker females by 4.1 kg on the average.
- Never smoker females weigh less than former smoker males by 12.3 kg on the average.
- Never smoker females weigh more than former smoker females by 0.56 kg on the average.

2.7. COMPARING ONE NUMERICAL AND ONE CATEGORICAL VARIABLE, GRAPHICAL METHODS

All plot types used for univariate continuous data (boxplot, histogram, density plot, violin plot) can be customized using the color or fill aesthetics to add 1 or 2 categorical variables.

2.7.1. Side by Side Boxplot

In regicor data, compare the age variable for females and males.

```
ggplot(data = regicor, aes(x = sex, y = age, col = sex))+geom_boxplot()
```

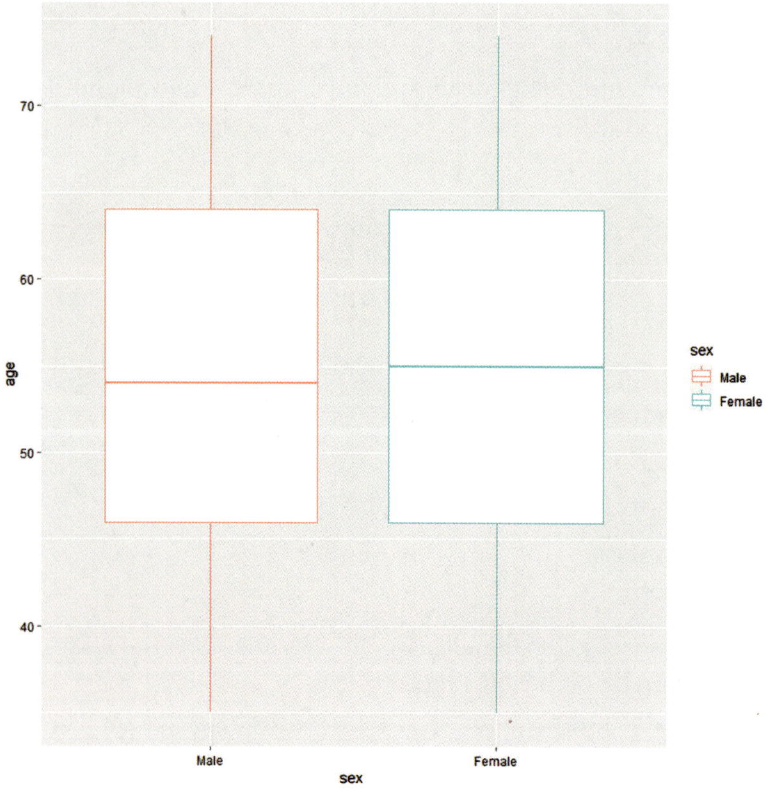

We see that the age variable is slightly shifted up for females, although females have lower mean age.

In regicor data, compare the weight variable for females and males.

```
ggplot(data = regicor, aes(x = sex, y = weight, col = sex))+geom_boxplot()
```

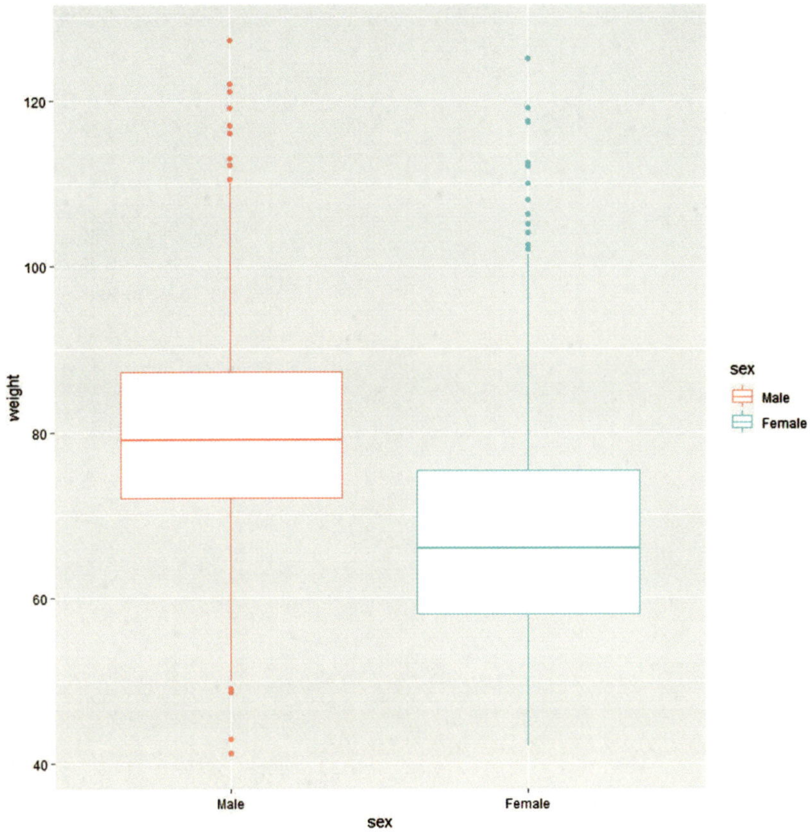

We see that the weight of females is shifted down compared to males.

In regicor data, compare the mean days to cardiovascular event for different smoker groups.

Assume the never smoker as our reference group.

```
ggplot(data = regicor, aes(x = smoker, y = tocv, col = smoker))+geom_
boxplot()
```

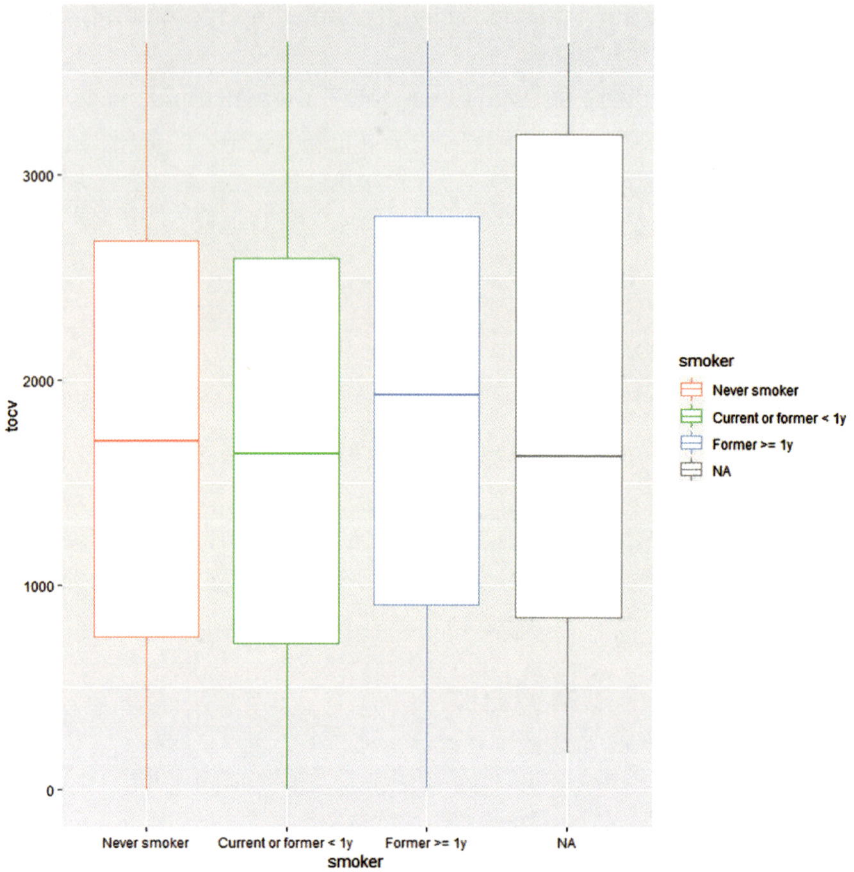

 # to remove NA from smoker column

```
dat<- regicor %>% filter(!is.na(smoker))
ggplot(data = dat, aes(x = smoker, y = tocv, col = smoker))+geom_boxplot()
```

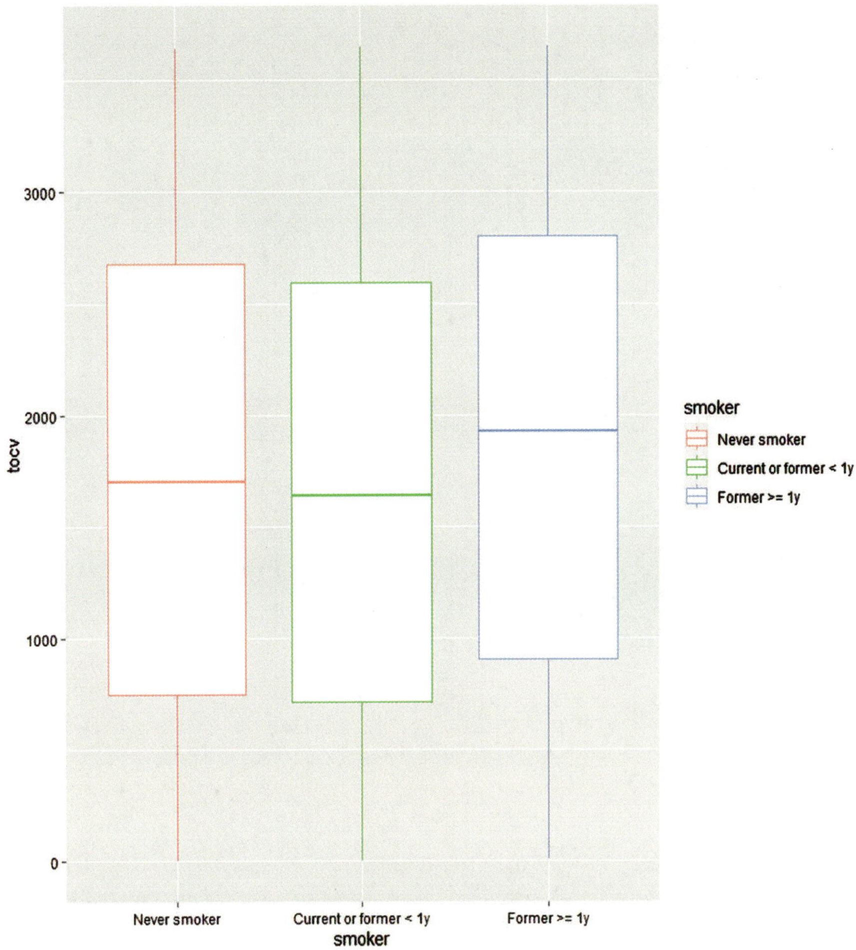

The days to cardiovascular event is longer for former smokers than other two smoker groups. The reference group is put as the first boxplot so the other groups can be compared to it.

In regicor data, compare the mean weight for different smoker and sex groups.

Assume that female never smokers are our reference group.

```
dat<- regicor %>% filter(!is.na(smoker))
ggplot(data = dat, aes(x = smoker, y = weight, fill = sex))+geom_boxplot()
```

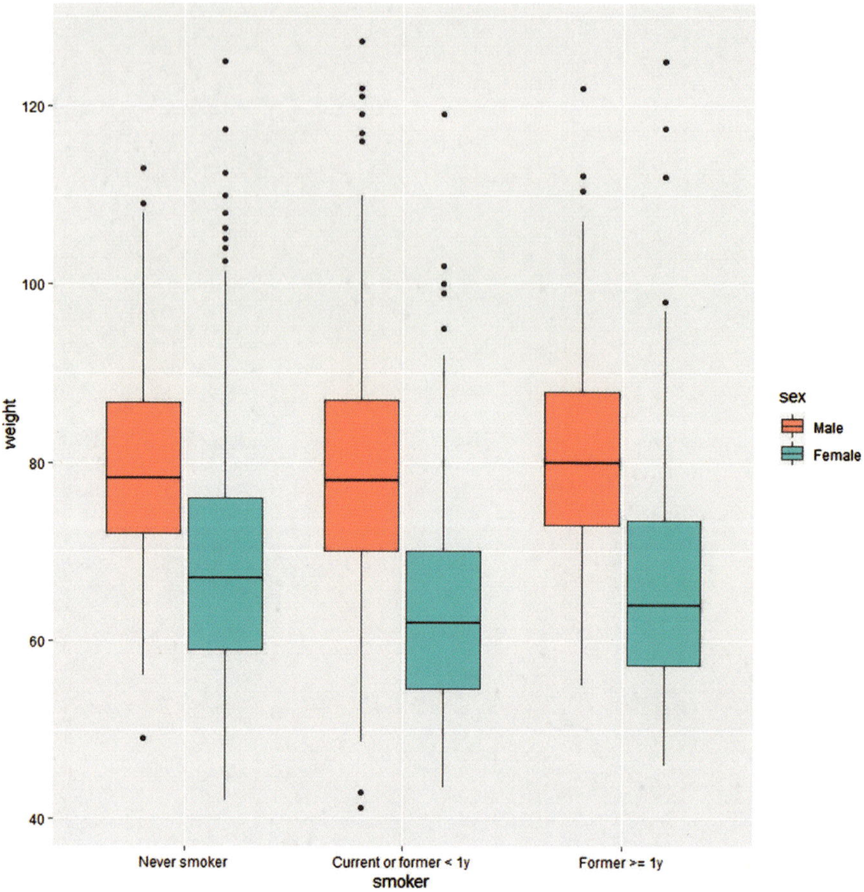


```
# to put female never smoker as first boxplot
dat$sex<-factor(dat$sex, levels = c("Female," "Male"))
ggplot(data = dat, aes(x = smoker, y = weight, fill = sex))+geom_boxplot()
```

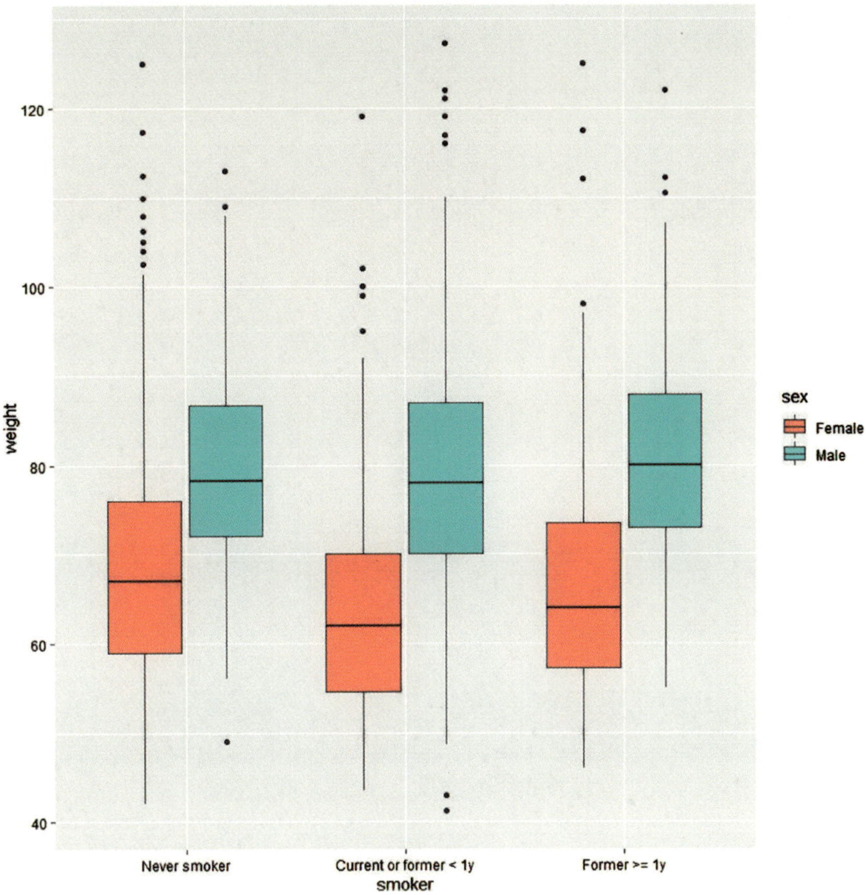

The weight of females is generally shifted down for the three smoker groups compared to males.

2.7.2. Side by Side Histograms

In regicor data, compare the age variable for females and males.

```
ggplot(data = regicor, aes(x = age, fill = sex))+geom_histogram()
```

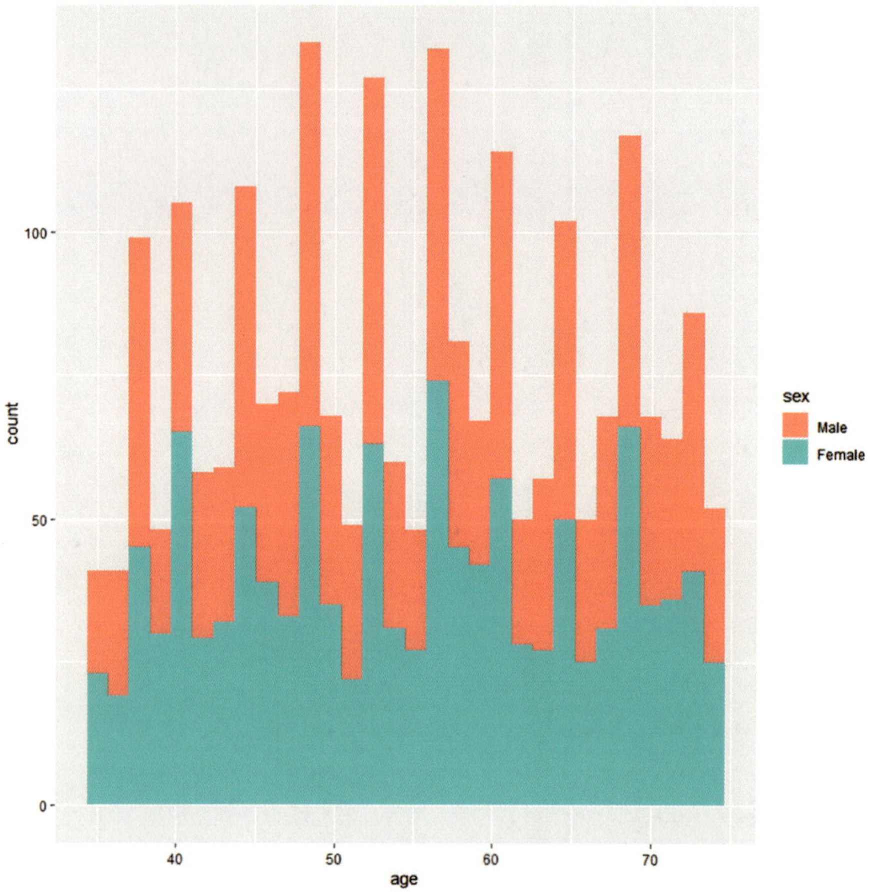

We see that the age variable is more represented for males than females.

In regicor data, compare the weight variable for females and males.

```
ggplot(data = regicor, aes(x = weight, fill = sex))+geom_histogram()
```

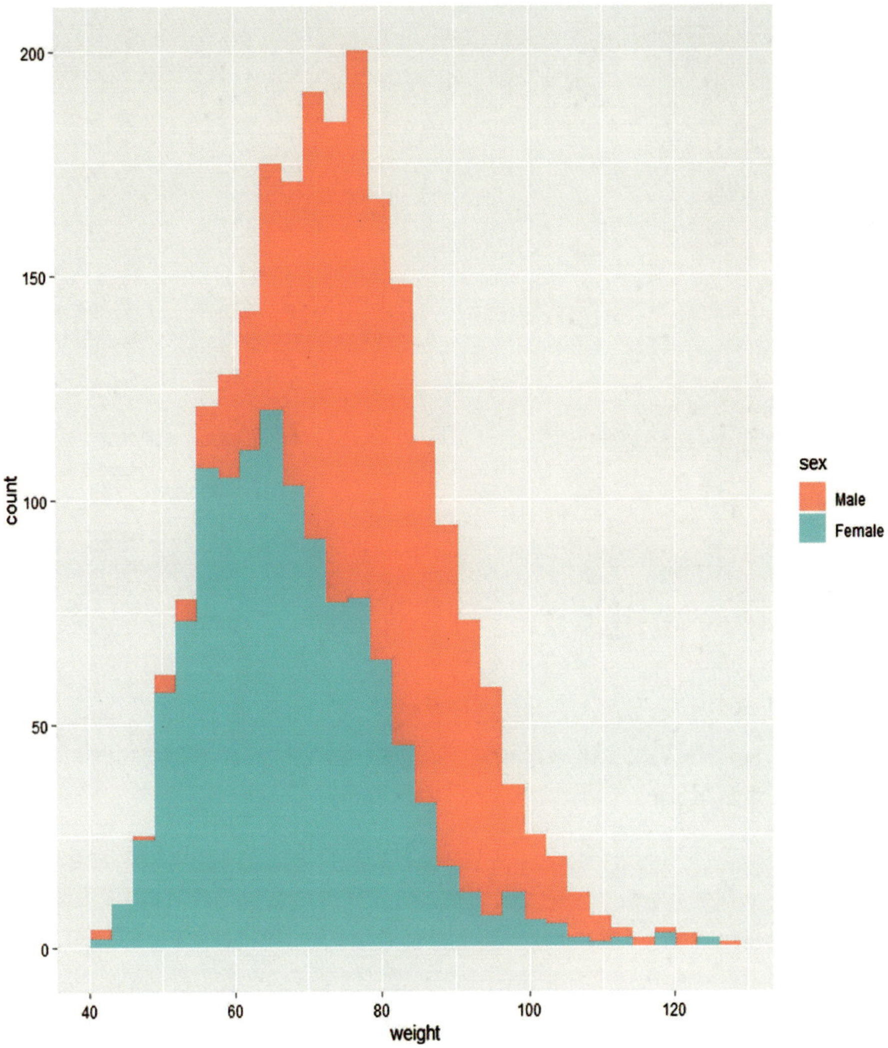

We see that the weight of females is shifted down compared to males.

In regicor data, compare the mean days to cardiovascular event for different smoker groups.

```
ggplot(data = regicor, aes(x = tocv, fill = smoker))+geom_histogram()
```

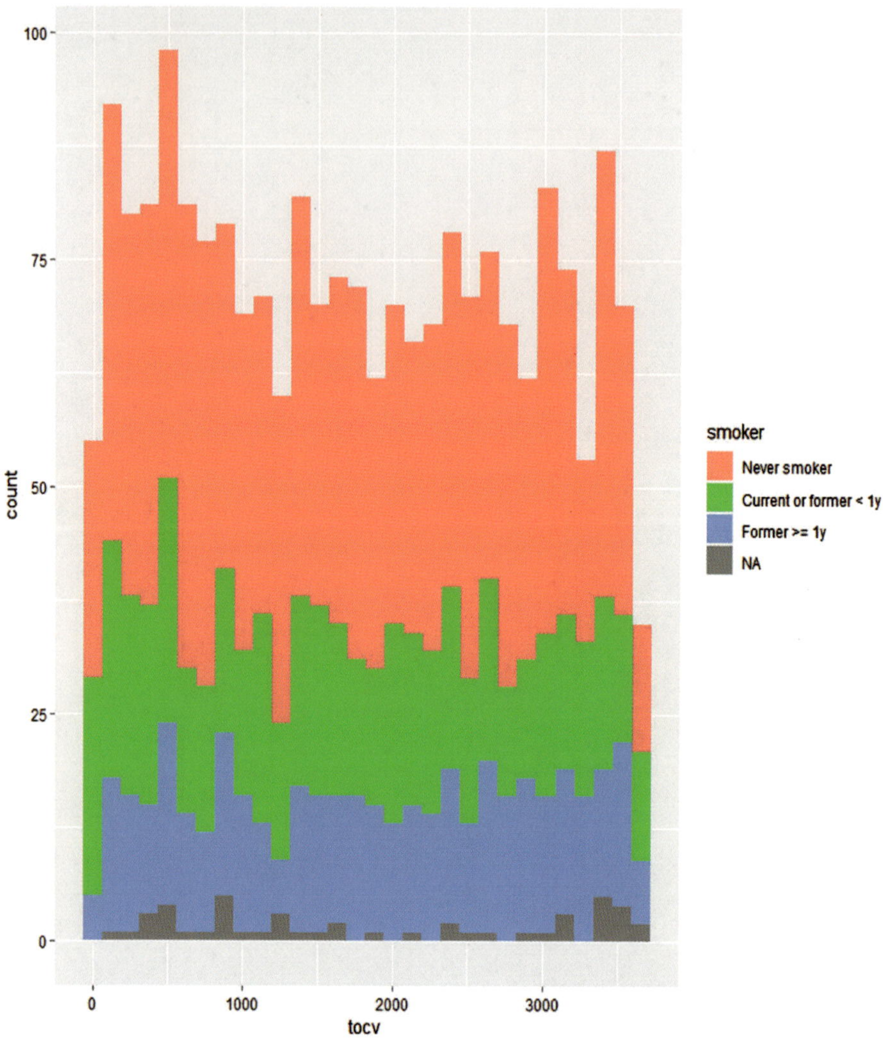


```
# to remove NA from smoker column
dat<- regicor %>% filter(!is.na(smoker))
ggplot(data = dat, aes(x = tocv, fill = smoker))+geom_histogram()
```

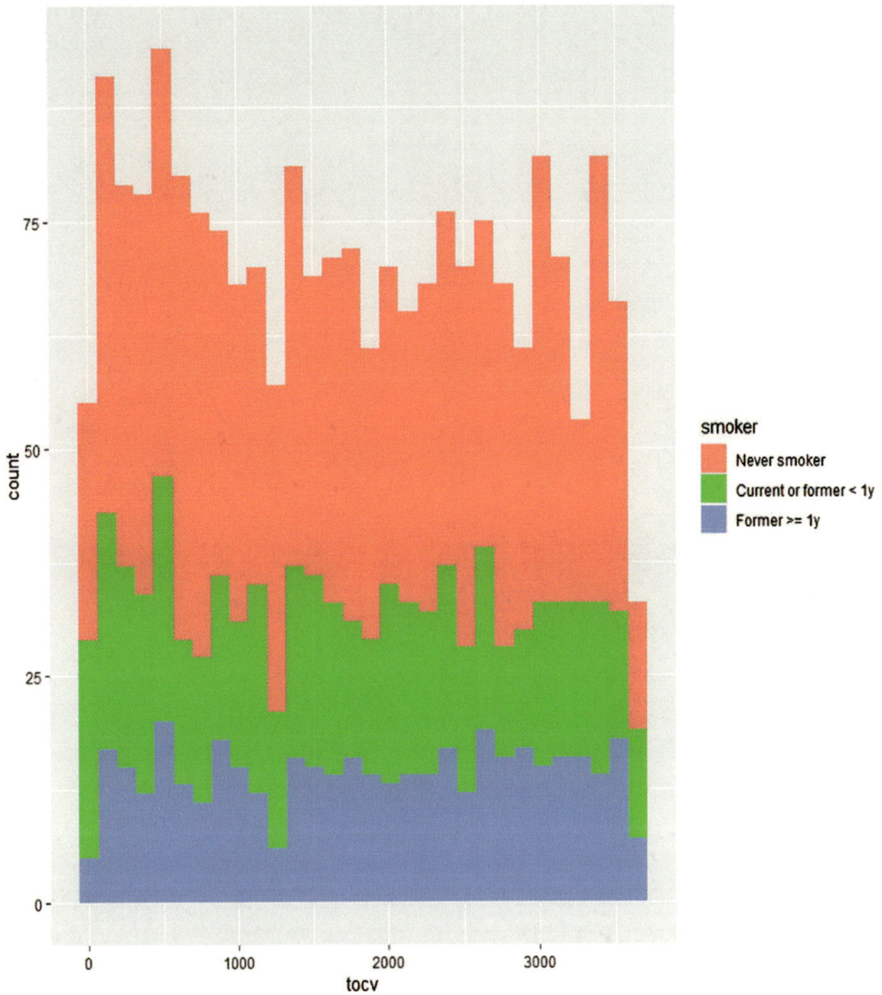

The days to cardiovascular event are more represented for never smokers than the other two smoker groups.

In regicor data, compare the mean weight for different smoker and sex groups.

```
dat<- regicor %>% filter(!is.na(smoker))
ggplot(data = dat, aes(x = weight, fill = smoker))+
 geom_histogram()+facet_grid(~sex)
```

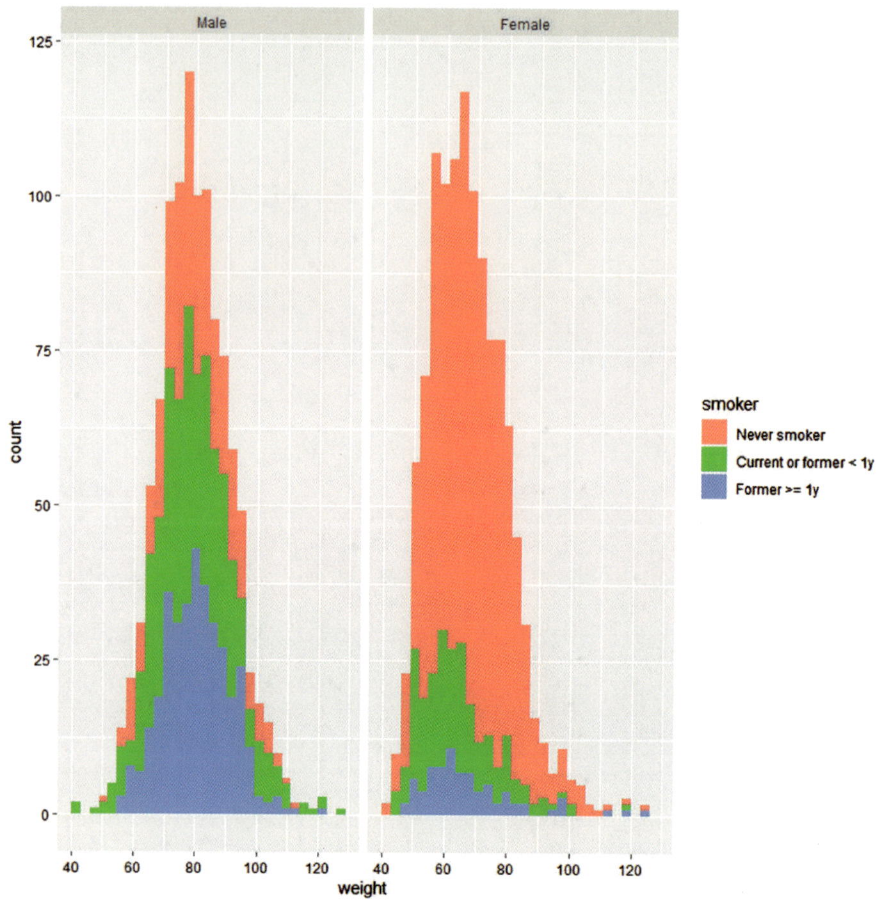

For females, never smokers weigh more than current or former smokers. On the other hand, the weight for the three smoker groups in males is nearly the same (same peak position for the three histograms).

2.7.3. Side by Side Density Plots

In regicor data, compare the age variable for females and males.

```
ggplot(data = regicor, aes(x = age, col = sex))+geom_density()
```

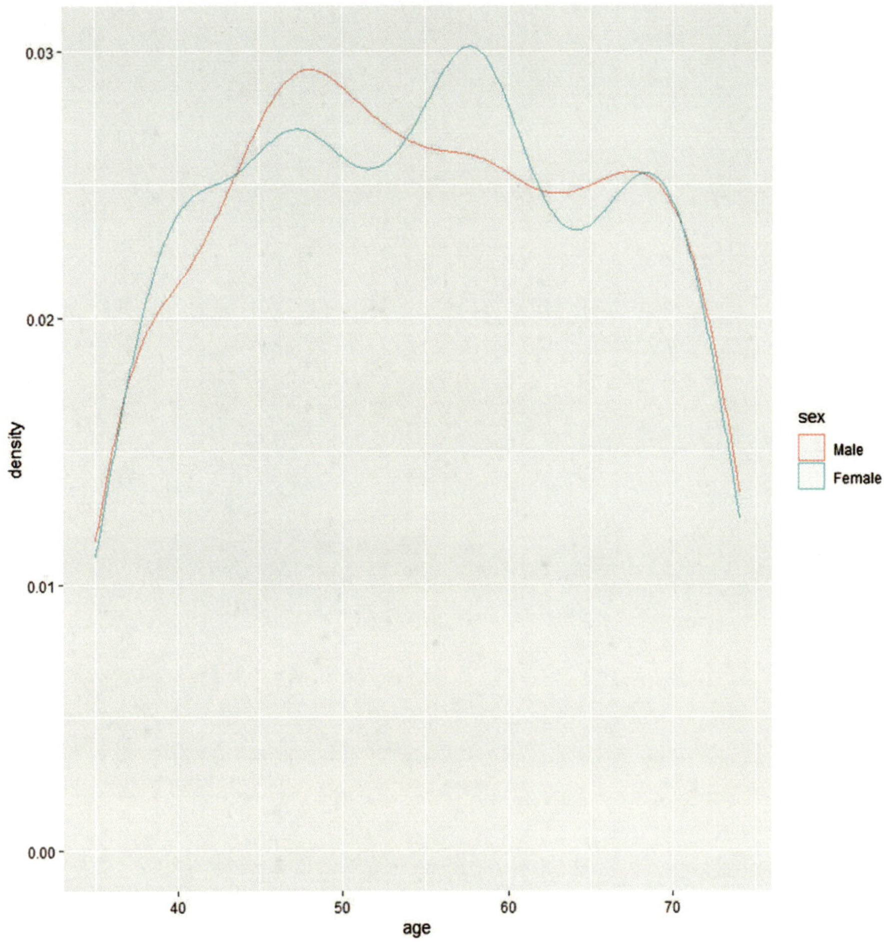

We see that females have a peak at age = 60 years while for males they have a peak at age = 50 years.

In regicor data, compare the weight variable for females and males.

```
ggplot(data = regicor, aes(x = weight, col = sex))+geom_density()
```

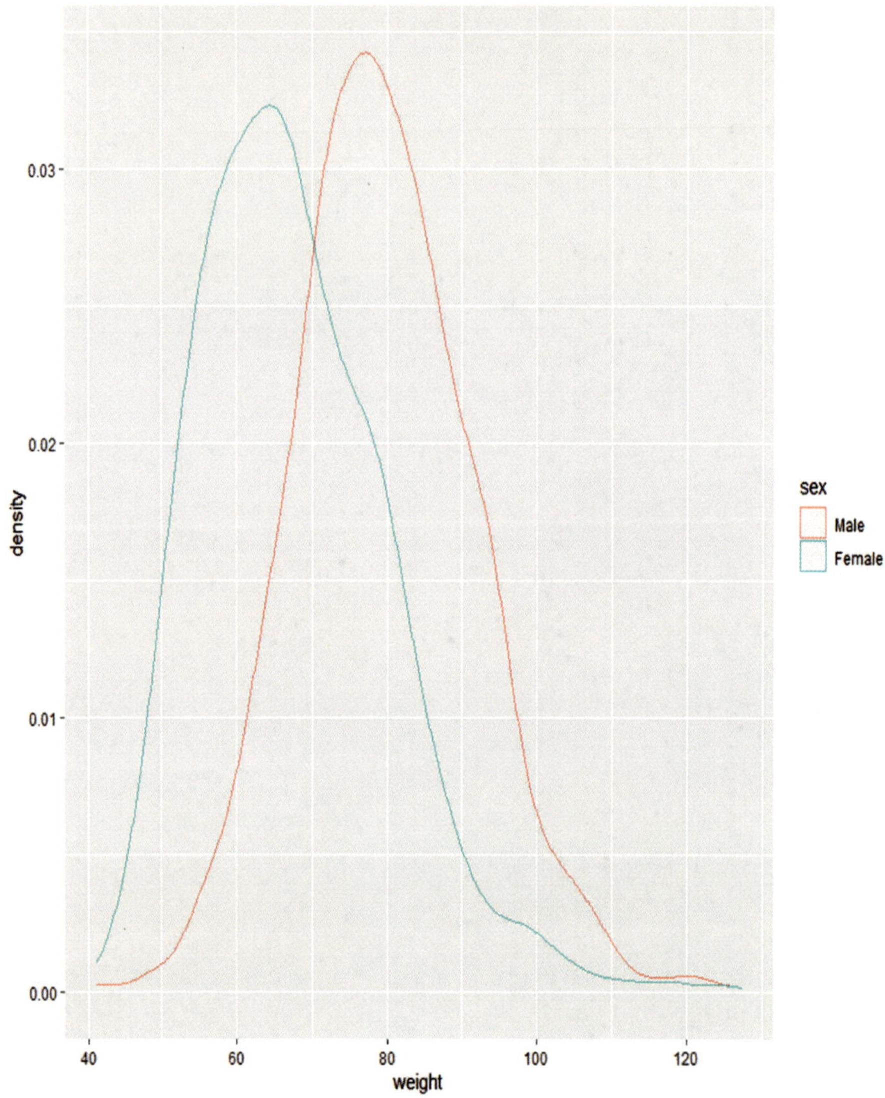

We see that the weight of females is shifted down compared to males.

In regicor data, compare the mean days to cardiovascular event for different smoker groups.

```
ggplot(data = regicor, aes(x = tocv, col = smoker))+geom_density()
```

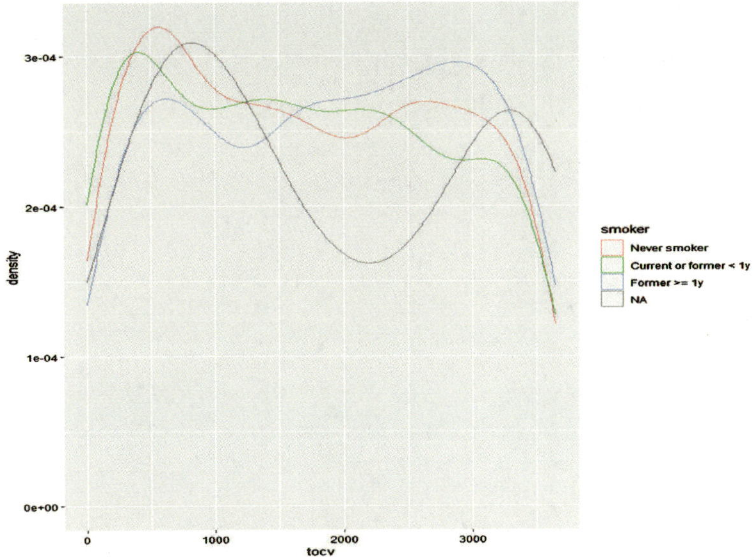


```
# to remove NA from smoker column
dat<- regicor %>% filter(!is.na(smoker))
ggplot(data = dat, aes(x = tocv, col = smoker))+geom_density()
```

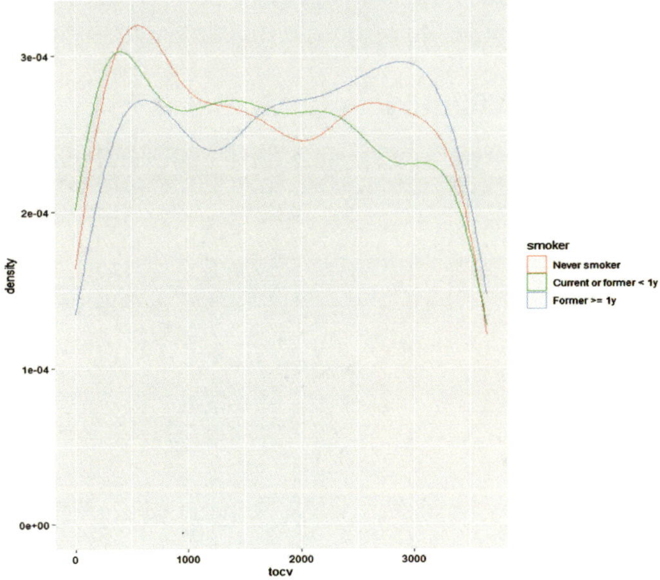

The days to cardiovascular event have a peak at 3000 days for never smokers, while the other two smoker groups have a peak at about 500 days.

In regicor data, compare the mean weight for different smoker and sex groups.

```
dat<- regicor %>% filter(!is.na(smoker))
ggplot(data = dat, aes(x = weight, col = smoker))+
 geom_density()+facet_grid(~sex)
```

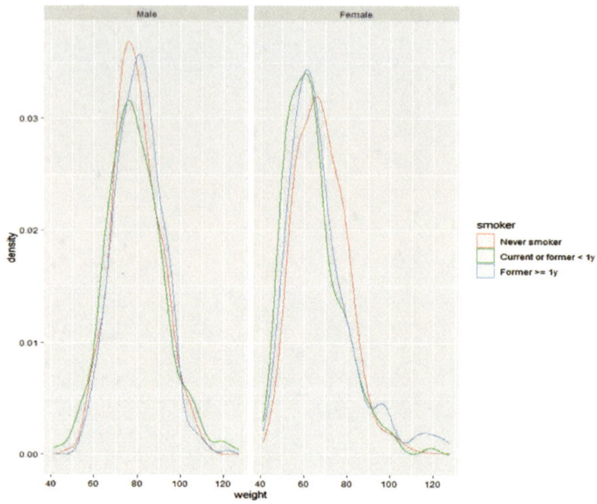

For females, never smokers weigh slightly more than current or former smokers. On the other hand, the weight for the three smoker groups in males is nearly the same (same peak position for the three density plots).

2.7.4. Side by Side Violin Plots

In regicor data, compare the age variable for females and males.

```
ggplot(data = regicor, aes(x = sex, y = age, fill = sex))+geom_violin()
```

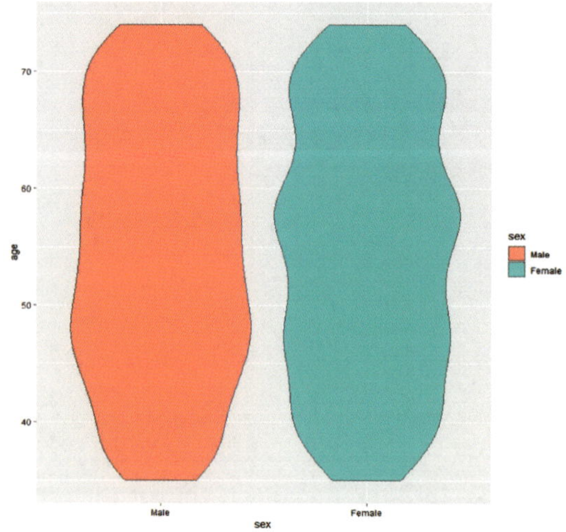

We see that the age variable is slightly shifted up for females, although females have lower mean age. The peak for female violin plot is at older age than males.

In regicor data, compare the weight variable for females and males.

```
ggplot(data = regicor, aes(x = sex, y = weight, fill = sex))+geom_violin()
```

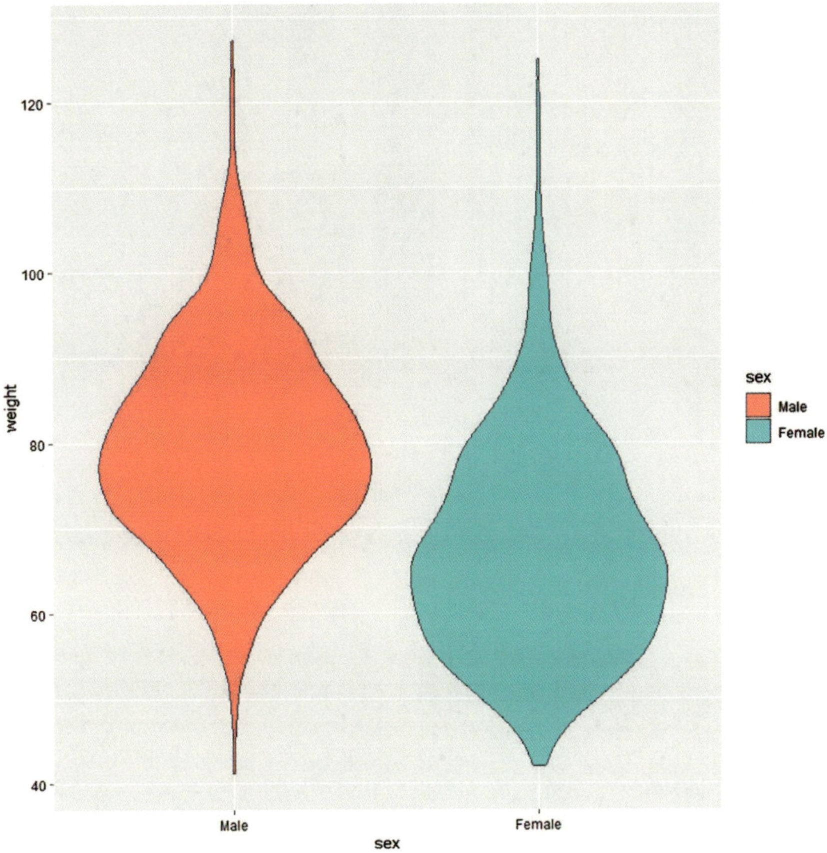

We see that the weight of females is shifted down compared to males. The peak for male violin plot is at higher weight than females.

In regicor data, compare the mean days to cardiovascular event for different smoker groups.

Assume the never smoker as our reference group.

```
ggplot(data = regicor, aes(x = smoker, y = tocv, fill = smoker))+geom_
violin()
```

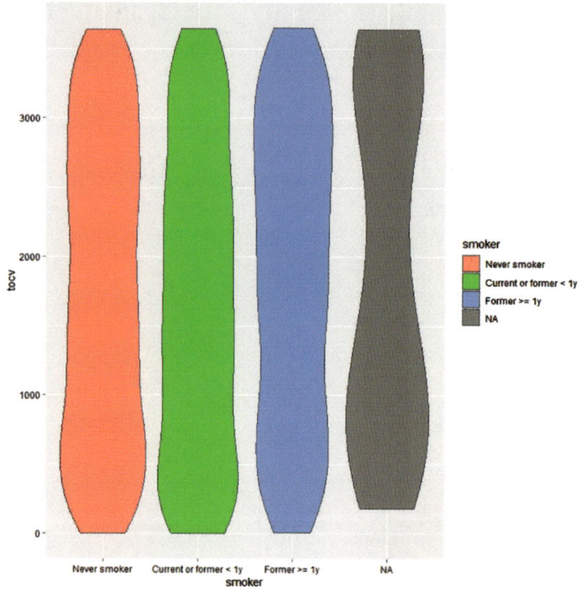

 # to remove NA from smoker column

```
dat<- regicor %>% filter(!is.na(smoker))
ggplot(data = dat, aes(x = smoker, y = tocv, fill = smoker))+geom_violin()
```

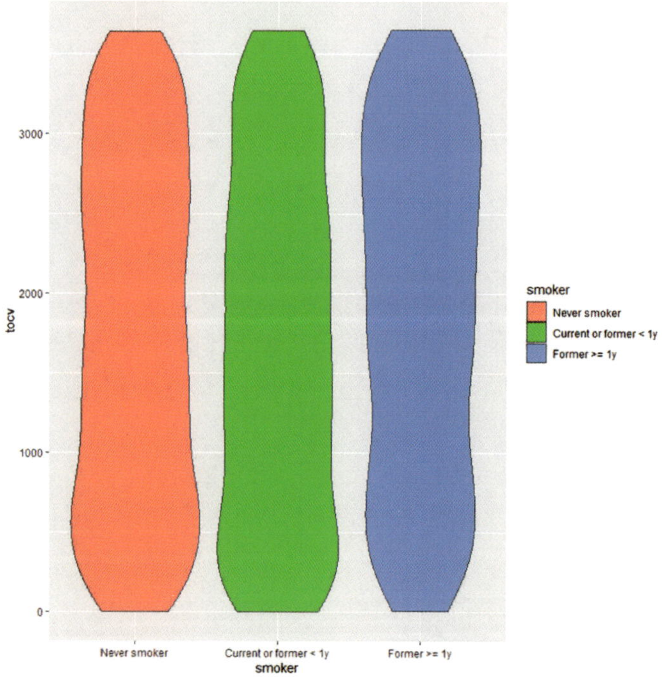

The days to cardiovascular event is longer for former smokers than other two smoker groups. The reference group is put as the first violin plot so the other groups can be compared to it.

In regicor data, compare the mean weight for different smoker and sex groups.

Assume that female never smokers are our reference group.

```
dat<- regicor %>% filter(!is.na(smoker))
ggplot(data = dat, aes(x = smoker, y = weight, fill = sex))+geom_violin()
```

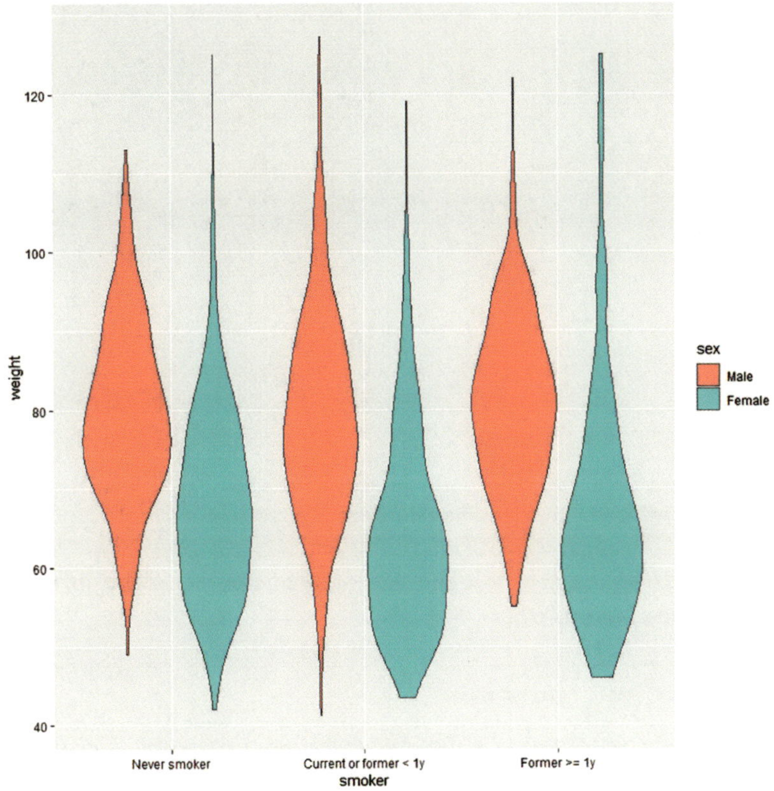


```
# to put female never smoker as first violin plot
dat$sex<-factor(dat$sex, levels = c("Female," "Male"))
ggplot(data = dat, aes(x = smoker, y = weight, fill = sex))+geom_violin()
```

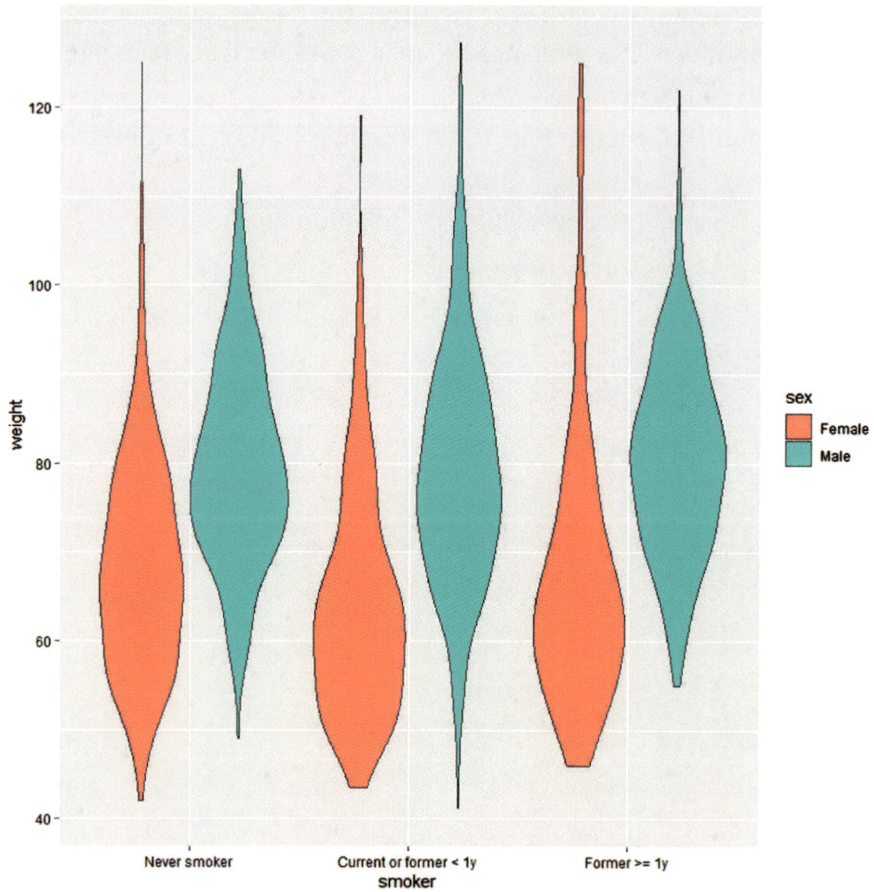

The weight of females is generally shifted down for the three smoker groups compared to males.

2.7.5. Scatter and Bar Plots

If you have one continuous value for each category, you can use geom_point() to draw scatterplot or geom_bar (with stat = "identity," position = "dodge") to draw a bar plot.

For example, the mean weight of participants in each year (1995, 2000, 2005).

```
dat<-regicor %>% group_by(year) %>% summarise(mean_wt = mean(weight, na.rm = TRUE))
dat
## # A tibble: 3 x 2
```

```
## year mean_wt
## <fct> <dbl>
## 1 1995 72.3
## 2 2000 73.8
## 3 2005 73.6
# plot the data
ggplot(data = dat, aes(x = year, y = mean_wt))+geom_point()
```

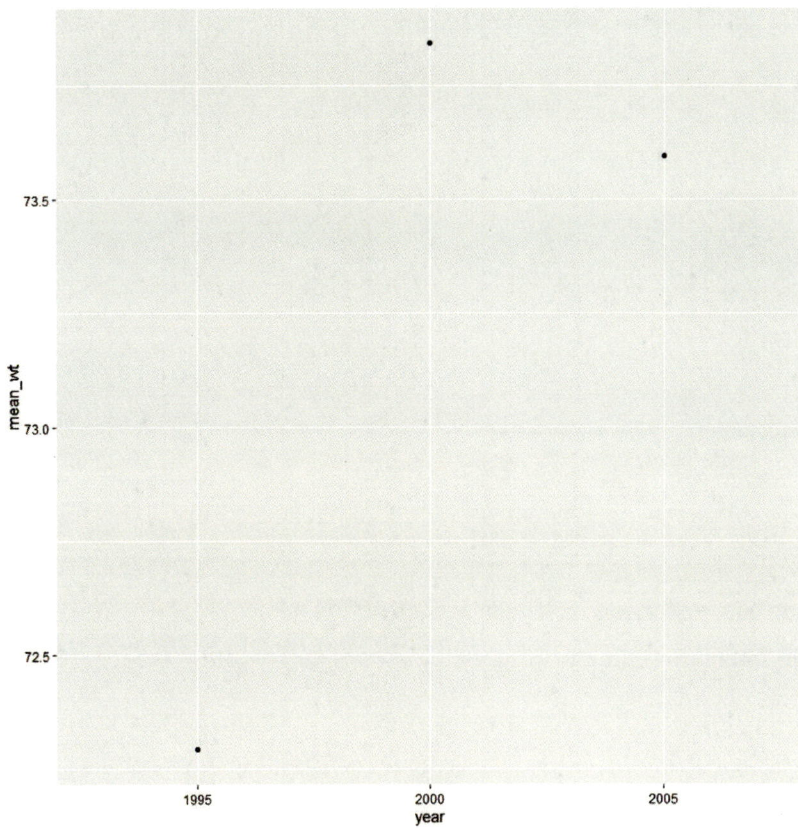


```
# Add lines and group = 1 to indicate that all points are from 1 group
ggplot(data = dat, aes(x = year, y = mean_wt))+
 geom_point()+geom_line(group=1)
```

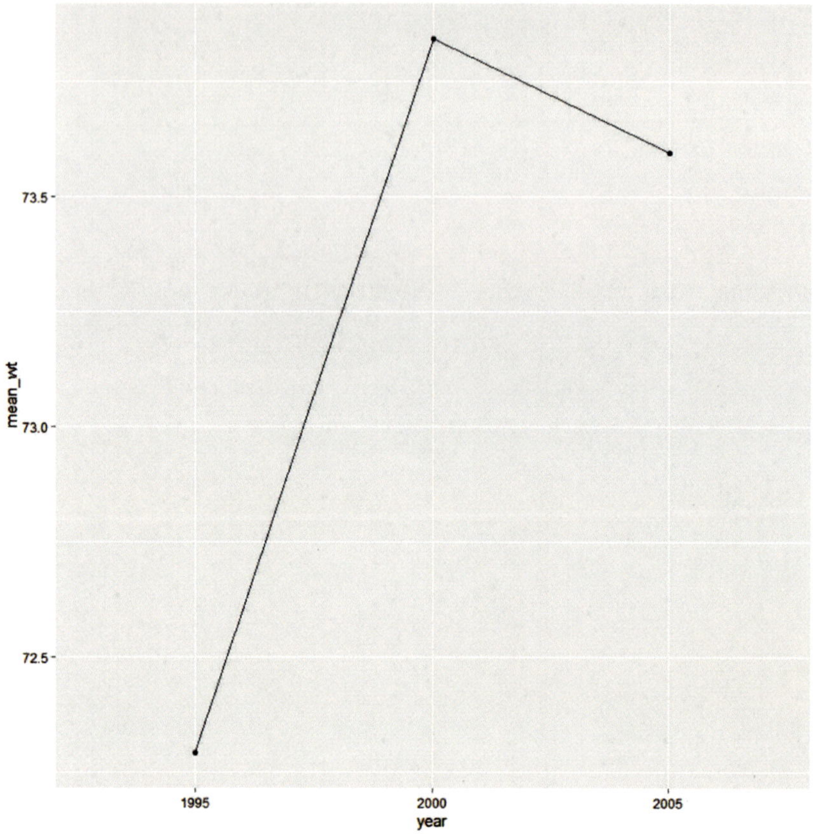


```
# plot bar graphs
ggplot(data = dat, aes(x = year, y = mean_wt))+
geom_bar(stat = "identity," position = "dodge," width = 0.5)
```

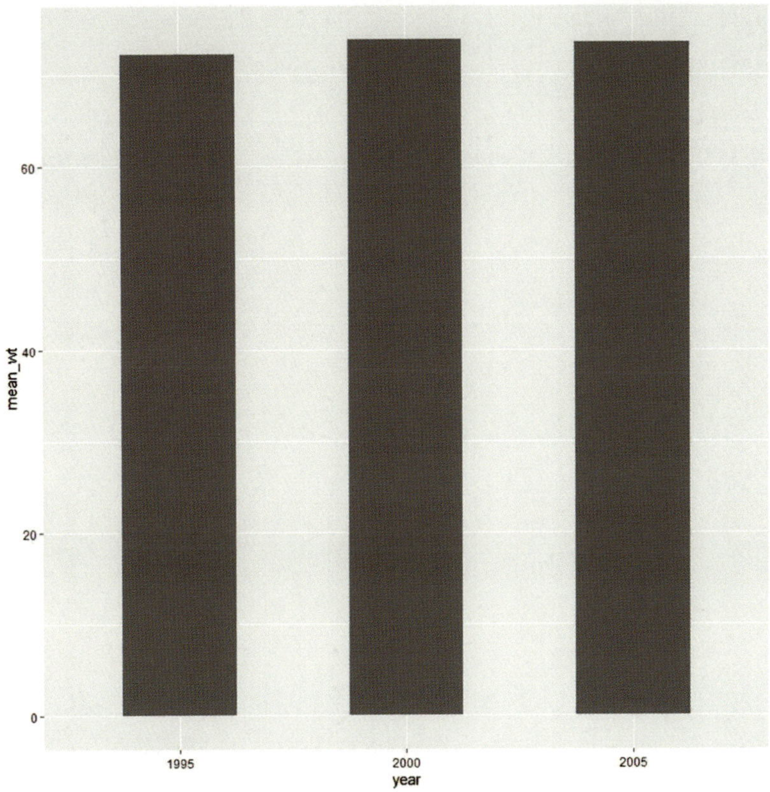

If you have another categorical variable as sex, you can add another aesthetic, group = sex, then add geom_line() to connect points of the same group.

```
dat<-regicor %>% group_by(year, sex) %>% summarise(mean_wt = mean(weight, na.rm = TRUE))
dat
## # A tibble: 6 x 3
## # Groups:  year [3]
##   year sex  mean_wt
##   <fct> <fct>  <dbl>
## 1 1995 Male   77.6
## 2 1995 Female  67.3
## 3 2000 Male   80.1
## 4 2000 Female  67.6
```

```
## 5 2005 Male   80.2
## 6 2005 Female  67.7
# plot the data
ggplot(data = dat, aes(x = year, y = mean_wt, color = sex))+geom_point()
```

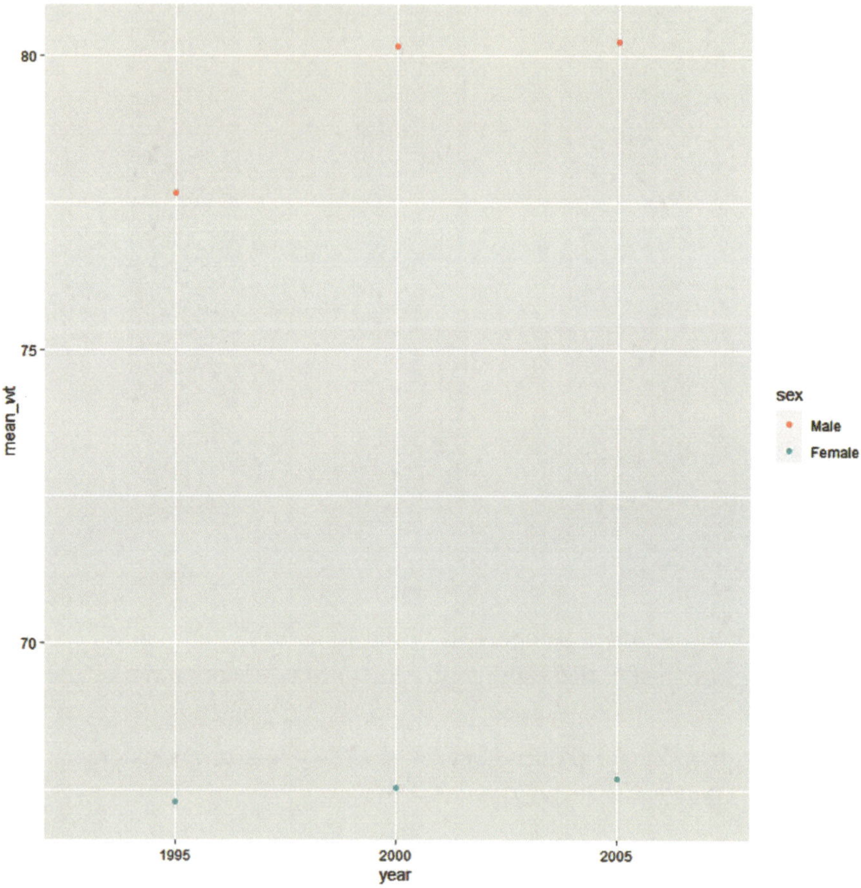


```
# Add lines
ggplot(data = dat, aes(x = year, y = mean_wt, color = sex, group = sex))+
 geom_point()+geom_line()
```

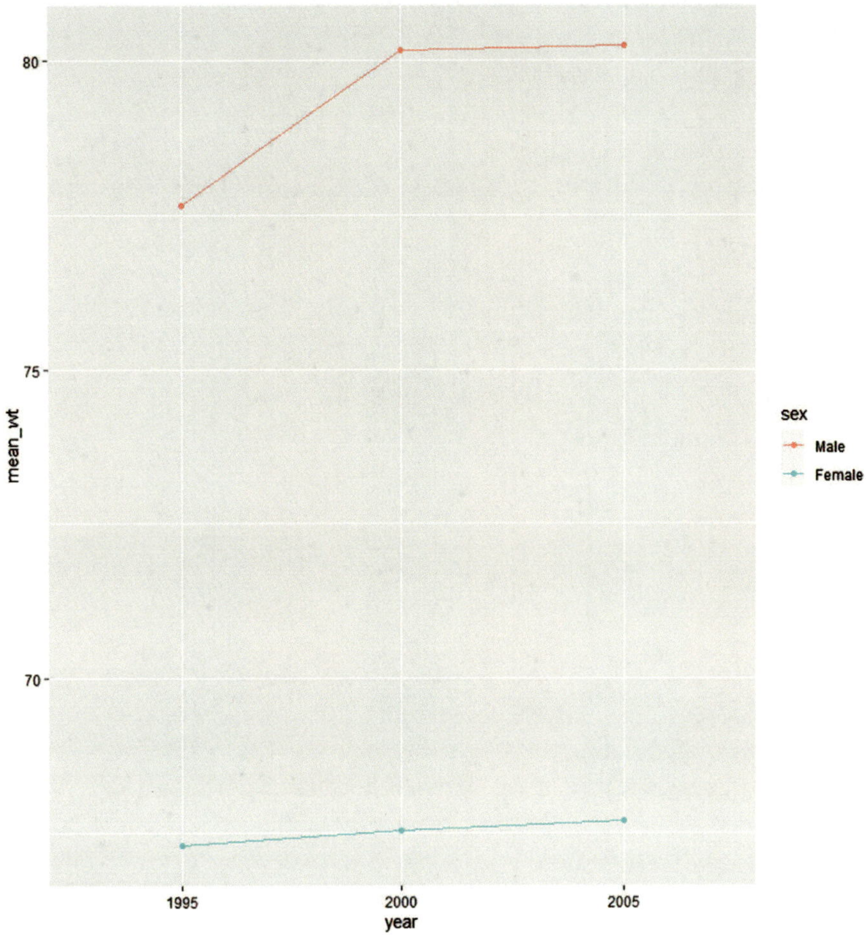


```
# plot bar graphs
ggplot(data = dat, aes(x = year, y = mean_wt, fill = sex))+
geom_bar(stat = "identity," position = "dodge")
```

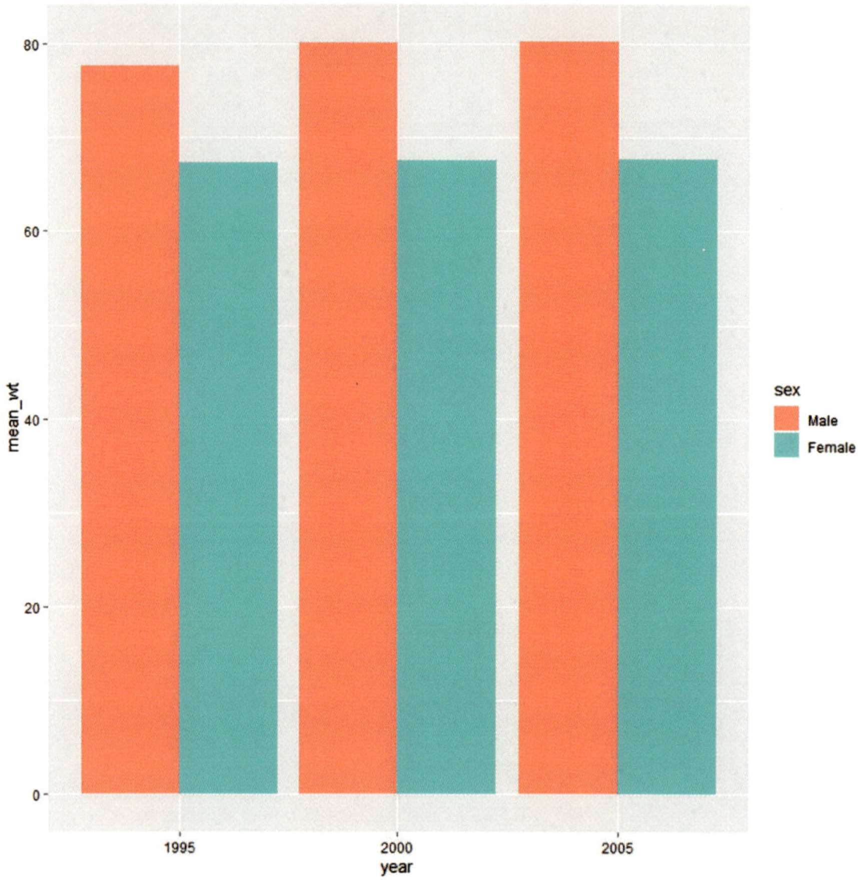

We see that in all years, the mean weight of males is larger than that of females.

CHAPTER 3

THE NORMAL DISTRIBUTION

CONTENTS

3.1. INTRODUCTION TO NORMAL DISTRIBUTION

Normal distribution is a theoretical probability distribution that is perfectly symmetric around its mean and has a bell-shape. The normal distribution is also called the Bell-curve or the Gaussian distribution after the name of the German mathematician, Carl Friedrich Gauss, who discovered it. The face of Carl Friedrich Gauss and the shape of the normal distribution was on the old German Mark currency.

3.1.1. Characters of the Normal Distribution

1. Bell-shaped distribution and symmetric around its mean.
2. The mean = median = mode (most frequent data value).
3. Values closer to the mean are more frequent than values far from the mean.
4. The limits of the normal distribution is from negative infinity to positive infinity.
5. Any normal distribution is completely defined by its mean () and standard deviation ().

3.1.2. Examining the Shape of Normal Distribution

We will use the dnorm function that generates a normal distribution for a given mean and standard deviation.

Example: Plotting a normal distribution curve with mean = 3 and sd (standard deviation) = 1.

```
#Activating the necessary packages
library(tidyverse)
#Note that the tidyverse packages include ggplot2
#Create a dataframe for a reasonable range of values plotted on the x-axis

df<-data.frame(x = c(–10,10))
# plotting the data
ggplot(data = df,aes(x = x))+
 stat_function(fun = dnorm, args = list(3,1),color=”red”)+
 scale_x_continuous(name = “values,” breaks = seq(–10,10,1))+
 ylab(“probability”)
```

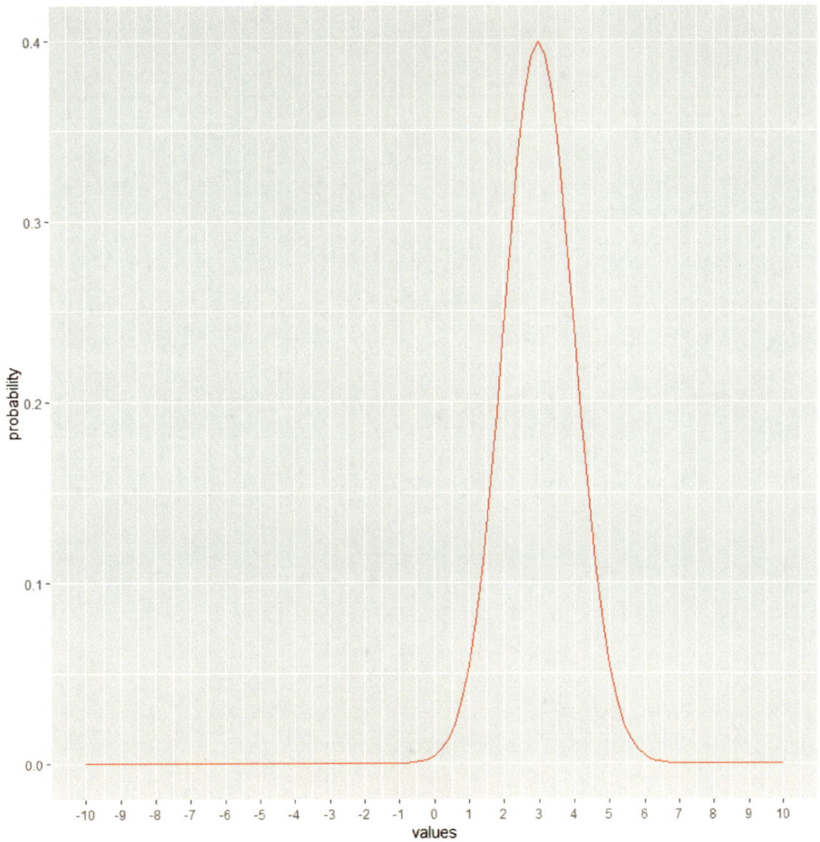

stat_function is used to compute a normal density function for every x value. args = list(3,1) means that the calculated normal density function should have a mean = 3 and sd = 1. scale_x_continuous to define the name of x-axis and its values and ylab to define a label for y-axis.

We see that the distribution has a peak at x = mean = 3 and tails dies off quickly. We notice that for values greater or less than 3 sd from the mean (i.e., values > 6 or values < 0) have a probability of nearly zero. The total Area Under the Curve (AUC) is = 1.

Example: Plotting different normal distributions with different means, 3, 5, –3 and the same sd = 1.

```
ggplot(data = df,aes(x = x))+
 stat_function(fun = dnorm, args = list(3,1),color="red")+
 stat_function(fun = dnorm,args = list(5,1),color="black")+
 stat_function(fun = dnorm,args = list(–3,1),color="blue")+
```

```
scale_x_continuous(name = "values," breaks = seq(–10,10,1))+
ylab("probability")
```

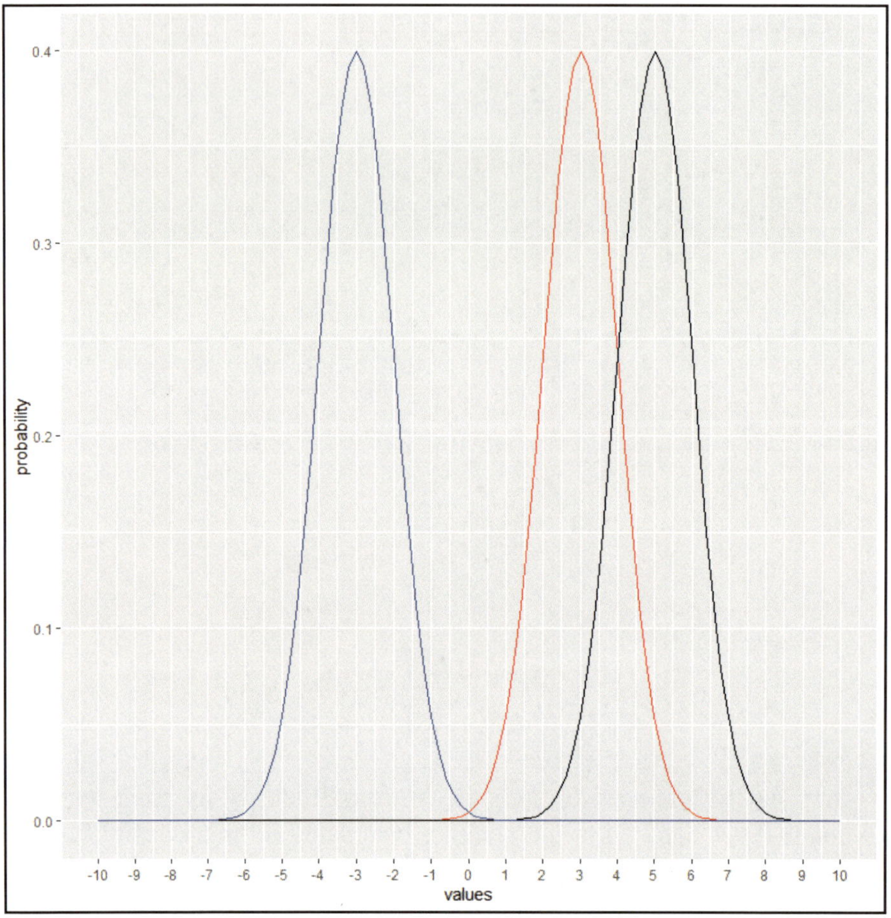

The normal distribution curve with a mean = 3 has a red color, a mean = 5 has a black color, and a mean = –3 has a blue color. Note how the location of each distribution has changed according to its mean.

Example: Plotting different normal distributions with different sd, 1, 0.5, 0.25 and the same mean = 3.

```
ggplot(data = df,aes(x = x))+
stat_function(fun = dnorm, args = list(3,1),color="red")+
stat_function(fun = dnorm,args = list(3,0.5),color="black")+
stat_function(fun = dnorm,args = list(3,0.25),color="blue")+
scale_x_continuous(name = "values," breaks = seq(–10,10,1))+
```

```
ylab("probability")
```

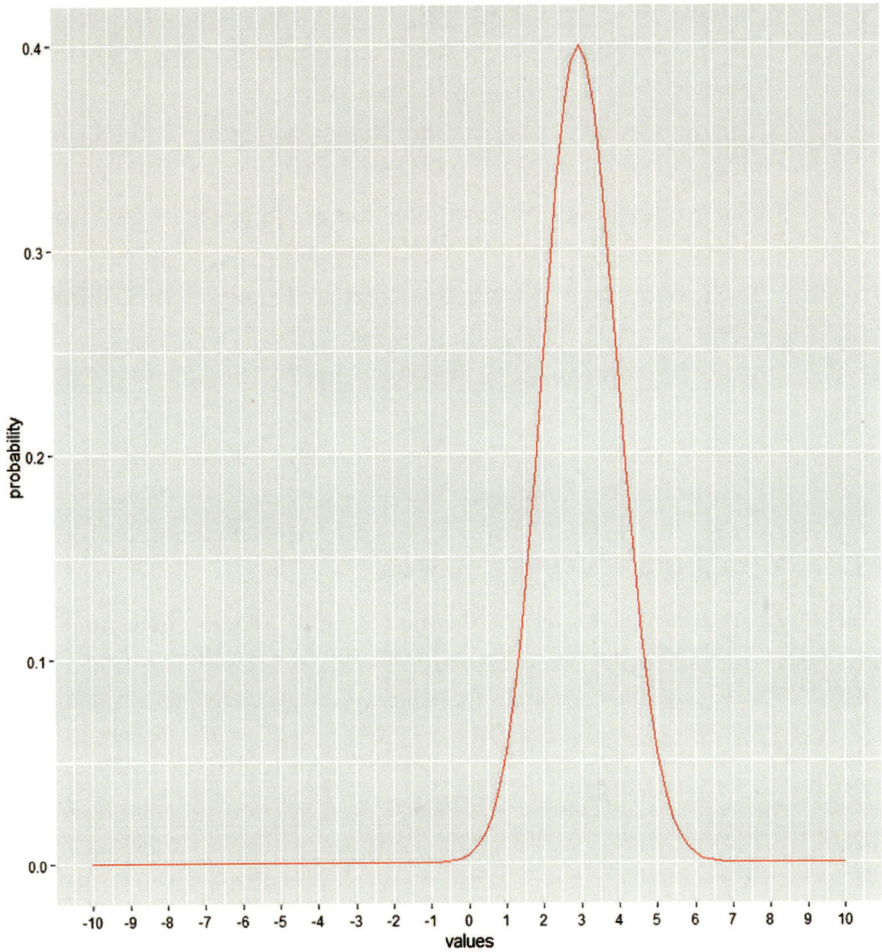

The normal distribution curve with a sd = 1 has a red color, a sd = 0.5 has a black color, and a sd = 0.25 has a blue color. Note that by decreasing the sd, the normal distribution curve becomes more thinner but the peaks of all curves are at the same position at 3 which is the mean value.

3.2. THE 68-95-99.7% RULE

3.2.1. Explaining the Rule

For any normal distribution, 68% of the data are within 1 sd from the mean, 95% of the data are within 2 sd (or more accurately 1.96) from the mean, and 99.7% of the data are within 3 sd from the mean.

As the normal distribution is symmetric around its mean and AUC =1. So, different conclusions can be extracted from that:

1. The proportion of observations that are larger than 1 sd from the mean = 1–0.68 = 0.32/2 = 0.16 or 16%.
2. The proportion of observations that are smaller than 1 sd from the mean = 1–0.68 = 0.32/2 = 0.16 or 16%.
3. The proportion of observations that are larger than 2 sd from the mean = 1–0.95= 0.05/2 = 0.025 or 2.5%.
4. The proportion of observations that are smaller than 2 sd from the mean = 1–0.95 = 0.05/2 = 0.025 or 2.5%.
5. The 2.5% = mean–2*sd. It is the value by which 2.5% of the data values are less than.
6. The 97.5% = mean+2*sd. It is the value by which 97.5% of the data values are less than.

3.2.2. Plotting the 68-95-99.7% Rule

Plotting the normal distribution with mean = 3 and sd = 1. Then shading the area within 1 sd from the mean that will occupy 68% of the total AUC from 2 to 4.

```
# creating the shading function
funcShaded <- function(x) {
  y <- dnorm(x, mean = 3, sd = 1)
  y[x < 2 | x > 4] <- NA
  return(y)}
# plotting the curve and the shaded area
ggplot(data = df,aes(x = x))+
 stat_function(fun = dnorm, args = list(3,1),color=”red”)+
 stat_function(fun= funcShaded, geom =”area,”fill=”green”) + scale_x_co
ntinuous(breaks=seq(–10,10,1))+ ylab(“probability”)
```

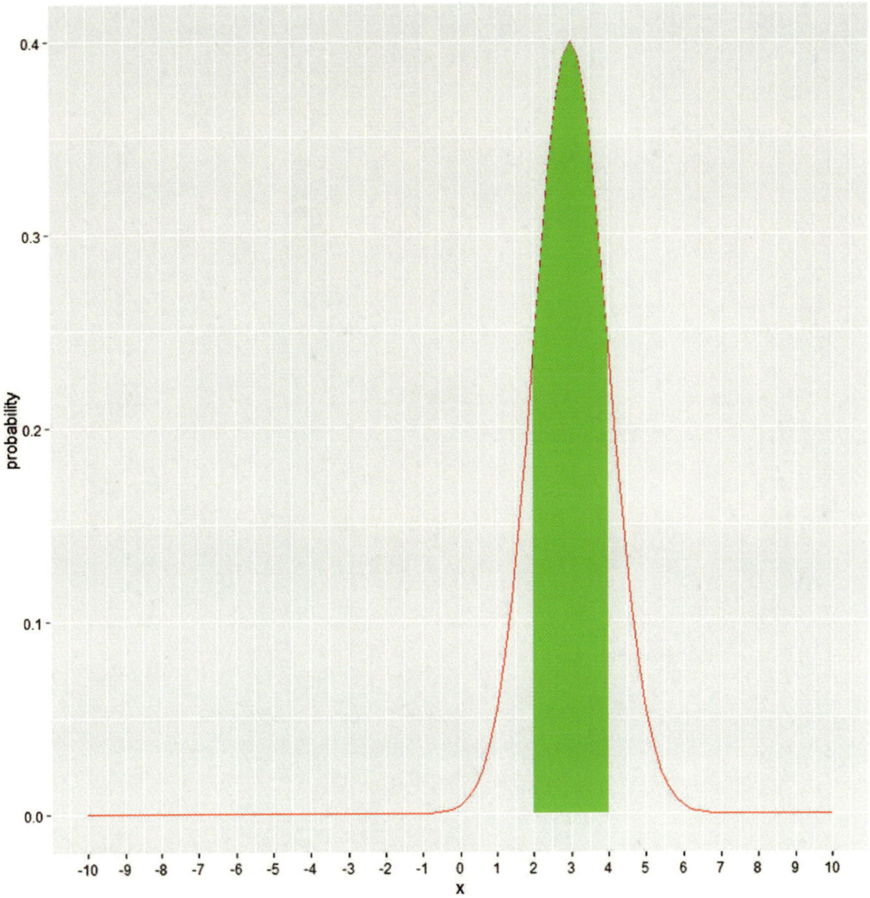

In ggplot, the plot is drawn in layers. We first plot the curve in red color then plot the shaded area with funcShaded function in green color.

Plotting the normal distribution with mean = 3 and sd = 1. Then shading the area within 2 sd from the mean that will occupy 95% of the total AUC from 1 to 5.

```
funcShaded2 <- function(x) {
  y <- dnorm(x, mean = 3, sd = 1)
  y[x < 1 | x > 5] <- NA

  return(y)}
ggplot(data = df,aes(x = x))+
 stat_function(fun = dnorm, args = list(3,1),color="red")+
```

```
stat_function(fun= funcShaded2, geom = "area,"fill="green")+ scale_x_continuous(breaks=seq(–10,10,1))+ ylab("probability")
```

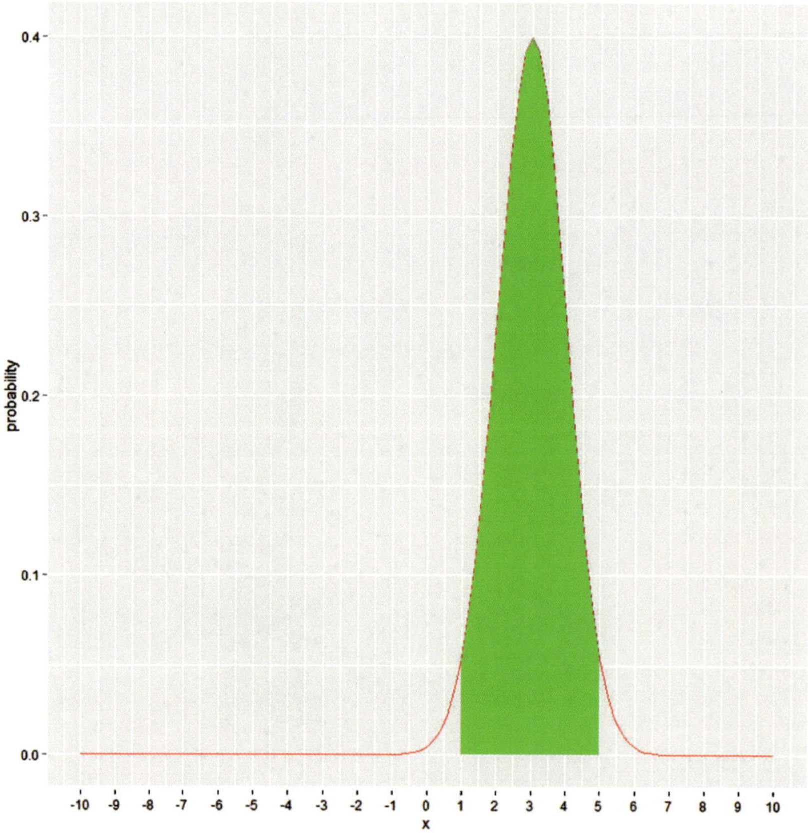

Plotting the normal distribution with mean = 3 and sd = 1. Then shading the area within 3 sd from the mean that will occupy 99.7 % of the total AUC from 0 to 6.

```
funcShaded3 <- function(x) {
  y <- dnorm(x, mean = 3, sd = 1)
  y[x < 0 | x > 6] <- NA
  return(y)}
ggplot(data = df,aes(x = x))+
 stat_function(fun = dnorm, args = list(3,1),color="red")+
 stat_function(fun= funcShaded3, geom = "area,"fill="green")+
 scale_x_continuous(breaks=seq(–10,10,1))+ylab("probability")
```

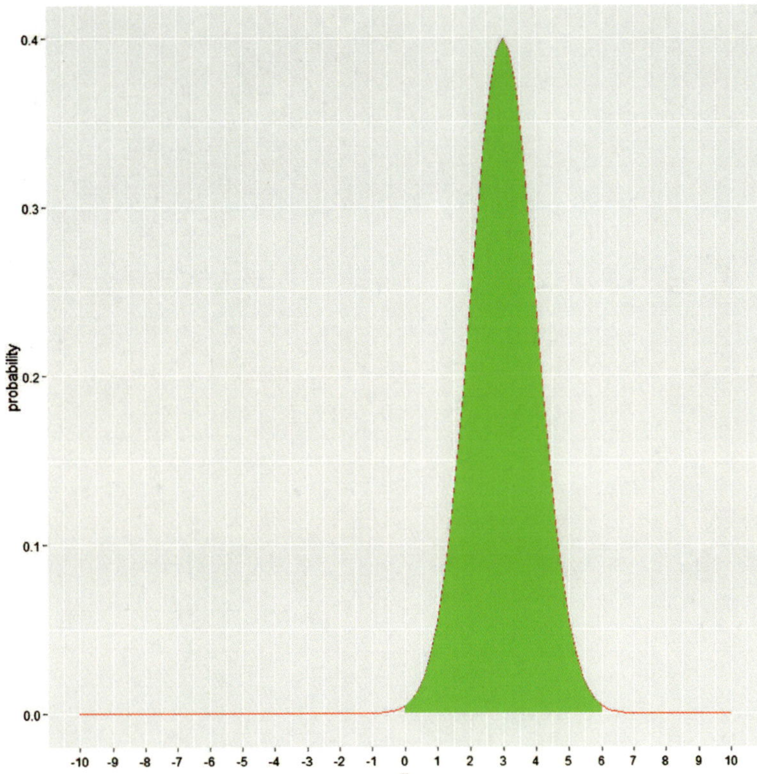

Plotting the normal distribution with mean = 3 and sd = 1. Then shading the area from the mean to 2 sd from the mean that will occupy 47.5 % of the total AUC from 3 to 5.

```
funcShaded4 <- function(x) {
  y <- dnorm(x, mean = 3, sd = 1)
  y[x < 3 | x > 5] <- NA
  return(y)}
ggplot(data = df,aes(x = x))+
 stat_function(fun = dnorm, args = list(3,1),color="red")+
 stat_function(fun= funcShaded4, geom = "area,"fill="green")+
 scale_x_continuous(breaks=seq(–10,10,1))+ ylab("probability")
```

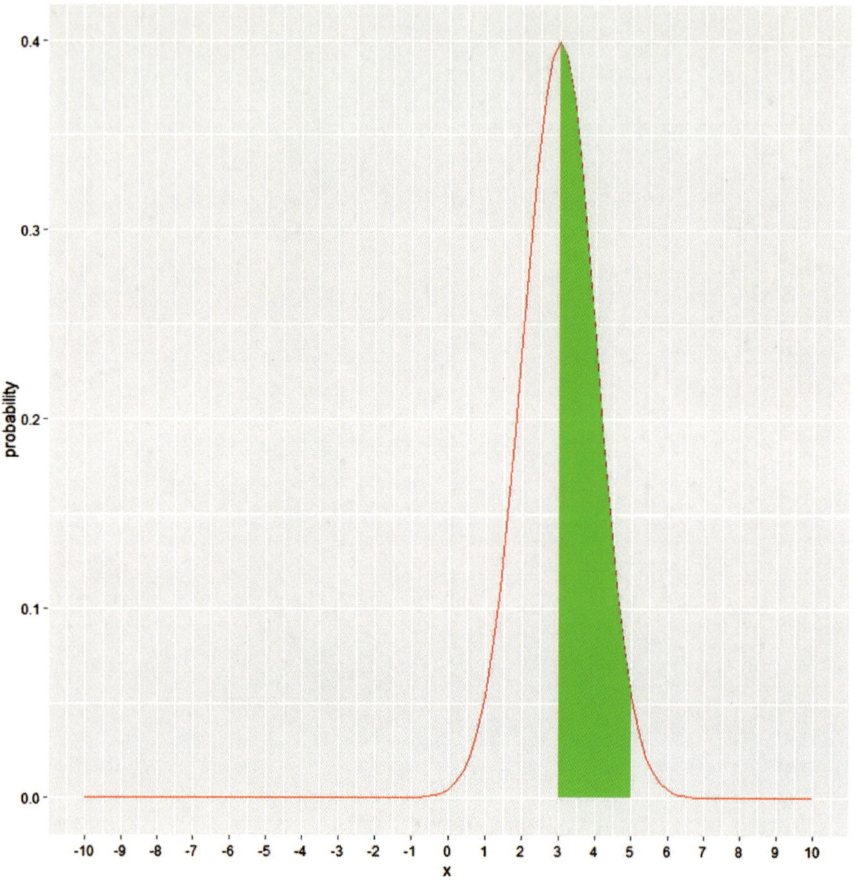

Plotting the normal distribution with mean = 3 and sd = 1. Then shading the area larger than 2 sd above the mean that will occupy 2.5 % of the total AUC or larger than 5 which is the 97.5%.

```
funcShaded5 <- function(x) {
  y <- dnorm(x, mean = 3, sd = 1)
  y[x < 5 ] <- NA
  return(y)}
ggplot(data = df,aes(x = x))+
 stat_function(fun = dnorm, args = list(3,1),color="red")+
 stat_function(fun= funcShaded5, geom = "area,"fill="green")+
 scale_x_continuous(breaks=seq(-10,10,1))+ylab("probability")
```

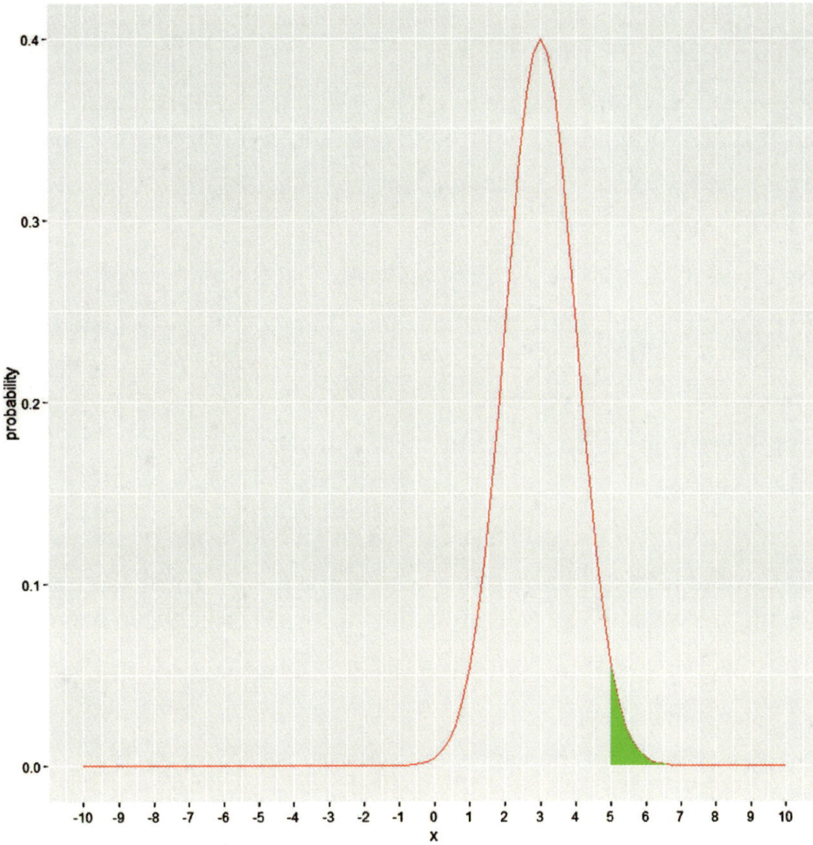

3.2.3. Examining the 68-95-99.7% Rule with Built-in R Functions

For a normal distribution with mean = 3 and sd = 1, calculate the % of values that are less than 1 sd (=1) above the mean(=3) or less than 4 with pnorm function (p for probability)

```
pnorm(4,mean = 3,sd = 1)
## [1] 0.8413447
```

So 84% of the values are less than 4 and 4 is the 84%.

For a normal distribution with mean = 3 and sd = 1, calculate the % of values that are less than 1.96 sd (=1.96) above the mean(=3) or less than 4.96 with pnorm function (p for probability).

```
pnorm(4.96,mean = 3,sd = 1)
## [1] 0.9750021
```

So 97.5% of the values are less than 4.96 and 2.5% of the values are larger than 4.96. 4.96 is the 97.5%.

For a normal distribution with mean = 3 and sd = 1, calculate the % of values that are within 1 sd (=1) from the mean(=3) or between 4 and 2.

```
pnorm(4,mean = 3,sd = 1)-pnorm(2,mean = 3,sd = 1)
## [1] 0.6826895
```

So about 68% of the values are between 1 sd from the mean

For a normal distribution with mean = 3 and sd = 1, calculate the % of values that are within 1.96 sd (=1.96) from the mean(=3) or between 4.96 and 1.04

```
pnorm(4.96,mean = 3,sd = 1)-pnorm(1.04,mean = 3,sd = 1)
## [1] 0.9500042
```

So about 95% of the values are between 1.96 sd from the mean.

For a normal distribution with mean = 3 and sd = 1, calculate the % of values that are within 3 sd (=3) from the mean(=3) or between 0 and 6

```
pnorm(6,mean = 3,sd = 1)-pnorm(0,mean = 3,sd = 1)
## [1] 0.9973002
```

So about 99.7 % of the values are between 3 sd from the mean.

To calculate the percentile for any normal distribution, use the qnorm function

```
qnorm(0.95,mean = 3,sd = 1)
## [1] 4.644854
#4.64 is 95%
qnorm(0.975,mean = 3,sd = 1)
## [1] 4.959964
#4.96 is 97.5%
qnorm(0.2,mean = 3,sd = 1)
## [1] 2.158379
#2.16 is 20%
```

From the results, we see that 4.64 is 95%, 4.96 is 97.5%, and 2.16 is 20%.

3.3. APPLYING NORMAL DISTRIBUTION TO SAMPLE DATA

No real data is perfectly described by the normal distribution because for any normal distribution, the range of the data goes from negative infinity to positive infinity. However, the distribution of some data can be approximated by the normal distribution if they are symmetric and bell-shaped. In this case, the properties of the normal distribution (as the 68-95-99.7% rule) can be used to characterize the aspects of the underlying population data distribution if our sample were randomly selected or representative of these population(s).

3.3.1. Applying the Normal Distribution to Regicor Data

The regicor dataframe, from the package compareGroups, contains data from three different cross-sectional surveys of individuals representative of the population from a north-west Spanish province (Girona), REGICOR study. Visit [www.regicor.org].

Examining the weight column,

```
# loading the package
library(compareGroups)
library(ggpubr)
data("regicor")
#Examining the weight column of this dataset
# a histogram
ggplot(data = regicor, aes(x = weight))+geom_histogram()
```

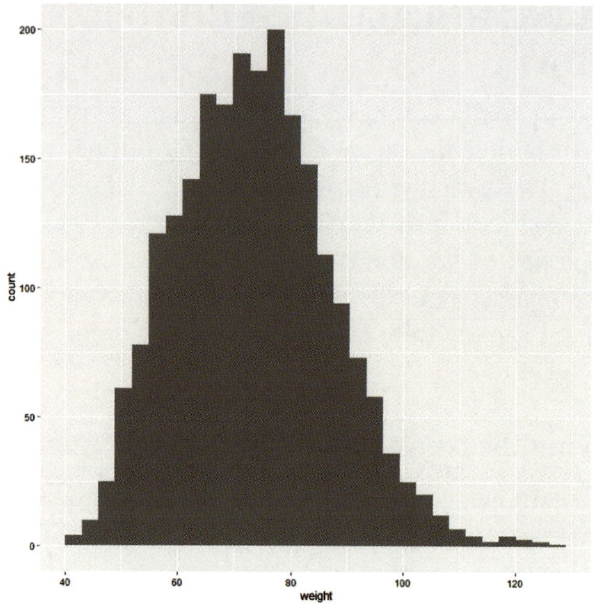

 # a boxplot

```
ggplot(data = regicor, aes(y=weight))+geom_boxplot()
```

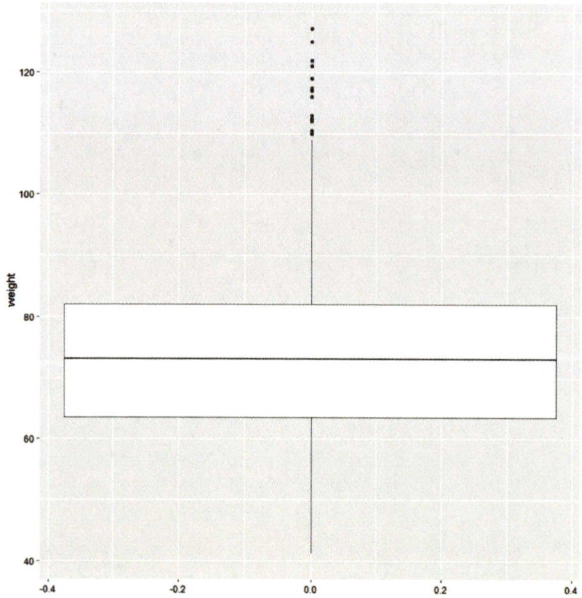

 # density plot

```
ggplot(data = regicor, aes(x = weight))+geom_density()
```

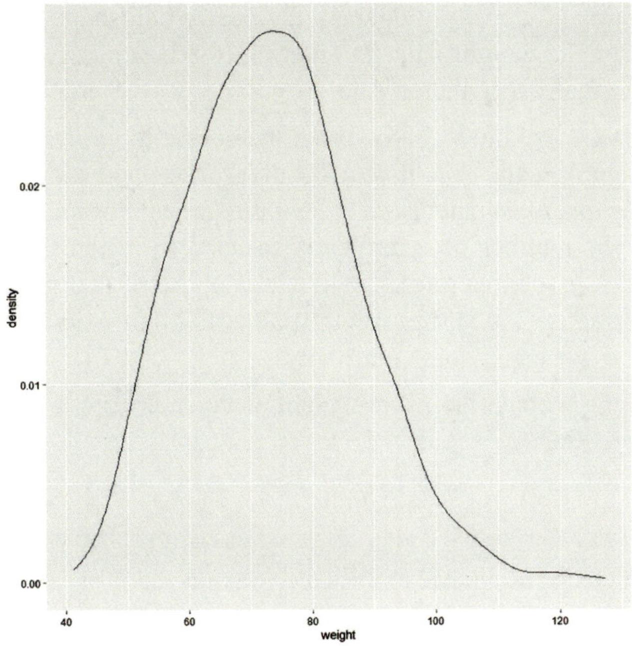

 # QQ plot

ggqqplot(data = regicor, x = "weight")

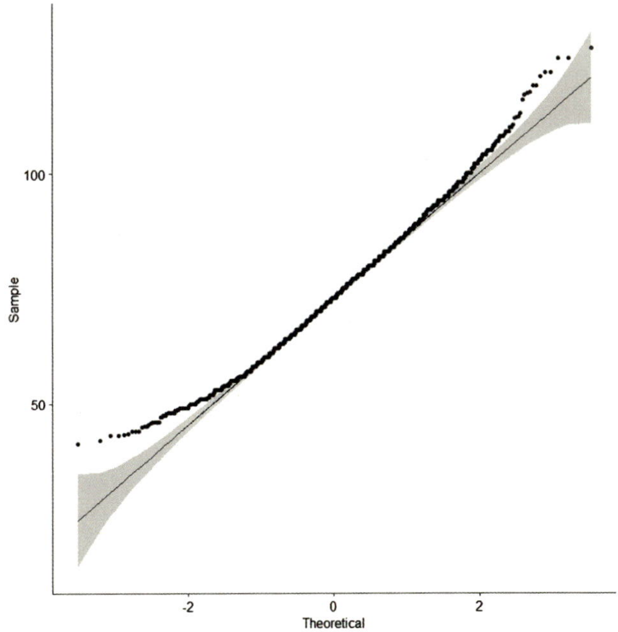

The histogram and density plot show the nearly bell shape and symmetric distribution of the weight data. In addition, the box plot has equal halves indicating normally distributed data.

QQ plots are used to visually check the normality of the data. QQ plots take your sample data, sort it in ascending order, and then plot them on y-axis versus quantiles calculated from a theoretical normal distribution on the x-axis. The number of quantiles is selected to match the size of your sample data. A 45-degree reference line is also plotted that indicates perfect fitting to normality. As all the weight points fall approximately along this reference line, we can assume normality.

Calculating percentile from weight data and from a corresponding normal distribution,

```
# summary statistics of weight column
summary(regicor$weight)
## Min. 1st Qu. Median Mean 3rd Qu. Max. NA's
## 41.20 63.50 73.00 73.44 82.00 127.20 35
sd(regicor$weight, na.rm = TRUE)
## [1] 13.67845
#2.5% according to normal distribution =mean–2*sd
mean(regicor$weight, na.rm = TRUE)–1.96*sd(regicor$weight, na.rm=
TRUE)
## [1] 46.6261
#The observed 2.5% using the quantile function
quantile(regicor$weight,0.025, na.rm = TRUE)
## 2.5%
## 50
#97.5% according to normal distribution =mean+2*sd
mean(regicor$weight, na.rm = TRUE)+1.96*sd(regicor$weight, na.rm =
TRUE)
## [1] 100.2456
#The observed 97.5% using the quantile function
quantile(regicor$weight,0.975, na.rm = TRUE)
## 97.5%
## 102
```

The calculated percentiles are nearly equal to the observed percentiles.

From the different plots and calculated and observed percentiles, it can be assumed that the weight data came from a population that is nearly normally distributed. Therefore, without looking at the data and knowing that its mean = 73.44 kg and its sd = 13.68 kg and based on the sample data, 95% of individuals from this north-west Spanish province (Girona) have weight between 46.6 and 100.2 kg and only 5% of individuals from this Spanish province have weight below 46.6 kg or more than 100.2 kg.

Examining the height column,

```
#Examining the height column of this dataset
# a histogram
ggplot(data = regicor, aes(x = height))+geom_histogram()
```

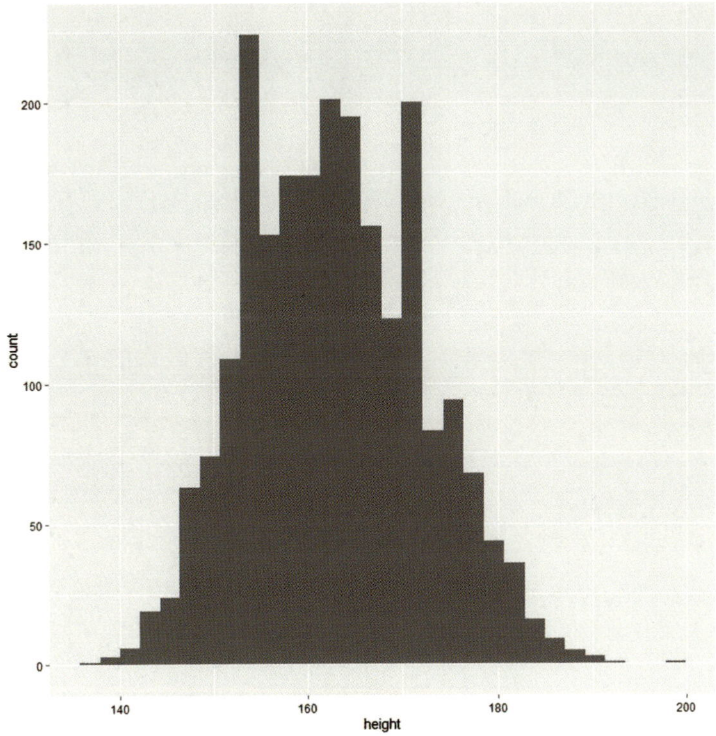


```
# a boxplot
ggplot(data = regicor, aes(y=height))+geom_boxplot()
```

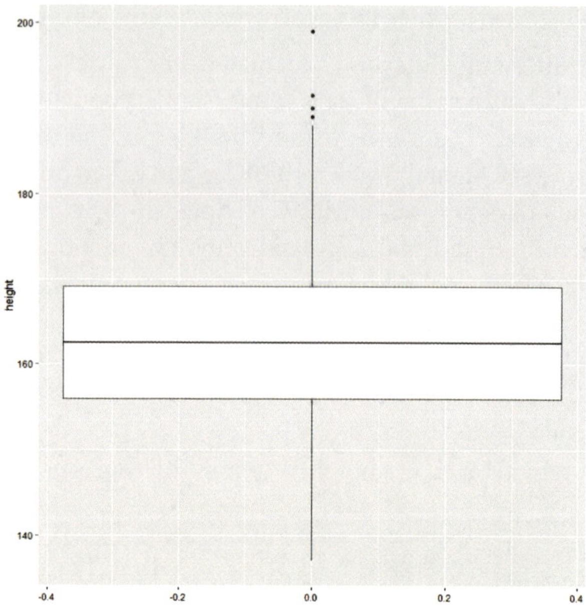


```
# density plot
ggplot(data = regicor, aes(x = height))+geom_density()
```

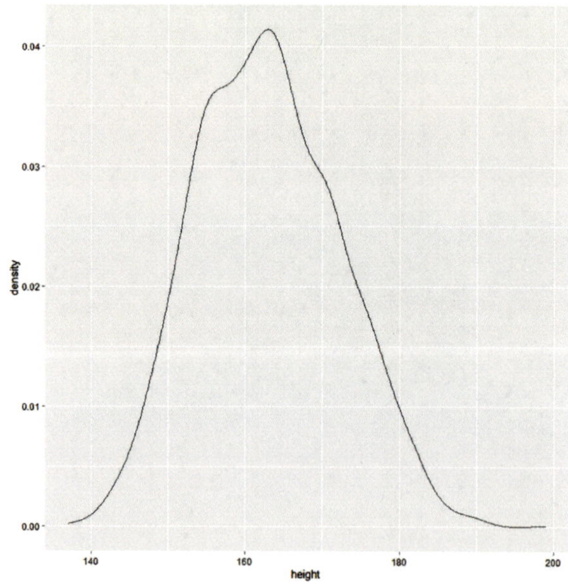


```
# QQ plot
ggqqplot(data = regicor, x = "height")
```

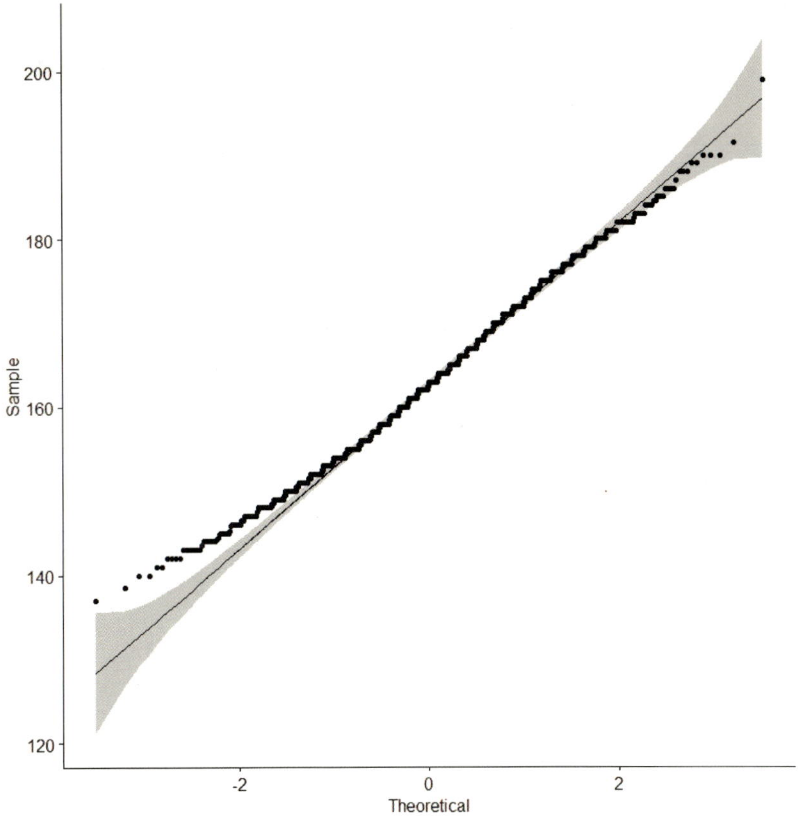

The histogram and density plot show the nearly bell shape and symmetric distribution of the weight data. In addition, the box plot has equal halves indicating normally distributed data. Furthermore, all the height points fall approximately along the reference line of QQ plot, so we can assume normality.

Calculating percentile from height data and from a corresponding normal distribution,

```
# summary statistics of height column
summary(regicor$height)
## Min. 1st Qu. Median  Mean 3rd Qu.  Max.  NA's
## 137.0  156.0  162.5  162.9  169.2  199.0   35
sd(regicor$height, na.rm = TRUE)
## [1] 9.216404
#2.5% according to normal distribution =mean–2*sd
```

```
mean(regicor$height, na.rm = TRUE)-1.96*sd(regicor$height, na.rm= TRUE)
## [1] 144.8515
#The observed 2.5% using the quantile function
quantile(regicor$height,0.025, na.rm = TRUE)
## 2.5%
## 146.5
#97.5% according to normal distribution =mean+2*sd
mean(regicor$height, na.rm = TRUE)+1.96*sd(regicor$height, na.rm = TRUE)
## [1] 180.9798
#The observed 97.5% using the quantile function
quantile(regicor$height,0.975, na.rm = TRUE)
## 97.5%
## 181
```

The calculated percentiles are nearly equal to the observed percentiles.

From the different plots and calculated and observed percentiles, it can be assumed that the height data came from a population that is nearly normally distributed. Therefore, without looking at the data and knowing that its mean = 162.9 cm and its sd = 9.2 cm and based on the sample data, 95% of individuals from this north-west Spanish province (Girona) have heights between 144.9 cm and 181 cm and only 5% of individuals from this Spanish province have heights below 144.9 cm or more than 181 cm.

3.4. THE Z SCORE "STATISTICAL MILE"

It is a measure of the relative distance and direction of a single data value in a data distribution relative to the mean of that distribution. This distance is converted into units of sd.

z score = (value-mean)/sd

This z score has a normal distribution with mean = 0 and sd = 1.

3.4.1. Calculating z Score for Single Continuous Variables

Example: calculate the z score for two participants' weight with id 2265 and 2259.

```
regicor %>% filter(id == 2265)
##   id year age  sex    smoker sbp dbp histhtn txhtn chol hdl triglyc
## 1 2265 2005 70 Female Never smoker 138 75   No  No 294 57   93
##   ldl histchol txchol height weight bmi phyact  pcs  mcs cv    tocv
## 1 218.4    No    No  160   64 25 304.2 54.455 58.918 No 1024.882
##  death todeath
## 1  Yes 1299.163
regicor %>% filter(id== 2259)
##   id year age sex    smoker sbp dbp histhtn txhtn chol hdl triglyc  ldl
## 1 2259 2005 54 Male Former >= 1y 130 66   No  No 268 37    128 205.4
##  histchol txchol height weight   bmi phyact  pcs  mcs cv   tocv death
## 1     No    No  163    79 29.7339 284.7 52.026 38.175 No 3054.901  No
##  todeath
## 1 2855.711
z_2265<-(64 – mean(regicor$weight, na.rm = TRUE))/sd(regicor$weight,
na.rm = TRUE)
z_2259<-(79 – mean(regicor$weight, na.rm = TRUE))/sd(regicor$weight,
na.rm = TRUE)
```

The weight of id 2265 participant = 64 kg while the weight of id 2259 participant = 79 kg.

The z score of id 2265 participant = –0.69. This means that his weight is 0.69 standard deviation less than the mean weight in this population.

The z score of id 2259 participant = 0.41. This means that his weight is 0.41 standard deviation more than the mean weight in this population.

Assuming that weight of individuals in Girona follows a normal distribution, we can calculate the % (or proportion) of individuals that have weight less than the weight of id 2265 participant in this population

```
pnorm(64,mean = mean(regicor$weight, na.rm = T),sd =sd(regicor$weight,
na.rm = T))
## [1] 0.2451493
#using the z-score where the default values for pnorm is mean = 0 and sd
= 1
```

```
pnorm(z_2265)
## [1] 0.2451493
```

From the results, it is estimated that about 25% of individuals in Girona have weights less than 64 kg which is the weight of id 2265 participant (64 is 25%) and 75% have weights more than 64 kg.

We can calculate the % (or proportion) of individuals that have weight less than the weight of id 2259 participant in this population

```
pnorm(79,mean = mean(regicor$weight, na.rm = T),sd =sd(regicor$weight, na.rm = T))
## [1] 0.6579159
#using the z-score
pnorm(z_2259)
## [1] 0.6579159
```

From the results, it is estimated that about 65.8% of individuals in Girona have weights less than 79 kg which is the weight of id 2259 participant (79 is 65.8%) and 34.2% have weights more than 79 kg.

To calculate the % (or proportion or probability) of individuals that have weights between 64 and 79 kg in this population.

```
pnorm(79,mean = mean(regicor$weight, na.rm = T),sd =sd(regicor$weight, na.rm = T)) –
pnorm(64,mean = mean(regicor$weight, na.rm = T),sd =sd(regicor$weight, na.rm = T))
## [1] 0.4127665
#or using the z-scores
pnorm(z_2259)-pnorm(z_2265)
## [1] 0.4127665
```

From the results, it is estimated that about 41.3% of individuals in Girona have weights 64 and 79 kg.

All these results are our first steps in statistical inference where we take the information from a single sample to make conclusions about the larger population from which this sample was taken. More solid statistical inference tests will come in the next chapters.

3.4.2. Calculating z Score for Two Continuous Variables

z score can also be used for comparing different continuous variables.

Example: calculate the z score for weight and height of participant with id 2265.

```
regicor %>% filter(id == 2265)
##   id year age  sex    smoker sbp dbp histhtn txhtn chol hdl triglyc
## 1 2265 2005 70 Female Never smoker 138 75   No  No 294 57   93
##   ldl histchol txchol height weight bmi phyact  pcs  mcs cv   tocv
## 1 218.4    No    No  160    64 25 304.2 54.455 58.918 No 1024.882
##  death todeath
## 1  Yes 1299.163
        z_2265_wt<-(64 – mean(regicor$weight, na.rm = TRUE))/
                    sd(regicor$weight, na.rm = TRUE)
        z_2265_ht<-(160 – mean(regicor$height, na.rm = TRUE))/
                    sd(regicor$height, na.rm = TRUE)
```

Assuming that weight and height of individuals in Girona follows a normal distribution, we can calculate the % (or proportion) of individuals that have weight and height less than the weight and height of id 2265 participant in this population.

```
pnorm(64,mean = mean(regicor$weight, na.rm = T),sd =sd(regicor$weight,
na.rm = T))
## [1] 0.2451493
#using the z-score where the default values for pnorm is mean = 0 and sd
= 1
pnorm(z_2265_wt)
## [1] 0.2451493
pnorm(160,mean = mean(regicor$height, na.rm = T),sd =sd(regicor$height,
na.rm = T))
## [1] 0.3758677
#using the z-score
pnorm(z_2265_ht)
## [1] 0.3758677
```

From the results, it is estimated that about 25% of individuals in Girona have weights less than 64 kg which is the weight of id 2265 participant (64 is 25%) and 75% have weights more than 64 kg.

On the other hand, it is estimated that about 37.6% of individuals in Girona have heights less than 160 cm which is the height of id 2265 participant (160 is 37.6%) and 62.4% have heights more than 160 cm.

3.5. APPLYING THE NORMAL DISTRIBUTION TO SKEWED DATA

z scores and the use of pnorm() or qnorm() functions produce values that do not align with the observed percentiles for skewed data.

Examining the physical activity column

```
# a histogram
ggplot(data = regicor, aes(x = phyact))+geom_histogram()
```

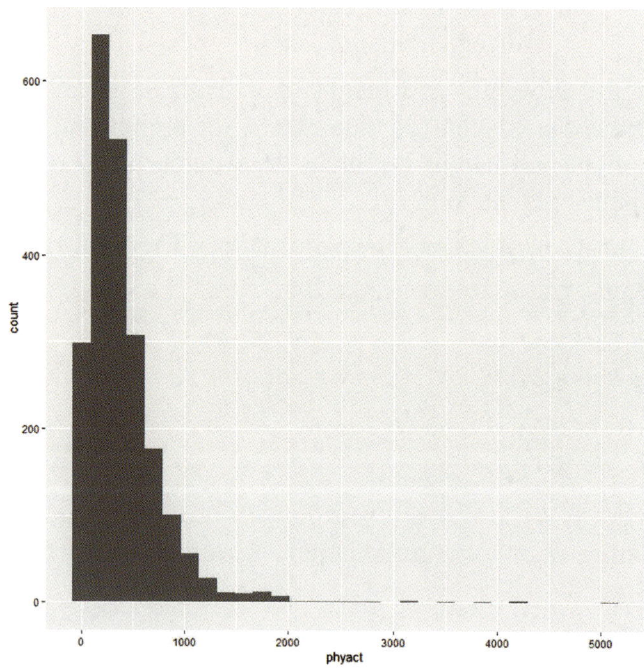


```
# a boxplot
ggplot(data = regicor, aes(y=phyact))+geom_boxplot()
```

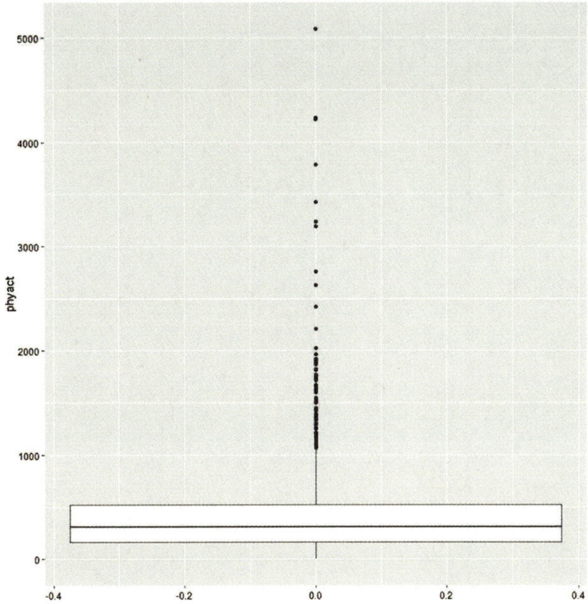

 # density plot

```
ggplot(data = regicor, aes(x = phyact))+geom_density()
```

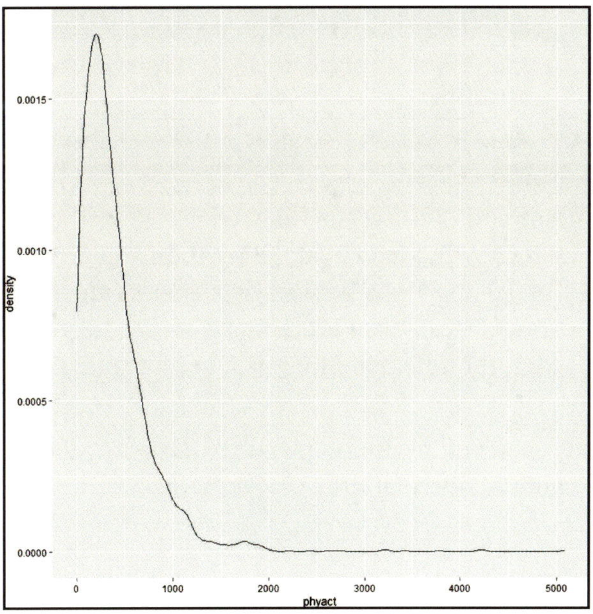

 # QQ plot

```
ggqqplot(data = regicor, x = "phyact")
```

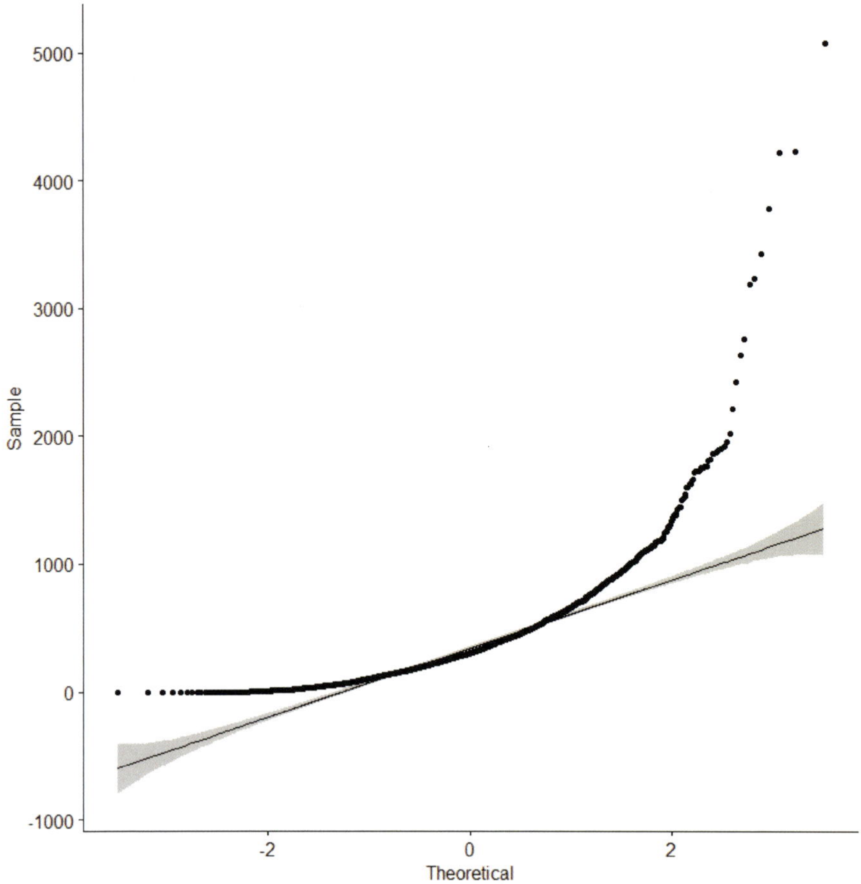

The histogram and density plot show right skewed distribution of the physical activity data with long tail at the right. In addition, the box plot has unequal halves indicating right skewed data (longer upper half and longer line extending from the 75%). Furthermore, many large physical activity points do not along the reference line of QQ plot indicating right skewed data.

Calculating percentile from physical activity data and from a corresponding normal distribution,

```
# summary statistics of phyact column
summary(regicor$phyact)
```

```
## Min. 1st Qu. Median Mean 3rd Qu. Max. NA's
## 0.0 159.5 303.7 398.8 521.9 5083.2 88
sd(regicor$phyact, na.rm = TRUE)
## [1] 388.1642
#2.5% according to normal distribution =mean–2*sd
mean(regicor$phyact, na.rm = TRUE)–1.96*sd(regicor$phyact, na.rm= TRUE)
## [1] –361.9705
#The observed 2.5% using the quantile function
quantile(regicor$phyact,0.025, na.rm = TRUE)
## 2.5%
## 11
#97.5% according to normal distribution =mean+2*sd
mean(regicor$phyact, na.rm = TRUE)+1.96*sd(regicor$phyact, na.rm = TRUE)
## [1] 1159.633
#The observed 97.5% using the quantile function
quantile(regicor$phyact,0.975, na.rm = TRUE)
## 97.5%
## 1284.703
```

When the normal distribution principles are applied, the calculated percentiles are either not real (2.5%, negative number) or far from the real observed numbers (97.5%= 1159 while the observed number= 1284). For non-normal data, use the observed percentiles of the data as estimates of percentiles for the underlying population and do not rely on the normal distribution principles.

CHAPTER 4

BINARY AND CATEGORICAL DATA

CONTENTS

4.1. DEFINITIONS

Binary data are data that have only two values. For example, sleeping status (sleeping, not sleeping), sex (Male or Female). They are examples of nominal categorical data.

Categorical data are extensions of the binary data, where it takes more than two values. Example of nominal categorical data, race (White, Black, Hispanic, etc.) where the values have no inherent ordering. Example of ordinal categorical data, degree of agreement (disagree, neutral, agree), academic ranks (professor, assistant professors, lecturers, assistant lecturers) where the values have a certain order. For more information, see the first chapter.

4.2. SUMMARIZING CATEGORICAL DATA

The sample proportion (along with the sample size) is the only measure that is used to describe categorical data. The sample proportion is denoted by (p-hat) as it estimates the true population proportion (p).

4.2.1. Examining the Regicor Data

The full description of this dataset is showed in Chapter 2. The regicor dataframe contains data from three different cross-sectional surveys of individuals representative of the population from a north-west Spanish province (Girona)

```
library(tidyverse)
library(compareGroups)
data("regicor")
str(regicor)
## 'data.frame': 2294 obs. of 25 variables:
## $ id   : num 2.26e+03 1.88e+03 3.00e+09 3.00e+09 3.00e+09 ...
##  ..- attr(*, "label")= Named chr "Individual id"
##  .. ..- attr(*, "names")= chr "id"
## $ year  : Factor w/ 3 levels "1995,""2000,"..: 3 3 2 2 2 2 2 1 3 1 ...
##  ..- attr(*, "label")= Named chr "Recruitment year"
##  .. ..- attr(*, "names")= chr "year"
## $ age   : int 70 56 37 69 70 40 66 53 43 70 ...
```

```
## ..- attr(*, "label")= Named chr "Age"
## .. ..- attr(*, "names")= chr "age"
## $ sex   : Factor w/ 2 levels "Male,""Female": 2 2 1 2 2 2 1 2 2 1 ...
## ..- attr(*, "label")= chr "Sex"
## $ smoker : Factor w/ 3 levels "Never smoker,"..: 1 1 2 1 NA 2 1 1 3 3 ...
## ..- attr(*, "label")= Named chr "Smoking status"
## .. ..- attr(*, "names")= chr "smoker"
## $ sbp   : int 138 139 132 168 NA 108 120 132 95 142 ...
## ..- attr(*, "label")= Named chr "Systolic blood pressure"
## .. ..- attr(*, "names")= chr "sbp"
## $ dbp   : int 75 89 82 97 NA 70 72 78 65 78 ...
## ..- attr(*, "label")= Named chr "Diastolic blood pressure"
## .. ..- attr(*, "names")= chr "dbp"
## $ histhtn : Factor w/ 2 levels "Yes,""No": 2 2 2 2 2 2 1 2 2 2 ...
## ..- attr(*, "label")= Named chr "History of hypertension"
## .. ..- attr(*, "names")= chr "histbp"
## $ txhtn  : Factor w/ 2 levels "No,""Yes": 1 1 1 1 1 1 2 1 1 1 ...
## ..- attr(*, "label")= chr "Hypertension treatment"
## $ chol  : num 294 220 245 168 NA NA 298 254 194 188 ...
## ..- attr(*, "label")= Named chr "Total cholesterol"
## .. ..- attr(*, "names")= chr "chol"
## $ hdl   : num 57 50 59.8 53.2 NA ...
## ..- attr(*, "label")= Named chr "HDL cholesterol"
## .. ..- attr(*, "names")= chr "hdl"
## $ triglyc : num 93 160 89 116 NA 94 71 NA 68 137 ...
## ..- attr(*, "label")= Named chr "Triglycerides"
## .. ..- attr(*, "names")= chr "triglyc"
## $ ldl   : num 218.4 138 167.4 91.6 NA ...
## ..- attr(*, "label")= Named chr "LDL cholesterol"
## .. ..- attr(*, "names")= chr "ldl"
## $ histchol: Factor w/ 2 levels "Yes,""No": 2 2 2 2 NA 2 1 2 2 2 ...
## ..- attr(*, "label")= chr "History of hyperchol."
```

```
## $ txchol : Factor w/ 2 levels "No,""Yes": 1 1 1 1 NA 1 1 1 1 1 ...
## ..- attr(*, "label")= Named chr "Cholesterol treatment"
## .. ..- attr(*, "names")= chr "txchol"
## $ height : num 160 163 170 147 NA ...
## ..- attr(*, "label")= Named chr "Height (cm)"
## .. ..- attr(*, "names")= chr "height"
## $ weight : num 64 67 70 68 NA 43.5 79.2 45.8 53 62 ...
## ..- attr(*, "label")= Named chr "Weight (Kg)"
## .. ..- attr(*, "names")= chr "weight"
## $ bmi   : num 25 25.2 24.2 31.5 NA ...
## ..- attr(*, "label")= Named chr "Body mass index"
## .. ..- attr(*, "names")= chr "bmi"
## $ phyact : num 304 160 553 522 NA ...
## ..- attr(*, "label")= Named chr "Physical activity (Kcal/week)"
## .. ..- attr(*, "names")= chr "phyact"
## $ pcs   : num 54.5 58.2 43.4 54.3 NA ...
## ..- attr(*, "label")= Named chr "Physical component"
## .. ..- attr(*, "names")= chr "pcs"
## $ mcs   : num 58.9 48 62.6 57.9 NA ...
## ..- attr(*, "label")= chr "Mental component"
## $ cv   : Factor w/ 2 levels "No,""Yes": 1 1 1 1 NA 1 1 1 1 1 ...
## ..- attr(*, "label")= chr "Cardiovascular event"
## $ tocv  : num 1025 2757 1906 1055 NA ...
## ..- attr(*, "label")= chr "Days to cardiovascular event or end of follow-
up"
## $ death  : Factor w/ 2 levels "No,""Yes": 2 1 1 1 NA 1 2 1 1 1 ...
## ..- attr(*, "label")= chr "Overall death"
## $ todeath : num 1299.2 39.3 858.4 1833.1 NA ...
## ..- attr(*, "label")= chr "Days to overall death or end of follow-up"
head(regicor)
##     id year age  sex     smoker sbp dbp histhtn txhtn
```

```
## 6101     2265 2005 70 Female      Never smoker 138 75   No  No
## 5762     1882 2005 56 Female      Never smoker 139 89   No  No
## 2992 3000105616 2000 37  Male Current or former < 1y 132 82   No  No
## 2611 3000103485 2000 69 Female      Never smoker 168 97   No  No
## 2762 3000103963 2000 70 Female          <NA> NA NA   No  No
## 1516 3000100883 2000 40 Female Current or former < 1y 108 70   No
No
##    chol   hdl triglyc    ldl histchol txchol height weight   bmi
## 6101 294 57.00000   93 218.40000    No    No  160  64.0 25.00000
## 5762 220 50.00000   160 138.00000    No    No  163  67.0 25.21736
## 2992 245 59.80429   89 167.39571    No    No  170  70.0 24.22145
## 2611 168 53.17571   116 91.62429    No    No  147  68.0 31.46837
## 2762  NA    NA   NA     NA   <NA>  <NA>   NA   NA    NA
## 1516  NA 68.90000   94    NA    No    No  158  43.5 17.42509
##     phyact  pcs  mcs  cv   tocv death  todeath
## 6101 304.2000 54.455 58.918 No 1024.882 Yes 1299.16343
## 5762 160.3000 58.165 47.995 No 2756.849 No  39.32629
## 2992 552.7912 43.429 62.585 No 1905.969 No 858.42203
## 2611 522.0000 54.325 57.900 No 1055.380 No 1833.07619
## 2762    NA    NA    NA <NA>    NA <NA>     NA
## 1516 386.9505 57.315 47.869 No 3239.241 No 877.61155
```

Examples of binary variables or columns: sex (“Male,” “Female”), histhtn or History of hypertension(“Yes,” “No”), cv or Cardiovascular event (“Yes,” “No”).

Example of ordinal categorical variable: year (1995, 2000, 2005).

Example of nominal categorical variable: smoker (Never smoker, Current or former < 1y, Former >= 1y).

Examining the actual counts with table() function and proportions with prop.table() function

```
# sex
table(regicor$sex)
```

```
##
##  Male Female
##  1101  1193
prop.table(table(regicor$sex))
##
##    Male   Female
## 0.4799477 0.5200523
# history of hypertension
table(regicor$histhtn)
##
## Yes  No
## 723 1563
prop.table(table(regicor$histhtn))
##
##    Yes     No
## 0.316273 0.683727
# cardiovascular event
table(regicor$cv)
##
##   No Yes
## 2071  92
prop.table(table(regicor$cv))
##
##      No     Yes
## 0.95746648 0.04253352
# year
table(regicor$year)
##
## 1995 2000 2005
## 431 786 1077
prop.table(table(regicor$year))
```

```
##
## 1995 2000 2005
## 0.1878814 0.3426330 0.4694856
# smoker
table(regicor$smoker)
##
##     Never smoker Current or former < 1y     Former >= 1y
##        1201          593          439
prop.table(table(regicor$smoker))
##
##     Never smoker Current or former < 1y     Former >= 1y
##      0.5378415      0.2655620      0.1965965
```

- **Interpretation:**

1. There are 1101 males in our data and 1193 females. The for males is nearly 0.48 or 48% and the for females is 0.52 or 52%.
2. There are 723 participants with hypertension history and 1563 participants with no such history. The for participants with hypertension history is nearly 31.6% and the for no such history is 68.4%.
3. There are 92 participants with cardiovascular event and 2071 participants with no such event. The for participants with cardiovascular event is nearly 4% and the or no such event is 96%.
4. There are 431 participants in 1995, 786 in 2000, and 1077 in 2005. These corresponding to percentages of 18.8%, 34.3%, and 46.9%, respectively.
5. There are 1201 non-smokers, 593 current smokers, and 439 former smokers in our data. The for non-smokers is nearly 53.8%, the for current smokers is nearly 26.6% and the for former smokers is 19.6%.

The proportion or percentage is also useful to learn the distribution of data and the percentiles.

For example, binary data for the hypertension history, the hypertension history percentage is 31.6% and for no such history is 68.4%.

Therefore, the 10% is hypertension history, 20% is hypertension history also, and the 31% is hypertension history also. The 32%, 40%, 60%, 80%, 100%s are all no hypertension history.

4.3. VISUALIZING CATEGORICAL DATA

4.3.1. Bar Plot

geom_bar() is used to plot a bar graph where each bar has a height equal to the number of rows or observations at each level of the categorical variable.

geom_bar() requires only one aesthetic (x) or the categorical variable you want to visualize. You can add another aesthetic, fill = categorical variable so every level will take a different color. To decrease the width of bars, use the argument width.

To add labels of counts, you will create a separate dataframe with counts for each level and use geom_col() and geom_text().

The geom_col() requires two aesthetics, x for categorical levels and y for the count column. The geom_text() requires x and y which will be the same as geom_col() for the position of label and label for the text label.

geom_label() can also be used instead of geom_text() with the same aesthetics but it will add a rectangle behind the text.

For example, different bar graph for different columns of regicor data

```
# sex
ggplot(data = regicor, aes(x = sex))+geom_bar()
```

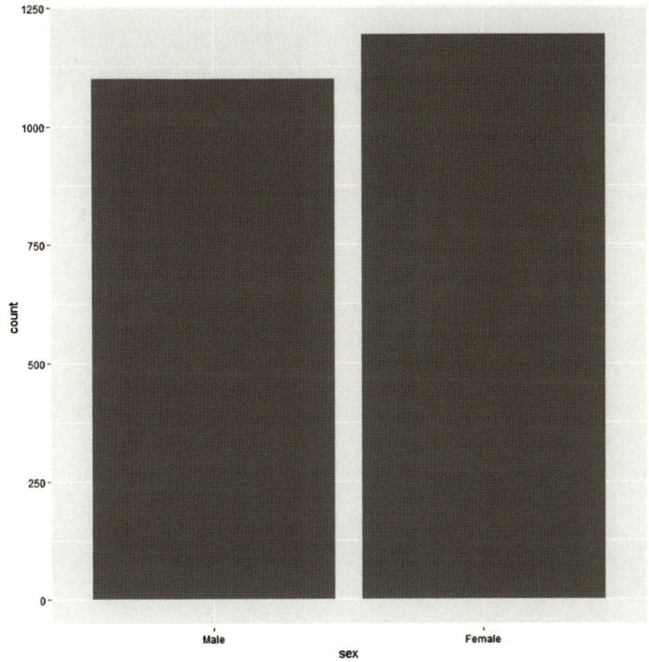

 # to reduce bar widths

```
ggplot(data = regicor, aes(x = sex))+geom_bar(width = 0.5)
```

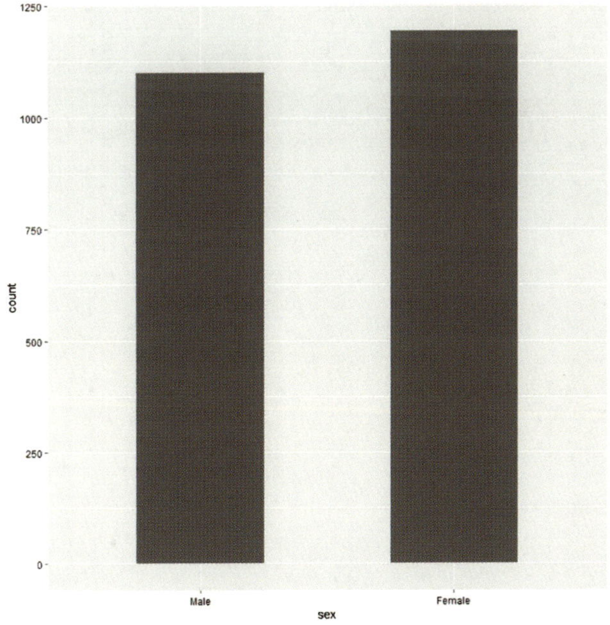

```
# to add color to each level
ggplot(data = regicor, aes(x = sex, fill = sex))+geom_bar(width = 0.5)
```

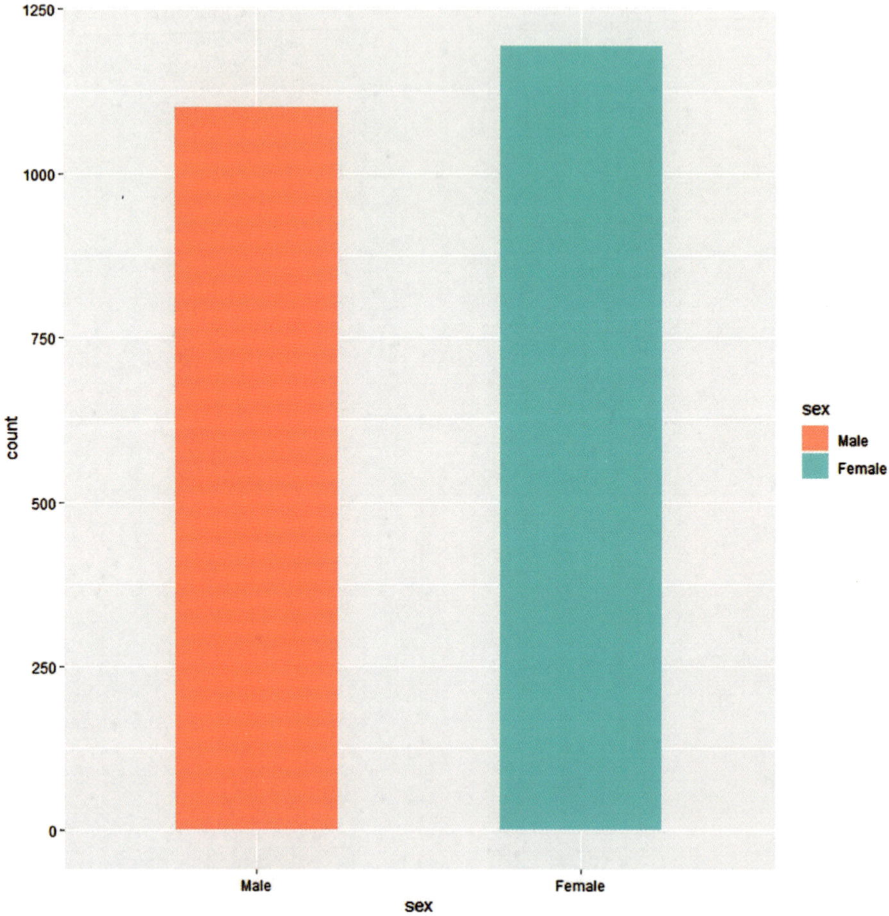


```
# to add labels of numbers
dat<-regicor %>% count(sex)
dat
## # A tibble: 2 x 2
##  sex    n
```

```
## <fct> <int>
## 1 Male 1101
## 2 Female 1193
ggplot(data = dat, aes(x = sex, y = n, fill = sex))+geom_col(width = 0.5)+
 geom_text(aes(label = n))
```

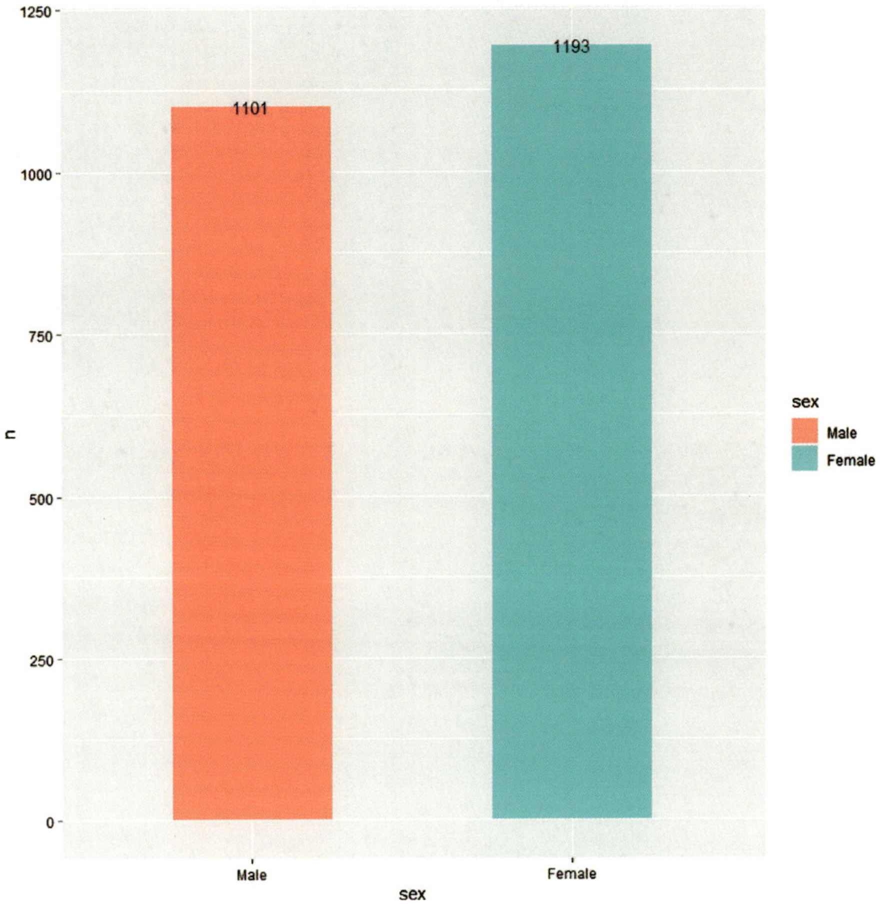


```
ggplot(data = dat, aes(x = sex, y = n, fill = sex))+geom_col(width = 0.5)+
 geom_label(aes(label = n))
```

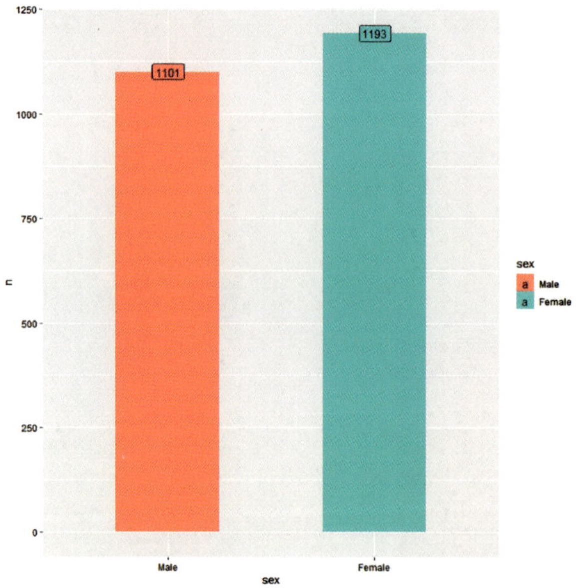

 # history of hypertension

```
ggplot(data = regicor, aes(x = histhtn, fill = histhtn))+geom_bar(width =
0.5)
```

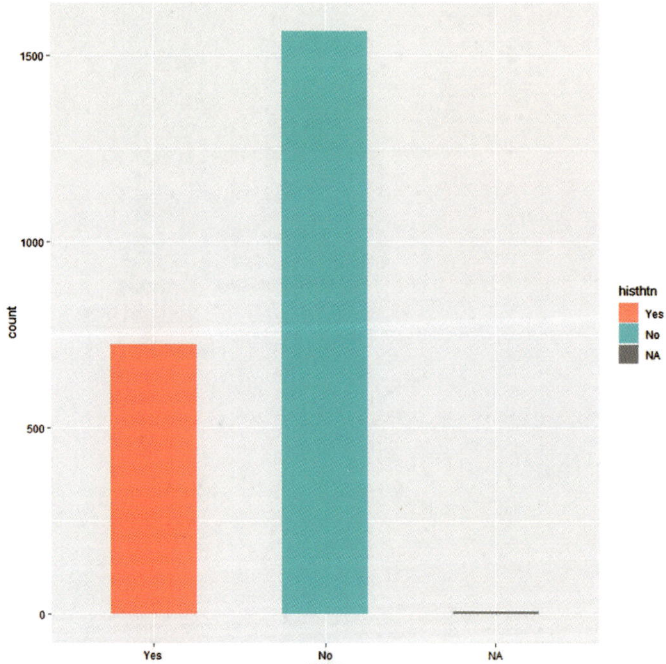

```
# to remove NA values
dat<- regicor %>% filter(!is.na(histhtn))
ggplot(data = dat, aes(x = histhtn, fill = histhtn))+geom_bar(width = 0.5)
```

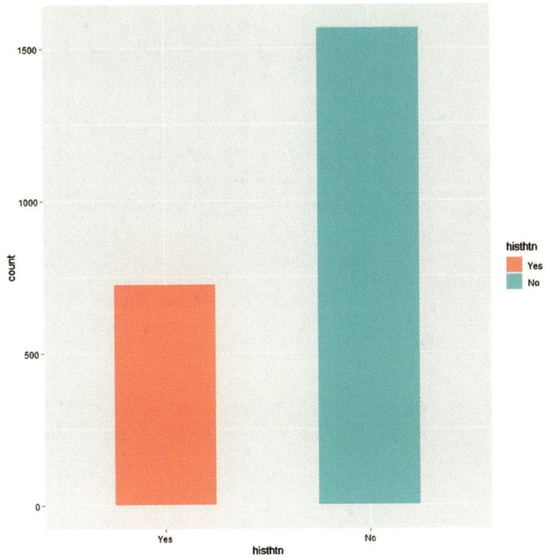


```
# cardiovascular event
ggplot(data = regicor, aes(x = cv, fill = cv))+geom_bar(width = 0.5)
```

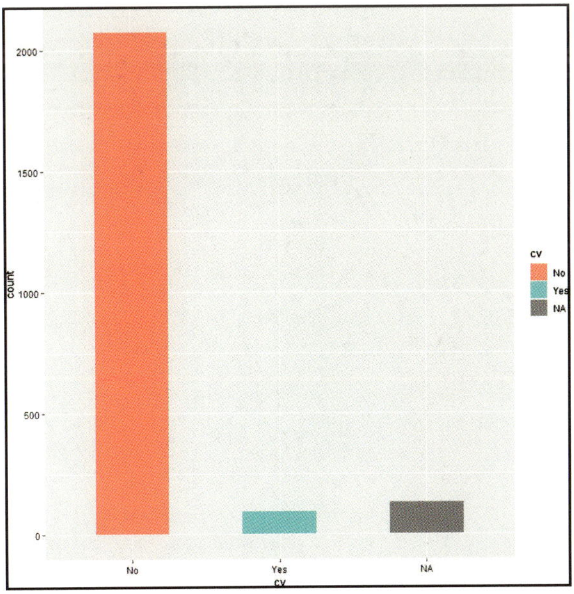

```
# to remove NA values
dat<- regicor %>% filter(!is.na(cv))
ggplot(data = dat, aes(x = cv, fill = cv))+geom_bar(width = 0.5)
```

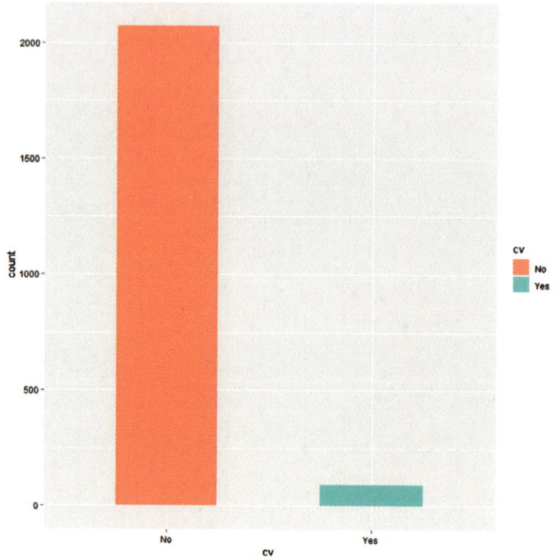


```
# year
ggplot(data = regicor, aes(x = year, fill = year))+geom_bar(width = 0.5)
```

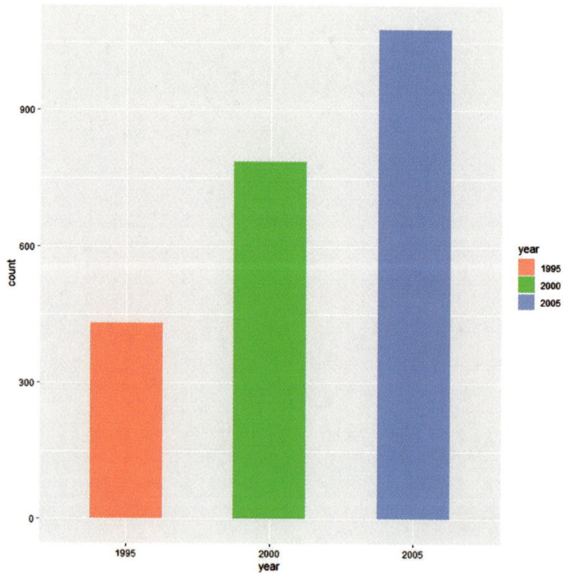

 # smoker

```
ggplot(data = regicor, aes(x = smoker, fill = smoker))+geom_bar(width =
0.5)
```

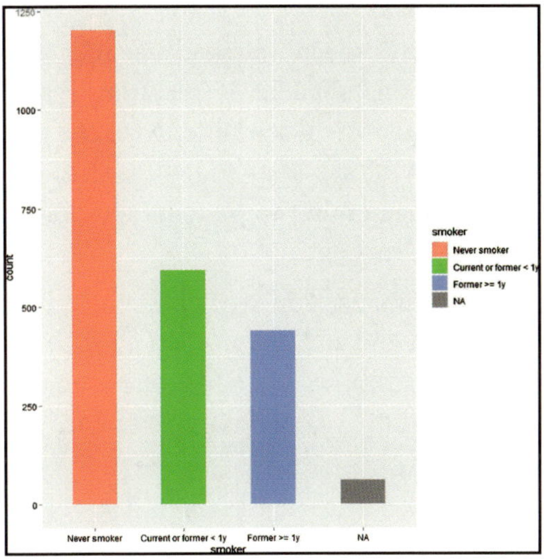

 # to remove NA values

```
dat<- regicor %>% filter(!is.na(smoker))
ggplot(data = dat, aes(x = smoker, fill = smoker))+geom_bar(width = 0.5)
```

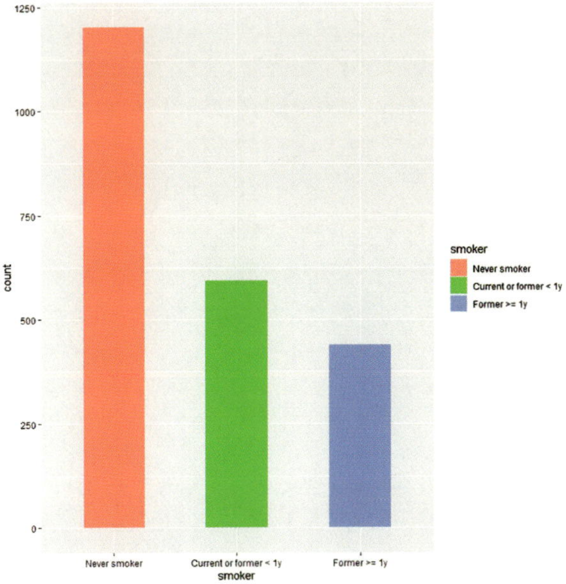

Notice how the bar heights are equal to counts from table() function in the previous code chunk.

4.3.2. Stacked Bar Plot

geom_bar() also can be used to plot stacked bar plot. geom_bar() requires two aesthetics, x to be a constant like 1 (meaning that all data points are from 1 group) and fill = categorical variable. Also the position argument of geom_bar is set to "fill." We have also added another layer using scale_y_continuous() to convert the count to percentage format using percent_format() function of scales package.

To add labels of percentages, you will create a separate dataframe with percentages for each level and use geom_bar() for the old dataframe and geom_text() for the new dataframe. We add some arguments to geom_text() to force the text to be in middle of each compartment.

```
library(scales)
# sex
ggplot(data = regicor, aes(x = 1, fill = sex))+geom_bar(position = "fill")+
 scale_y_continuous(name = "percentage,"labels = percent_format())
```

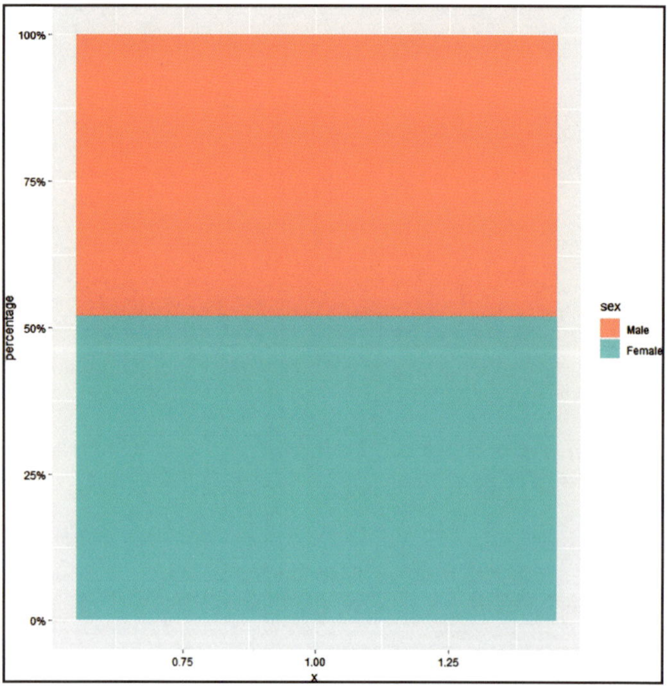

```
# to add percentages labels
dat<- regicor %>% count(sex) %>% mutate(ratio = percent(n/sum(n)))
ggplot(data = regicor, aes(x = 1, fill = sex))+geom_bar(position = "fill")+
 geom_text(data = dat, aes(x = 1, y=n, label=ratio),
     position = position_fill(vjust = 0.5))+
  scale_y_continuous(name = "percentage,"labels = percent_format())
```

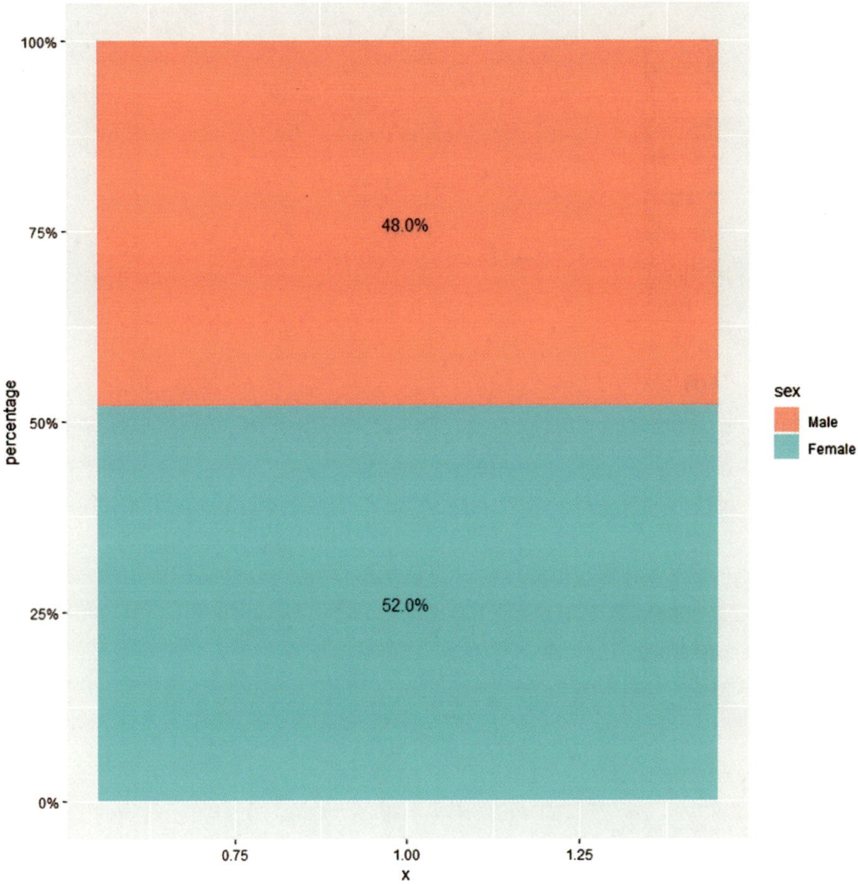


```
ggplot(data = regicor, aes(x = 1, fill = sex))+geom_bar(position = "fill")+
 geom_label(data = dat, aes(x = 1, y=n, label=ratio),
     position = position_fill(vjust = 0.5))+
  scale_y_continuous(name = "percentage,"labels = percent_format())
```

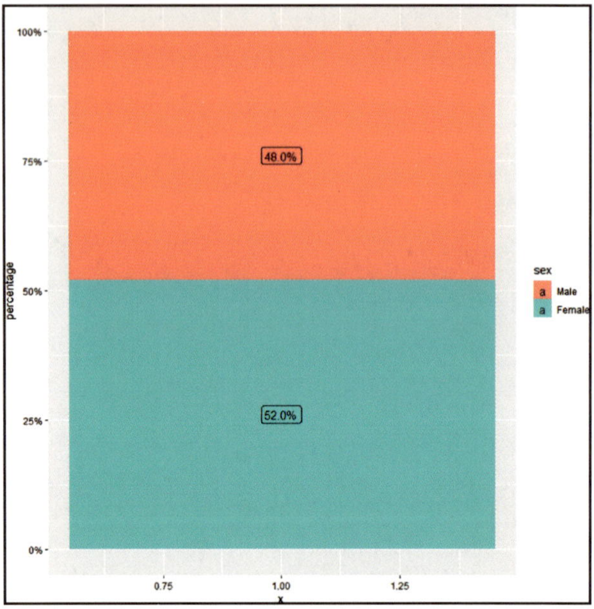

 # history of hypertension

```
ggplot(data = regicor, aes(x = 1, fill = histhtn))+geom_bar(position = "fill")+
 scale_y_continuous(name = "percentage,"labels = percent_format())
```

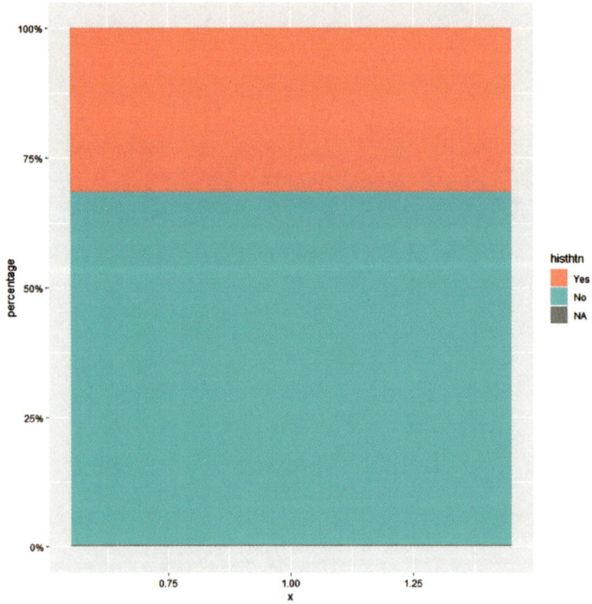

```
# to remove NA
dat<-regicor %>% filter(!is.na(histhtn))
ggplot(data = dat, aes(x = 1, fill = histhtn))+geom_bar(position = "fill")+
  scale_y_continuous(name = "percentage,"labels = percent_format())+
 scale_fill_discrete(name = "hypertension history")
```

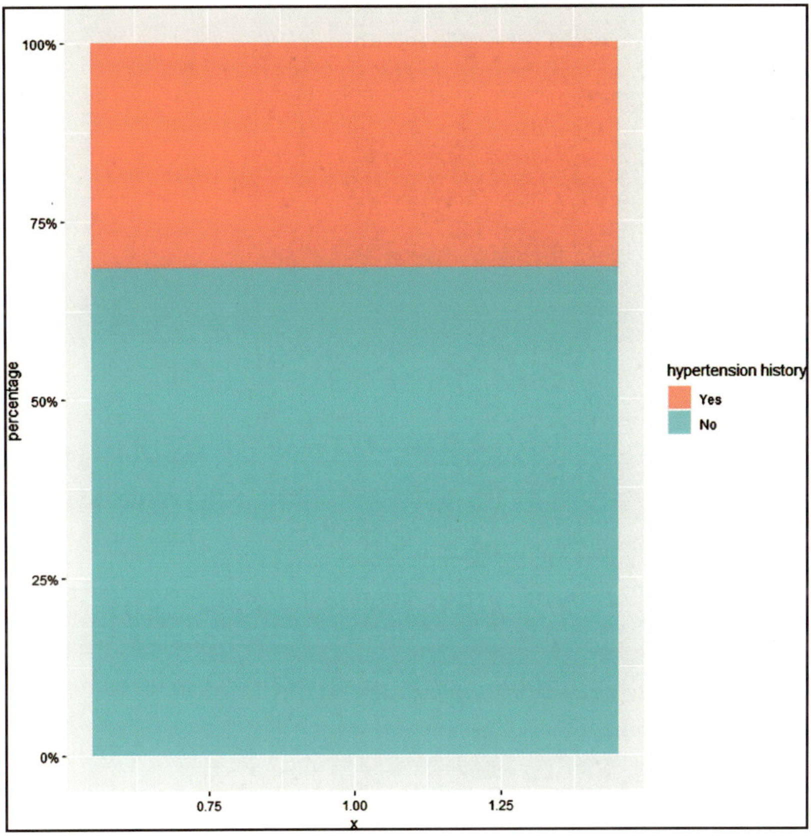


```
# to add percentages labels
dat2<- dat %>% count(histhtn) %>% mutate(ratio = percent(n/sum(n)))
ggplot(data = dat, aes(x = 1, fill = histhtn))+geom_bar(position = "fill")+
 geom_text(data = dat2, aes(x = 1, y=n, label=ratio),
     position = position_fill(vjust = 0.5))+
  scale_y_continuous(name = "percentage,"labels = percent_format())+
  scale_fill_discrete(name = "hypertension history")
```

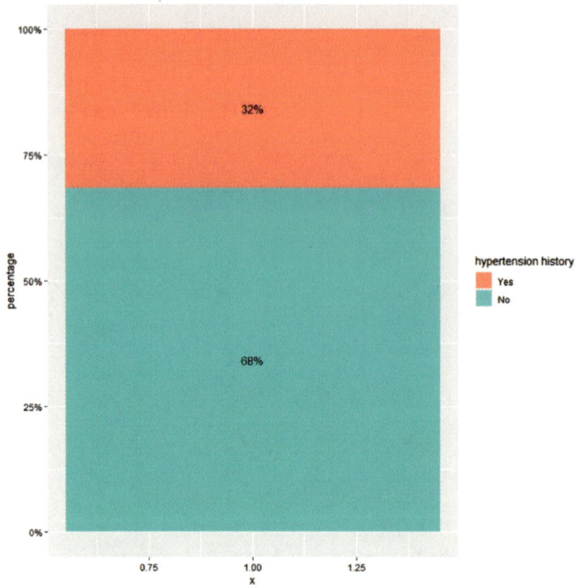


```
# cardiovascular event
ggplot(data = regicor, aes(x = 1, fill = cv))+geom_bar(position = "fill")+
  scale_y_continuous(name = "percentage,"labels = percent_format())
```

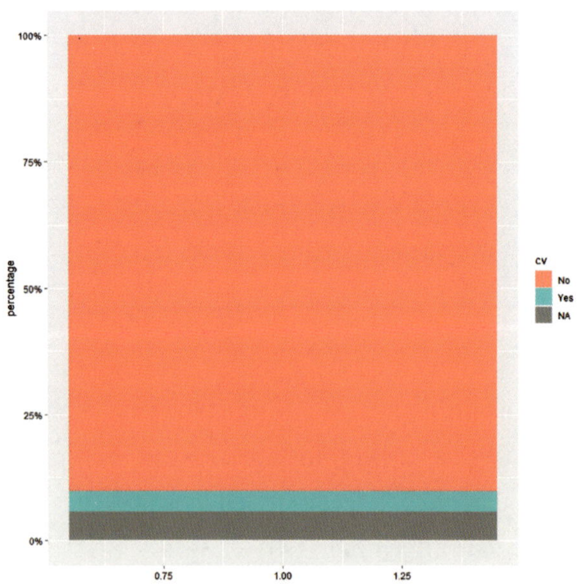


```
# to remove NA
```

```
dat<-regicor %>% filter(!is.na(cv))
ggplot(data = dat, aes(x = 1, fill = cv))+geom_bar(position = "fill")+
 scale_y_continuous(name = "percentage,"labels = percent_format())+
 scale_fill_discrete(name = "cardiovascular event")
```

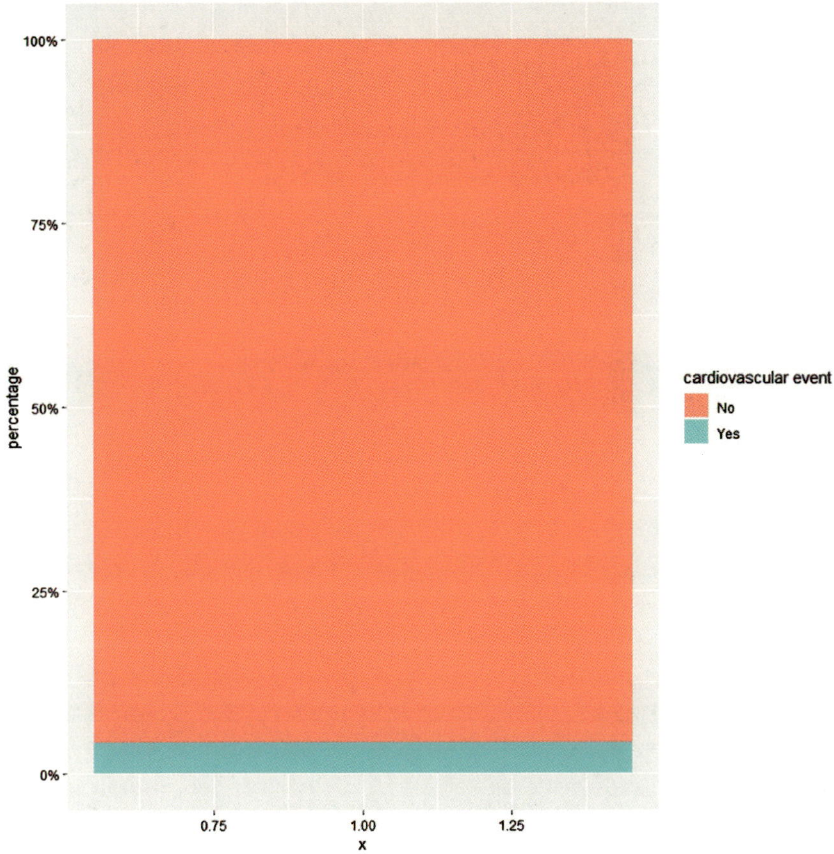


```
# to add percentages labels
dat2<- dat %>% count(cv) %>% mutate(ratio = percent(n/sum(n)))
ggplot(data = dat, aes(x = 1, fill = cv))+geom_bar(position = "fill")+
 geom_text(data = dat2, aes(x = 1, y=n, label=ratio),
     position = position_fill(vjust = 0.5))+
  scale_y_continuous(name = "percentage,"labels = percent_format())+
 scale_fill_discrete(name = "cardiovascular event")
```

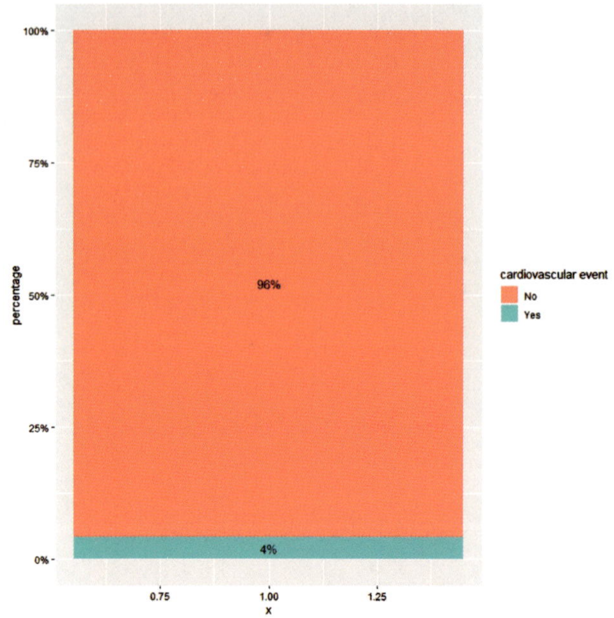


```
# year
ggplot(data = regicor, aes(x = 1, fill = year))+geom_bar(position = "fill")+
  scale_y_continuous(name = "percentage,"labels = percent_format())
```

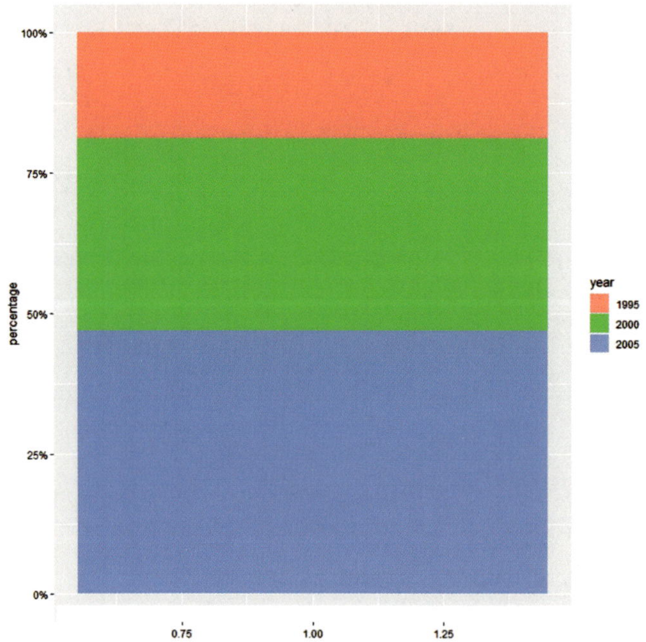

```
# to add percentages labels
dat<- regicor %>% count(year) %>% mutate(ratio = percent(n/sum(n)))
ggplot(data = regicor, aes(x = 1, fill = year))+geom_bar(position = "fill")+
 geom_text(data = dat, aes(x = 1, y=n, label=ratio),
    position = position_fill(vjust = 0.5))+
  scale_y_continuous(name = "percentage,"labels = percent_format())
```

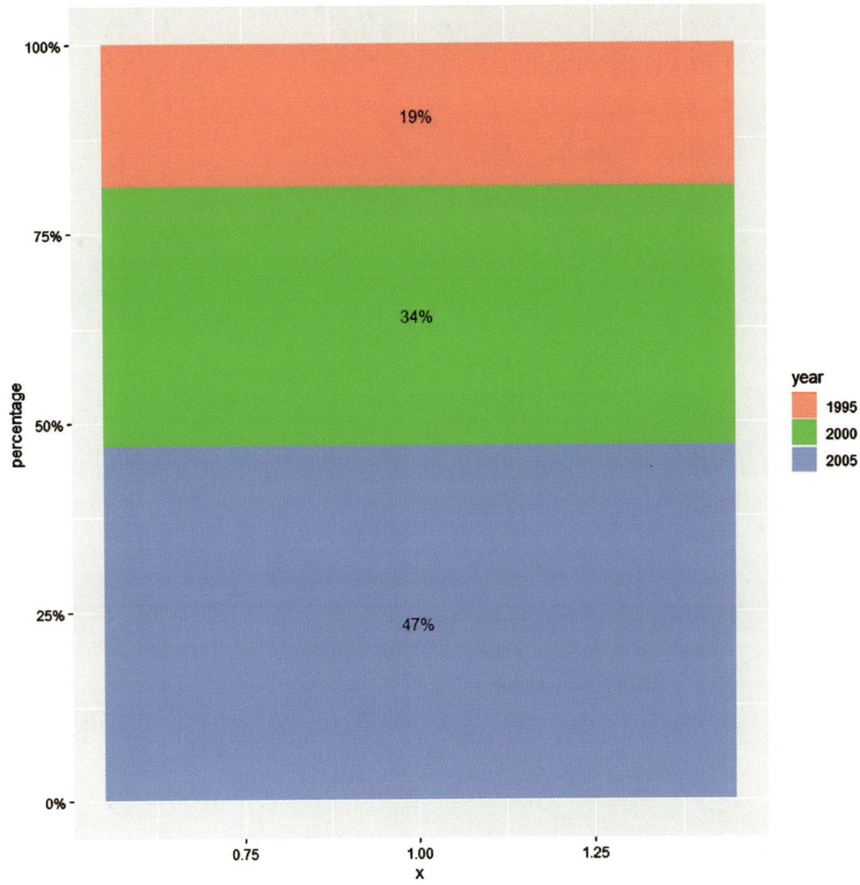


```
# smoker
ggplot(data = regicor, aes(x = 1, fill = smoker))+geom_bar(position =
"fill")+
  scale_y_continuous(name = "percentage,"labels = percent_format())
```

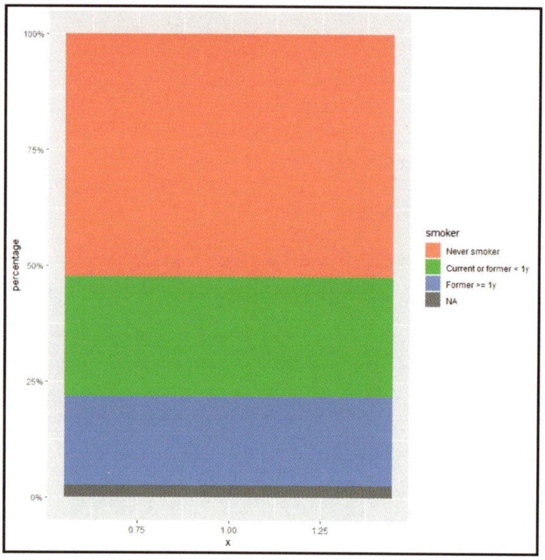

 # to remove NA

```
dat<-regicor %>% filter(!is.na(smoker))
ggplot(data = dat, aes(x = 1, fill = smoker))+geom_bar(position = "fill")+
 scale_y_continuous(name = "percentage,"labels = percent_format())+
 scale_fill_discrete(name = "smoking status")
```

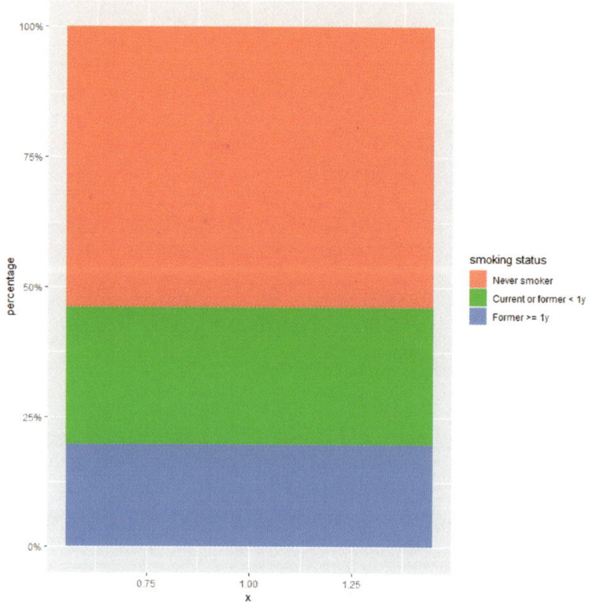

```
# to add percentages labels
dat2<- dat %>% count(smoker) %>% mutate(ratio = percent(n/sum(n)))
ggplot(data = dat, aes(x = 1, fill = smoker))+geom_bar(position = "fill")+
 geom_text(data = dat2, aes(x = 1, y=n, label=ratio),
    position = position_fill(vjust = 0.5))+
  scale_y_continuous(name = "percentage,"labels = percent_format())+
 scale_fill_discrete(name = "smoking status")
```

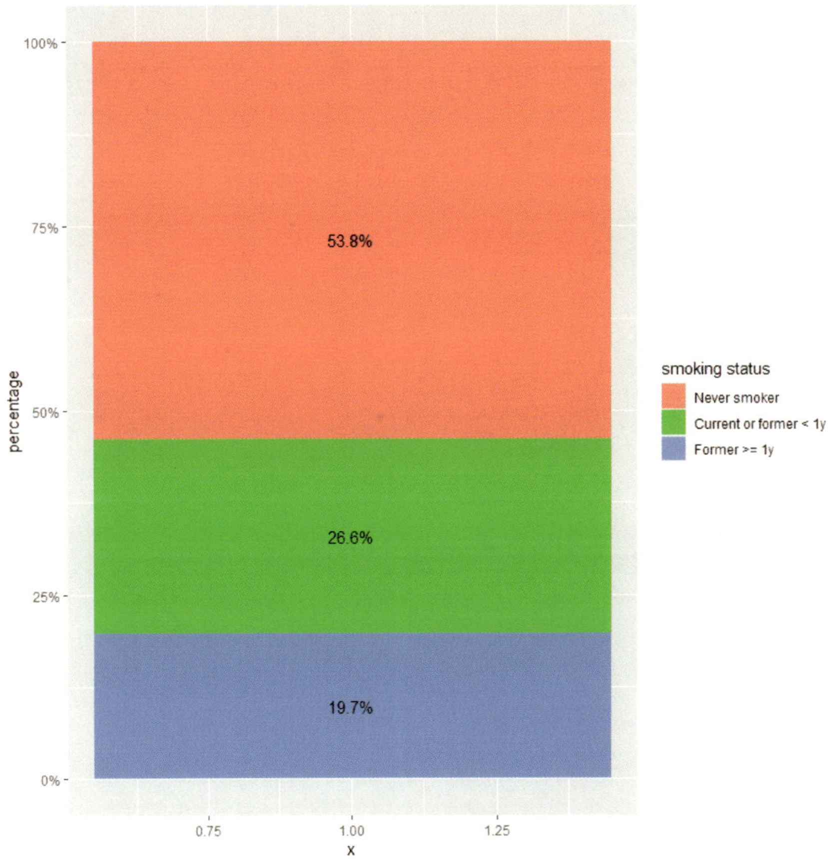

Note that we renamed the legend using scale_fill_discrete() function because the categorical variable hypertension history (histhtn) is passed to fill aesthetic. Also, notice how the percentages in stacked bar plot are equal to that of prop.table(table()) function.

4.4. COMPARING CATEGORICAL DATA ACROSS TWO OR MORE POPULATIONS, NUMERICAL MEASURES

There are three types of numerical measures (sample estimates) that are used for this purpose: risk difference (difference in proportions), relative risk (RR) (ratio between proportions), and odds ratios (ORs). Risk difference can be interpreted as impact (assuming causality) at the population level while the RR and OR can be interpreted as impact at the individual level (assuming causality).

4.4.1. Risk Difference (Proportion Difference or Attributable Risk)

For example, the relation between sex (binary predictor) and cardiovascular (cv) event (binary outcome)

```
# table() and prop.table()
prop.table(table(regicor$sex, regicor$cv),1)
##
##         No    Yes
## Male  0.95585413 0.04414587
## Female 0.95896521 0.04103479
# using tidyverse functions
regicor %>% group_by(sex) %>% summarise(cv_event = mean(cv=="Yes," na.rm = TRUE))
## # A tibble: 2 x 2
## sex  cv_event
## <fct>  <dbl>
## 1 Male   0.0441
## 2 Female  0.0410
```

The proportion of cv event for males = 4.41%

The proportion of cv event for females = 4.1%

The risk difference = 4.41–4.1 = 0.31%

So there is a 0.31% greater absolute risk of cardiovascular event for males compared to females. In a population of 1000 individuals from Girona, we expect to see (0.3/100 × 1,000)= 31 fewer cardiovascular events in females compared to males. In a population of 100,000 individuals from Girona, we expect to see (0.3/100X100000) = 3100 fewer cardiovascular events in females compared to males.

Using epi.2by2() function from the epiR package will allow us to calculate all the 3 types of numerical measures. The epi.2by2() function requires 2 by 2 table format as follows:

1. The first row for exposure +ve and second row for exposure –ve. The first row will be for males and second row for females;
2. The first column for disease +ve and second column for disease –ve. The first column will be for cv event and second column for absence of cv event.

We will use the table() function to generate this table in the required format. However, the ordering of levels of cv factor is not as we desired so we order the positive outcome first

```
library(epiR)
table(regicor$sex, regicor$cv)
##
##         No Yes
## Male   996  46
## Female 1075  46
regicor$cv<-factor(regicor$cv, levels = c("Yes,""No"))
table(regicor$sex, regicor$cv)
##
##         Yes  No
## Male    46 996
## Female  46 1075
dat<- table(regicor$sex, regicor$cv)
epi.2by2(dat = dat, method = "cross.sectional")
##            Outcome + Outcome –  Total  Prevalence *  Odds
## Exposed +      46       996      1042      4.41   0.0462
## Exposed –      46      1075      1121      4.10   0.0428
## Total          92      2071      2163      4.25   0.0444
```

```
##
## Point estimates and 95% CIs:
## ---------------------------------------------------------------
## Prevalence ratio                 1.08 (0.72, 1.60)
## Odds ratio                   1.08 (0.71, 1.64)
## Attrib prevalence *               0.31 (–1.39, 2.02)
## Attrib prevalence in population *     0.15 (–1.29, 1.59)
## Attrib fraction in exposed (%)      7.05 (–38.65, 37.68)
## Attrib fraction in population (%)     3.52 (–17.83, 21.01)
## ---------------------------------------------------------------
## Test that OR = 1: chi2(1) = 0.128 Pr>chi2 = 0.72
## Wald confidence limits
## CI: confidence interval
## * Outcomes per 100 population units
```

From the result, the Attribute risk or risk difference = 0.31. As this risk difference calculates the risk difference per 100 population units so it is 0.31%. This means that males

4.4.2. Relative Risk (Risk Ratio or Proportion Ratio)

For example, the relation between sex (binary predictor) and cardiovascular event (binary outcome)

The proportion of cv event for males = 4.41%, while the proportion of cv event for females = 4.1%

The risk ratio = 4.41/4.1= 1.08. So males have 1.08 times the chance or risk of cardiovascular events compared to females. In other words, males have (1.08–1 = 0.8) or 8% greater RR of cardiovascular events compared to females. By seeing the results of epi.2by2() function, it gives the same prevalence ratio = 1.08.

Risk difference and RR agree in direction but not magnitude. The 8% greater RR for males is an increase over the low percentage of females (4.1%) so both estimates are needed to tell the whole story.

4.4.3. The Odds Ratio (OR)

The estimated odds of an event (outcome) is the estimated probability(or proportion) of the event occurring divided by the estimated probability of the event not occurring. In all subsequent sections, we will use p instead of to denote the estimated probability or proportion.

The estimated odds = p/(1–p).

The estimated odds of an event is a function of the probability or risk of an event occurring.As p increases, so does the odds. If p =0, odds=0/(1–0)=0. If p= 0.25, odds=0.25/(1–0.25)=0.33.

If p =0.5, odds=0.5/(1–0.5)=1, so the risk of having the event is the same as the risk of not having the event. If p =0.75, odds=0.75/(1–0.75)=3. If p =0.95, odds=0.95/(1–0.95)=19. As p approaches 1, the odds will approach infinity.

OR is not a direct comparison of risks but a comparison of risk functions (odds). Risk difference, RR, and OR always agree in direction but not in value.

The OR is useful in two cases:

1. In case-control studies, the OR is the only measure of association that can be estimated. The RR and risk difference will provide meaningless results as the patients are not randomly selected from the population in these studies (see the first chapter); and
2. In logistic regression, the results are initially presented as OR and so frequently used in publications.

Example: the relation between sex (binary predictor) and cardiovascular (cv) event (binary outcome).

The proportion of cv event for males = 0.0441 or 4.41% and the proportion of no cv event for males = 1–0.0441 = 0.9559 or 95.59% so the odds of cv event for males = 0.0441/0.9559 = 0.0462.

The proportion of cv event for females = 0.041 or 4.1% and the proportion of no cv event for males = 1–0.041 = 0.959 or 95.9% so the odds of cv event for females = 0.041/0.959 = 0.0428.

OR for cv event for males compared to females = 0.0462/0.0428 = 1.08.

Males have 1.08 times the odds of cv event as females. In other words, males have (1.08 –1) × 100) 8% greater odds of cardiovascular events than females. The results of epi.2by2() function give the odds of males, females, total population, and the OR.

4.4.4. Example: The Relation Between Smoking (Categorical Predictor) and Cardiovascular (cv) Event (Binary Outcome)

When we compare more than two groups, design one group as a reference group and compare all other groups to it. Assuming that the never smoker group is our reference group.

Comparing current smokers to never smokers. The epi.2by2() function requires 2 by 2 table format as follows:

1. The first row for exposure +ve and second row for exposure –ve. The first row will be for males and second row for females;
2. The first column for disease +ve and second column for disease –ve. The first column will be for cv event and second column for absence of cv event

```
# filter for desired levels
dat<- regicor %>% filter(smoker %in% c("Never smoker,""Current or former < 1y"))
table(dat$smoker, dat$cv)
##
##            Yes  No
## Never smoker      37 1099
## Current or former < 1y  47 506
## Former >= 1y      0  0
# reordering levels of smoker
dat$smoker<- factor(dat$smoker, levels = c("Current or former < 1y,""Never smoker"))
# reordering levels of cv event
dat$cv<- factor(dat$cv, levels = c("Yes,""No"))
table(dat$smoker, dat$cv)
##
##            Yes  No
## Current or former < 1y  47 506
## Never smoker      37 1099
dat2<- table(dat$smoker, dat$cv)
epi.2by2(dat = dat2, method = "cross.sectional")
```

```
##      Outcome +  Outcome -   Total   Prevalence *   Odds
## Exposed +     47     506    553      8.50   0.0929
## Exposed -     37    1099   1136      3.26   0.0337
## Total         84    1605   1689      4.97   0.0523
##
## Point estimates and 95% CIs:
## -------------------------------------------------------------------
## Prevalence ratio                    2.61 (1.72, 3.97)
## Odds ratio                          2.76 (1.77, 4.30)
## Attrib prevalence *                 5.24 (2.70, 7.79)
## Attrib prevalence in population *   1.72 (0.25, 3.18)
## Attrib fraction in exposed (%)      61.68 (41.76, 74.79)
## Attrib fraction in population (%)   34.51 (17.13, 48.25)
## -------------------------------------------------------------------
## Test that OR = 1: chi2(1) = 21.626 Pr>chi2 = <0.001
## Wald confidence limits
## CI: confidence interval
## * Outcomes per 100 population units
```

- **Interpretation:**

  Current smokers have 524% (5.24*100) greater absolute risk of cardiovascular event compared to never smokers

  Current smokers have 2.61 times the risk of cardiovascular events compared to never smokers. In other words, current smokers have (2.61–1 = 1.61) or 161% greater RR of cardiovascular events compared to never smokers.

  Current smokers have 2.76 times the odds of cv event as never smokers. In other words, current smokers have (2.76 –1) × 100) 176% greater odds of cardiovascular events than never smokers.

Comparing former smokers to never smokers.

```
#filter for desired levels
dat<- regicor %>% filter(smoker %in% c("Never smoker,""Former >= 1y"))
table(dat$smoker, dat$cv)
```

```
##
##              Yes  No
## Never smoker        37 1099
## Current or former < 1y  0  0
## Former >= 1y        8 419
# reordering levels of smoker
dat$smoker<- factor(dat$smoker, levels = c("Former >= 1y,""Never smoker"))
# reordering levels of cv event
dat$cv<- factor(dat$cv, levels = c("Yes,""No"))
table(dat$smoker, dat$cv)
##
##          Yes  No
## Former >= 1y  8 419
## Never smoker  37 1099
# to get percentage by row
prop.table(table(dat$smoker, dat$cv),1)
##
##              Yes      No
## Former >= 1y 0.01873536 0.98126464
## Never smoker 0.03257042 0.96742958
dat2<- table(dat$smoker, dat$cv)
epi.2by2(dat = dat2, method = "cross.sectional")
##       Outcome +  Outcome –   Total   Prevalence *   Odds
## Exposed +      8     419    427        1.87   0.0191
## Exposed –      37     1099    1136        3.26   0.0337
## Total       45     1518    1563        2.88   0.0296
##
```

```
## Point estimates and 95% CIs:
## -------------------------------------------------------------
## Prevalence ratio                 0.58 (0.27, 1.23)
## Odds ratio                   0.57 (0.26, 1.23)
## Attrib prevalence *              –1.38 (–3.03, 0.27)
## Attrib prevalence in population *     –0.38 (–1.70, 0.95)
## Attrib fraction in exposed (%)     –73.84 (–270.27, 18.38)
## Attrib fraction in population (%)     –13.13 (–29.46, 1.14)
## -------------------------------------------------------------
## Test that OR = 1: chi2(1) = 2.124 Pr>chi2 = 0.14
## Wald confidence limits
## CI: confidence interval
## * Outcomes per 100 population units
```

- **Interpretation:**

Former smokers have:

Former smokers have 138% (1.38*100) smaller absolute risk of cardiovascular event compared to never smokers. This because the proportion of cardiovascular events in the former smokers group (0.019) is less than the proportion of cardiovascular events in the never smoker group (0.03).

Former smokers have 0.58 times the risk of cardiovascular events compared to never smokers. In other words, former smokers have (1–0.58 = 0.42) or 42% less RR of cardiovascular events compared to never smokers.

Former smokers have 0.57 times the odds of cv event as never smokers. In other words, former smokers have (1–0.57 = 0.43) 43% less odds of cardiovascular events compared to never smokers.

All these conclusions are strange. However, the next chapters will illustrate the use of statistical tests to show that all these conclusions are not real and are the results of sampling errors.

4.5. COMPARING CATEGORICAL DATA ACROSS TWO OR MORE POPULATIONS, GRAPHICAL METHODS

4.5.1. Bar Plot

geom_bar() is used to plot a bar graph where each bar has a height equal to the number of rows or observations at each level of the categorical variable. geom_bar() requires only one aesthetic (x) or the categorical variable you want to visualize. You can add another aesthetic, fill = categorical variable and position = "dodge" so every level will take a different color.

To add labels of counts, you will create a separate dataframe with counts for each level and use geom_col() and geom_text(). The geom_col() requires two aesthetics, x for one categorical variable, y for the count column, and fill for the other categorical variable. Also, position="dodge" to allow side by side bar plots. The geom_text() requires x and y which will be the same as geom_col() for the position of label and label for the text label. geom_label() can also be used instead of geom_text() with the same aesthetics but it will add a rectangle behind the text.

For example, bar graph for relation between smoking status and cardiovascular events

```
# frequency table
table(regicor$smoker, regicor$cv)
##
##                   Yes  No
## Never smoker       37 1099
## Current or former < 1y  47 506
## Former >= 1y       8 419
# plot bar graph
ggplot(data = regicor, aes(x = smoker, fill = cv))+geom_bar(width = 0.5, position = "dodge")
```

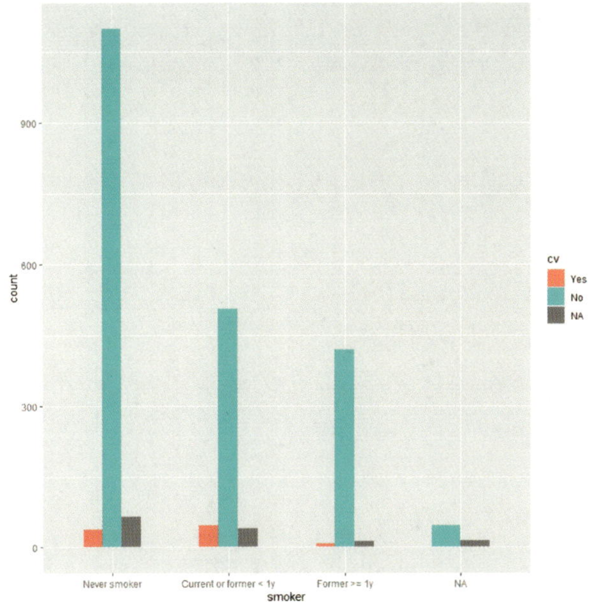

 # to remove NA

```
dat<- regicor %>% filter(!is.na(smoker) & !is.na(cv))
ggplot(data = dat, aes(x = smoker, fill = cv))+geom_bar(width = 0.5, position = "dodge")
```

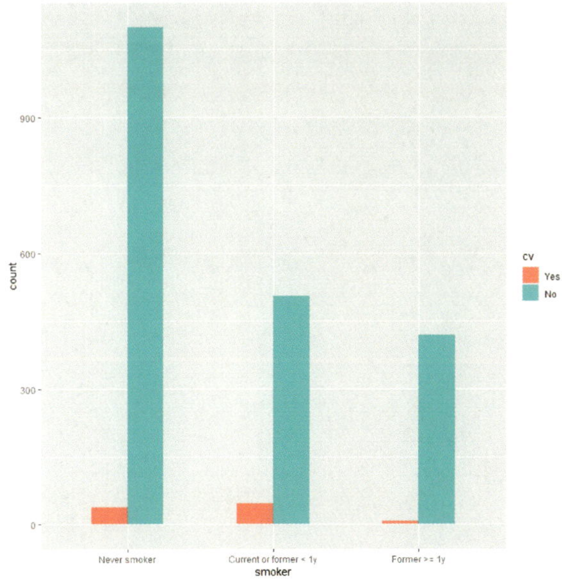

```
# to add labels of numbers
dat2<- dat %>% count(smoker, cv)
dat2
## # A tibble: 6 x 3
##   smoker             cv     n
##   <fct>              <fct> <int>
## 1 Never smoker       Yes    37
## 2 Never smoker       No   1099
## 3 Current or former < 1y Yes   47
## 4 Current or former < 1y No   506
## 5 Former >= 1y       Yes     8
## 6 Former >= 1y       No    419
ggplot(data = dat2, aes(x = smoker, y = n, fill = cv))+
 geom_col(width = 0.5, position = "dodge")+
 geom_text(aes(label = n))
```

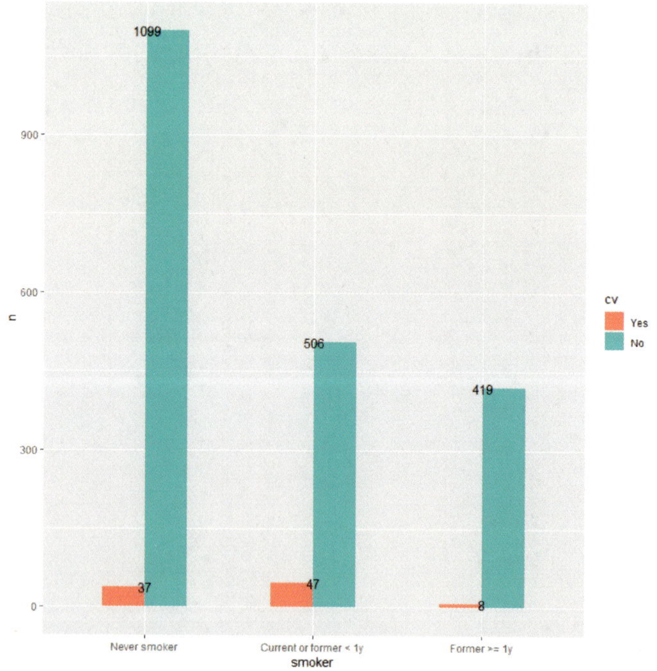


```
ggplot(data = dat2, aes(x = smoker, y = n, fill = cv))+
```

```
geom_col(width = 0.5, position = "dodge")+
geom_label(aes(label = n))
```

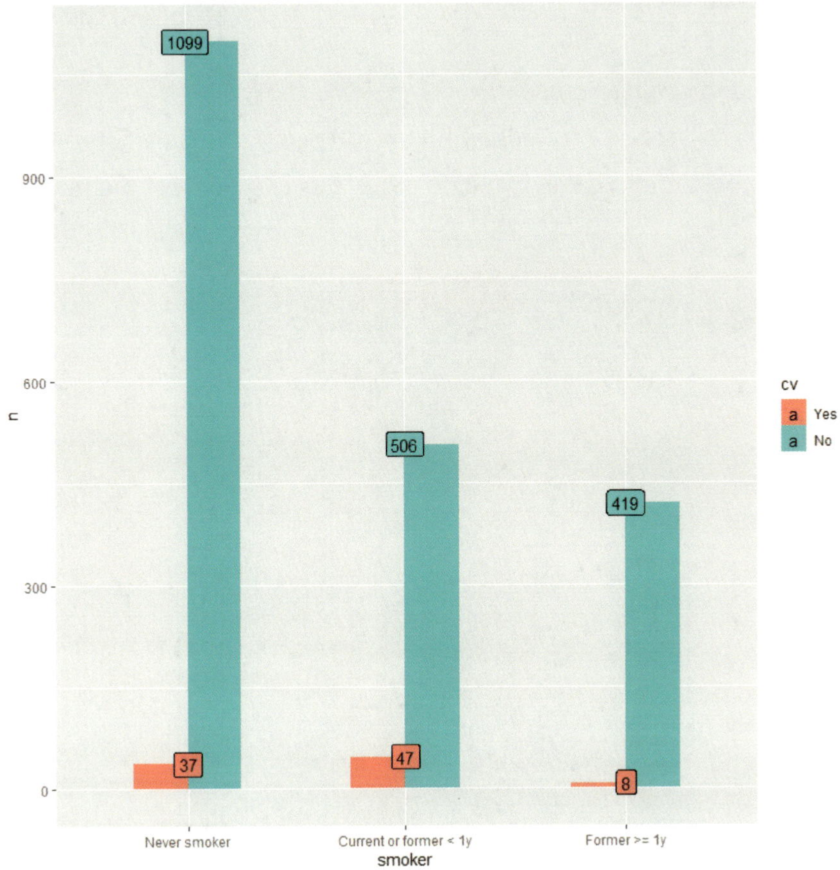

Notice how the bar heights are equal to counts from table() function.

4.5.2. Stacked Bar Plot

geom_bar() also can be used to plot stacked bar plot. geom_bar() requires two aesthetics, x = one categorical variable and fill = another categorical variable. Also the position argument of geom_bar is set to "fill." We have also added another layer using scale_y_continuous() to convert the count to percentage format using percent_format() function of scales package.

To add labels of percentages, you will create a separate dataframe with percentages for each level and use geom_bar() for the old dataframe and geom_text() for the new dataframe. We add some arguments to geom_text()

to force the text to be in middle of each compartment.

For example, bar graph for relation between smoking status and cardiovascular events

```
library(scales)
# bar graph for smoker and cv
ggplot(data = dat, aes(x = smoker, fill = cv))+geom_bar(position = "fill")+
 scale_y_continuous(name = "percentage,"labels = percent_format())
```

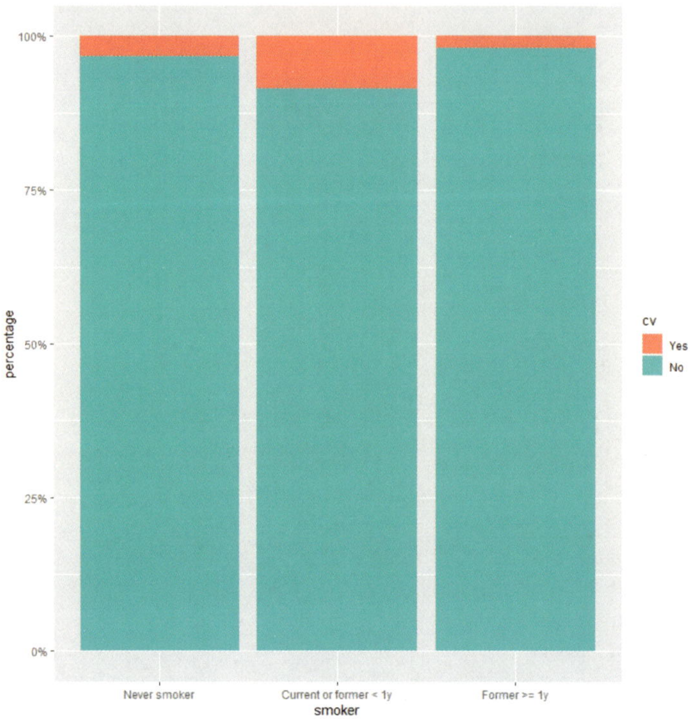


```
# to add percentages labels
dat2<- dat %>% group_by(smoker) %>% count(cv) %>%
 mutate(ratio = percent(n/sum(n)))
dat2
## # A tibble: 6 x 4
## # Groups:  smoker [3]
##  smoker        cv    n ratio
##  <fct>        <fct> <int> <chr>
```

```
## 1 Never smoker      Yes   37 3%
## 2 Never smoker      No    1099 97%
## 3 Current or former < 1y Yes   47 8%
## 4 Current or former < 1y No   506 92%
## 5 Former >= 1y      Yes    8 2%
## 6 Former >= 1y      No   419 98%
ggplot(data = dat, aes(x = smoker, fill = cv))+geom_bar(position = "fill")+
  geom_text(data = dat2, aes(x = smoker, y=n, label=ratio),

     position = position_fill(vjust = 0.5))+
  scale_y_continuous(name = "percentage,"labels = percent_format())
```

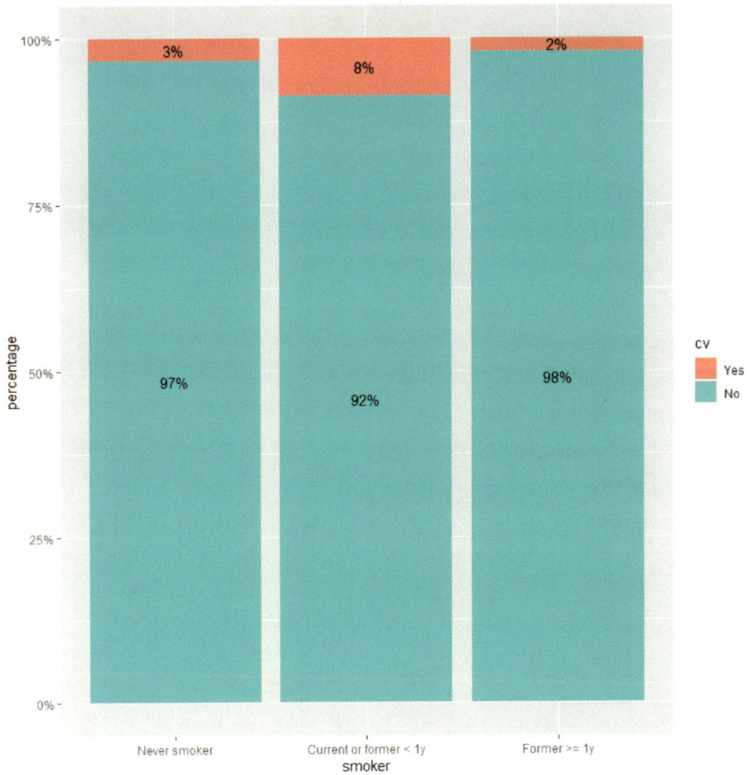


```
ggplot(data = dat, aes(x = smoker, fill = cv))+geom_bar(position = "fill")+
 geom_label(data = dat2, aes(x = smoker, y=n, label=ratio),
     position = position_fill(vjust = 0.5))+
```

```
scale_y_continuous(name = "percentage,"labels = percent_format())
```

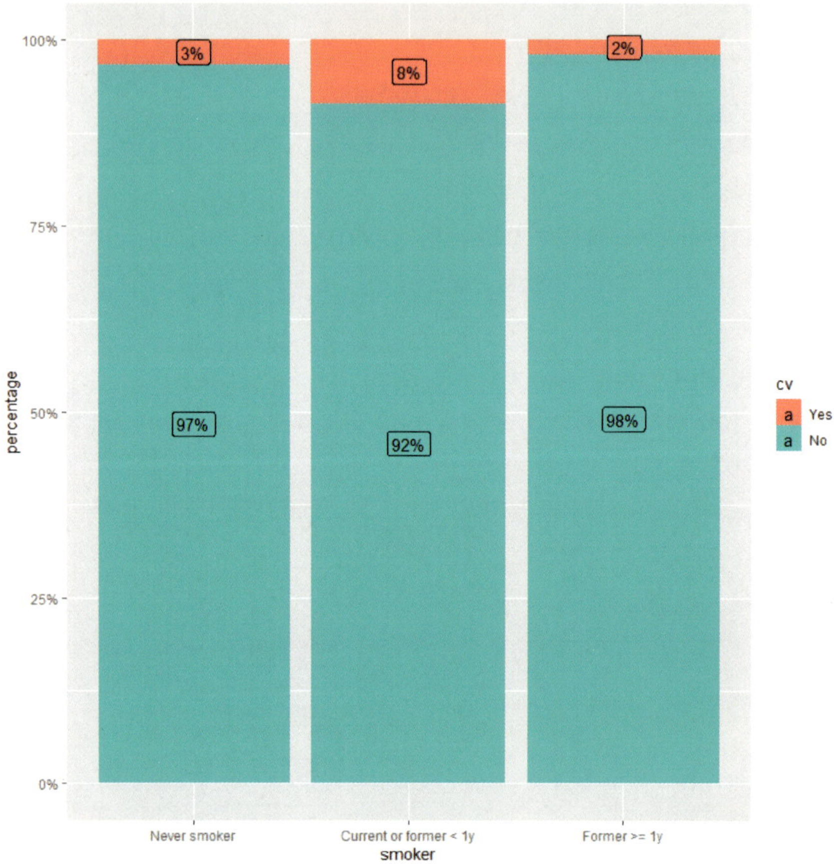

Note how the percentages in stacked bar plot are equal to that of group_ by() followed by the summary() function.

CHAPTER 5

TIME TO EVENT DATA = SURVIVAL DATA = FAILURE TIME DATA

CONTENTS

5.1. INTRODUCTION

Time to event data start with a cohort of observations and follow them over time till they have a well-defined binary outcome.

- Example 1: Subjects with primary biliary cirrhosis (PBC) followed for 10 years and the outcome of interest was death or alive.
- Example 2: Subjects given antiretroviral therapy (ART) and followed for 5 years till the outcome of interest occur or not which was partner to partner HIV transmission.
- Example 3: Smokers put into a counselling program and followed over a year to see rate of smoking quit which was the outcome of interest.

There are two dimensions of the data:

1. Number of events occurred which is a binary data; and
2. The total follow up time which is a continuous data.

5.2. NUMERICAL SUMMARIES

5.2.1. Incidence Rate (IR) for One Sample

Subjects with PBC followed for 10 years and the outcome of interest was death or alive.

We have three subjects:

1. Subject 1 enters at the start of the study and dies 7 years later. So, the follow up time for this subject is 7 years;
2. Subject 2 enters the study after 4 years of the study start and lost to follow up after 2 years. For this subject, we do not know if he is still alive or not and the follow up time for this subject is 2 years; and
3. Subject 3 enters the study after 3 years of the study start and ends the follow up without having the outcome. So, the follow up time for this subject is 7 years.

Subjects 2 and 3 are called censored data or partial information for subjects dropping out of the study or ending the study without having the outcome.

Incidence rate = IR = Total number of deaths (outcomes) in a sample/Total follow up time

IR in this sample of 3 subjects =1 death/(7+2+7 years)= 0.0625 death per year of follow up. This IR assumes that the rate of death is constant across the entire follow up period, i.e., the same death rate in the 1st year, 2nd year, …, 10th year.

- Interpretation:
  - If we follow 16 persons for a year, we expect to see 16 × 0.0625= 1 death.
  - If we follow 100 persons for a year, we expect to see 100 × 0.0625= 6.25 deaths.
  - If we follow 100 persons for 6 months, we expect to see 100 × 0.5 × 0.0625= 3.125 deaths.
  - If we follow 1000 persons for 1 month, we expect to see 1000 × 1/12 × 0.0625= 5.21 deaths.
  - If we follow 100 persons for 10 years, we expect to see 100 × 10 × 0.0625= 62.5 deaths.
  - If we follow 1000 persons for 10 years, we expect to see 1000 × 10 × 0.0625= 625 deaths.

Another example from [CDC] https://www.cdc.gov/csels/dsepd/ss1978/lesson3/section2.html.

The diabetes follow-up study included 218 diabetic women and 3,823 nondiabetic women. By the end of the study, 72 of the diabetic women and 511 of the nondiabetic women had died. The diabetic women were observed for a total of 1862 years; the nondiabetic women were observed for a total of 36653 years. Calculate the IRs of death for the diabetic and non-diabetic women.For diabetic women, Incidence rate =72/1862= 0.0386 deaths per year = 38.6 deaths per 1,000 person followed over a year.

For nondiabetic women, Incidence rate = 511/36653= 0.0139 deaths per year= 13.9 deaths per 1,000 person followed over a year.

5.2.2. Incidence Rate Ratio (IRR)

Used to compare time to event data between 2 or more groups numerically. In the above example from CDC

IR for diabetics = 0.0386 deaths per year and IR for nondiabetics = 0.0139 deaths per year

IRR = 0.0386/0.0139= 2.78

The risk of death in diabetic women in the follow up period is 2.78 times the risk in nondiabetic women OR the diabetic women had (2.78–1)=1.78 × 100= 178% greater risk of death compared to nondiabetic women.

When the outcome is death, the IR can be called the mortality rate.

When we have three or more groups, assign one group as the reference group and calculate IR ratios relative to this reference group.

For example, in a study of maternal vitamin supplementation and infant mortality where each woman was given vitamin A, vitamin B-carotene, or placebo.

$IR_{\text{vitamin A}} = 0.00041$ deaths per infant day

$IR_{\text{vitamin B-carotene}} = 0.00039$ deaths per infant day

$IR_{\text{placebo}} = 0.00039$ deaths per infant day

Making the placebo group the reference group,

$IRR_{\text{vitamin A}} = IR_{\text{vitamin A}} / IR_{\text{placebo}} = 1.05$

$IRR_{\text{vitamin B-carotene}} = IR_{\text{vitamin B}} / IR_{\text{placebo}} = 1.05$

Vitamin A had a 5% higher risk of death than the placebo group while vitamin B-carotene has the same death risk as that of the placebo. In other words, the estimated mortality rate in vitamin A group is 5% greater than that of the placebo.

5.3. GRAPHICAL SUMMARIES: KAPLAN-MEIER APPROACH

Kaplan Meier approach is a standard method for creating two dimensional graph of time to event data (time on x-axis and proportion event-free at y-axis)

This graph is called survival curve. The survival curve is a summary statistic that is estimated from a sample data and it estimates the cumulative survival of a population.

The cumulative survival is the proportion of the population, estimated from the sample, that remains event free at a certain time or beyond.

The estimated survival curve or S(t) at time=0 or start of the study is denoted by S(t) =1 as all our study subjects are event free at the start of the study. The S(t) remains at 1 till an events occur and drops in a step wise manner during the follow up time.

The lowest value of survival curve or S(t) is zero if every subject has the event or higher than zero indicating that this proportion is still event free.

The survival curve is based on all data, the complete one and the censored one.

5.3.1. Calculation of Kaplan Meier Curve

Generally done using the computer

S(t) = N(t) – E(t)/N(t) × S(previous t)

where; N(t) is the number at risk of having the event at time t; E(t) is the number who have had the event at time t; S(previous t) is the proportion survived beyond the previous event time.

5.3.2. Plotting Survival Curve for Whole Data

We will use survival package in R with its accompanying dataset veteran.

Veteran data is randomized trial of two treatment regimens for lung cancer. It consists of 137 rows or patients and eight columns which are:

1. trt: treatment type, 1=standard 2=test
2. celltype: cancer cell type, 1=squamous, 2=smallcell, 3=adeno, 4=large
3. time: survival time
4. status: censoring status,1=death was observed, 0=censoring observation
5. karno: Karnofsky performance score (100=good)
6. diagtime: months from diagnosis to randomization
7. age: in years
8. prior: prior therapy, 0=no, 10=yes

Loading and exploring the data:

```
library(survival)
library(survminer)
library(ggpubr)
data("veteran")
str(veteran)
## 'data.frame': 137 obs. of 8 variables:
```

```
## $ trt   : num 1 1 1 1 1 1 1 1 1 1 ...
## $ celltype: Factor w/ 4 levels "squamous,""smallcell,"..: 1 1 1 1 1 1 1 1 1 ...
## $ time  : num 72 411 228 126 118 10 82 110 314 100 ...
## $ status : num 1 1 1 1 1 1 1 1 1 0 ...
## $ karno  : num 60 70 60 60 70 20 40 80 50 70 ...
## $ diagtime: num 7 5 3 9 11 5 10 29 18 6 ...
## $ age   : num 69 64 38 63 65 49 69 68 43 70 ...
## $ prior  : num 0 10 0 10 10 0 10 0 0 0 ...
head(veteran,10)
##  trt celltype time status karno diagtime age prior
## 1  1 squamous  72   1  60    7 69   0
## 2  1 squamous 411   1  70    5 64  10
## 3  1 squamous 228   1  60    3 38   0
## 4  1 squamous 126   1  60    9 63  10
## 5  1 squamous 118   1  70   11 65  10
## 6  1 squamous  10   1  20    5 49   0
## 7  1 squamous  82   1  40   10 69  10
## 8  1 squamous 110   1  80   29 68   0
## 9  1 squamous 314   1  50   18 43   0
## 10  1 squamous 100   0  70    6 70   0
```

Building survival object:

Surv function takes the time and status column to build the survival object

```
km<-Surv(veteran$time,veteran$status)
head(km,10)
## [1] 72 411 228 126 118  10  82 110 314 100+
class(km)
## [1] "Surv"
#It is a survival object
head(veteran[,3:4],10)
##  time status
```

```
## 1  72   1
## 2  411   1
## 3  228   1
## 4  126   1
## 5  118   1
## 6  10   1
## 7  82   1
## 8  110   1
## 9  314   1
## 10 100   0
```

Note that a "+" after the time in the print out of km indicates a censoring observation. They are the same values as the time column but with "+" sign to indicate censoring.

Whole survival curve:

To see the whole Kaplan Meier curve for the proportion of survival, use the survfit function with km as the response variable and 1 as the predictor variable to see the whole curve. Then use ggsurvplot() to plot the curve.

```
km_fit<-survfit(km ~1, data = veteran)
ggsurvplot(km_fit)
```

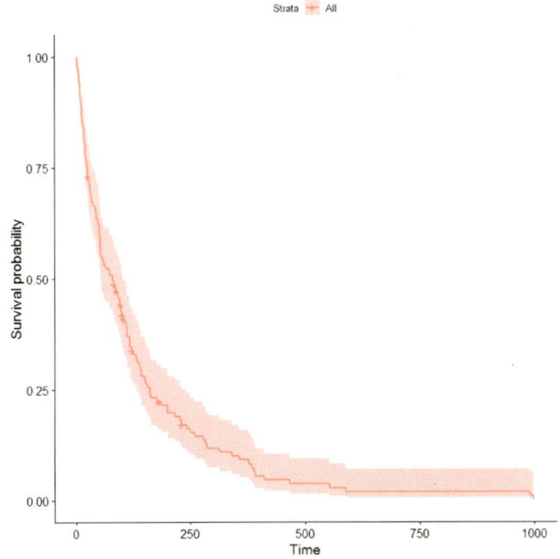

```
# without confidence interval
ggsurvplot(km_fit, conf.int = F)
```

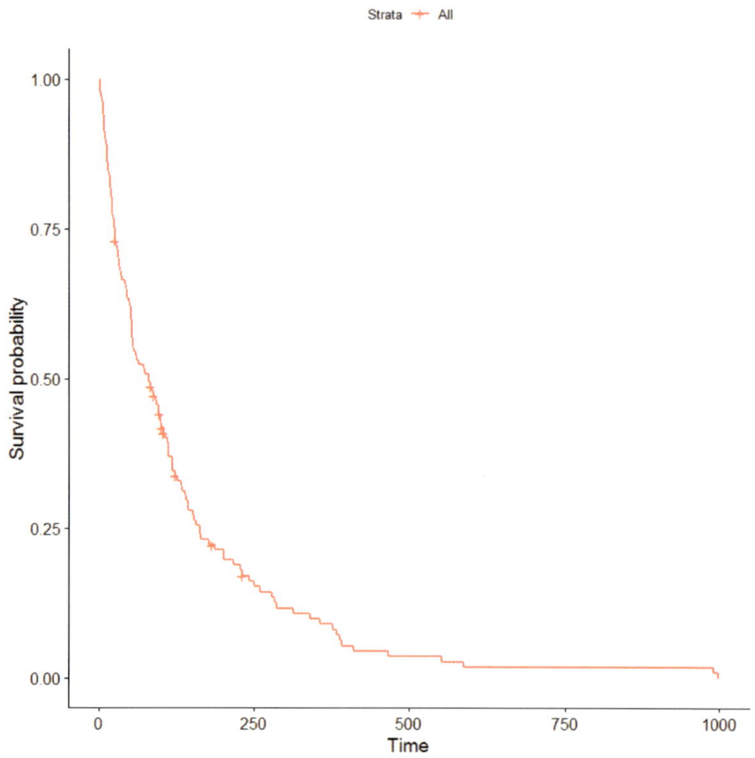

Summarizing the survival curve:

```
#To see the minimum time and maximum time in our data
summary(veteran$time)
## Min. 1st Qu. Median Mean 3rd Qu. Max.
## 1.0 25.0 80.0 121.6 144.0 999.0
#So it lasts for 999 days
summary(km_fit)
## Call: survfit(formula = km ~ 1, data = veteran)
##
## time n.risk n.event survival std.err lower 95% CI upper 95% CI
```

```
## 1 137 2 0.985 0.01025 0.96552 1.0000
## 2 135 1 0.978 0.01250 0.95390 1.0000
## 3 134 1 0.971 0.01438 0.94302 0.9994
## 4 133 1 0.964 0.01602 0.93261 0.9954
## 7 132 3 0.942 0.02003 0.90315 0.9817
## 8 129 4 0.912 0.02415 0.86628 0.9610
## 10 125 2 0.898 0.02588 0.84850 0.9500
## 11 123 1 0.891 0.02668 0.83973 0.9444
## 12 122 2 0.876 0.02817 0.82241 0.9329
## 13 120 2 0.861 0.02953 0.80534 0.9212
## 15 118 2 0.847 0.03078 0.78849 0.9092
## 16 116 1 0.839 0.03137 0.78013 0.9032
## 18 115 3 0.818 0.03300 0.75533 0.8848
## 19 112 2 0.803 0.03399 0.73900 0.8724
## 20 110 2 0.788 0.03490 0.72280 0.8598
## 21 108 2 0.774 0.03575 0.70674 0.8471
## 22 106 1 0.766 0.03615 0.69875 0.8407
## 24 105 2 0.752 0.03690 0.68286 0.8277
## 25 103 3 0.730 0.03793 0.65924 0.8082
## 27 99 1 0.723 0.03826 0.65133 0.8016
## 29 98 1 0.715 0.03857 0.64344 0.7949
## 30 97 2 0.700 0.03916 0.62774 0.7816
## 31 95 2 0.686 0.03970 0.61213 0.7681
## 33 93 1 0.678 0.03995 0.60436 0.7613
## 35 92 1 0.671 0.04019 0.59661 0.7545
## 36 91 1 0.664 0.04042 0.58889 0.7477
## 42 90 1 0.656 0.04064 0.58119 0.7409
## 43 89 1 0.649 0.04085 0.57351 0.7340
## 44 88 1 0.641 0.04104 0.56585 0.7272
## 45 87 1 0.634 0.04123 0.55821 0.7203
## 48 86 1 0.627 0.04140 0.55059 0.7133
```

```
## 49  85  1 0.619 0.04157  0.54299  0.7064
## 51  84  3 0.597 0.04200  0.52032  0.6855
## 52  81  3 0.575 0.04234  0.49782  0.6644
## 53  78  1 0.568 0.04243  0.49036  0.6573
## 54  77  2 0.553 0.04259  0.47549  0.6431
## 56  75  1 0.546 0.04266  0.46808  0.6360
## 59  74  1 0.538 0.04272  0.46070  0.6288
## 61  73  1 0.531 0.04276  0.45333  0.6216
## 63  72  1 0.523 0.04280  0.44598  0.6145

## 72  71  1 0.516 0.04283  0.43864  0.6073
## 73  70  1 0.509 0.04284  0.43133  0.6000
## 80  69  2 0.494 0.04285  0.41675  0.5855
## 82  67  1 0.487 0.04284  0.40949  0.5783
## 84  65  1 0.479 0.04283  0.40212  0.5709
## 87  64  1 0.472 0.04281  0.39478  0.5635
## 90  62  1 0.464 0.04279  0.38731  0.5560
## 92  61  1 0.456 0.04276  0.37986  0.5484
## 95  60  2 0.441 0.04267  0.36504  0.5333
## 99  57  2 0.426 0.04255  0.35000  0.5179
## 100  55  1 0.418 0.04248  0.34251  0.5101
## 103  53  1 0.410 0.04240  0.33488  0.5022
## 105  51  1 0.402 0.04233  0.32711  0.4942
## 110  50  1 0.394 0.04224  0.31936  0.4861
## 111  49  2 0.378 0.04201  0.30395  0.4700
## 112  47  1 0.370 0.04188  0.29628  0.4618
## 117  46  2 0.354 0.04158  0.28103  0.4455
## 118  44  1 0.346 0.04140  0.27345  0.4372
## 122  43  1 0.338 0.04121  0.26589  0.4290
## 126  41  1 0.329 0.04102  0.25815  0.4206
## 132  40  1 0.321 0.04082  0.25044  0.4121
```

```
## 133 39 1 0.313 0.04059 0.24277 0.4036
## 139 38 1 0.305 0.04035 0.23513 0.3951
## 140 37 1 0.297 0.04009 0.22752 0.3865
## 143 36 1 0.288 0.03982 0.21994 0.3779
## 144 35 1 0.280 0.03952 0.21240 0.3693
## 151 34 1 0.272 0.03921 0.20490 0.3606
## 153 33 1 0.264 0.03888 0.19743 0.3519
## 156 32 1 0.255 0.03852 0.18999 0.3432
## 162 31 2 0.239 0.03776 0.17525 0.3256
## 164 29 1 0.231 0.03734 0.16793 0.3168
## 177 28 1 0.222 0.03691 0.16066 0.3079
## 186 26 1 0.214 0.03647 0.15310 0.2987
## 200 25 1 0.205 0.03600 0.14560 0.2895
## 201 24 1 0.197 0.03550 0.13814 0.2802
## 216 23 1 0.188 0.03497 0.13075 0.2709
## 228 22 1 0.180 0.03441 0.12341 0.2615
## 231 21 1 0.171 0.03382 0.11613 0.2520
## 242 19 1 0.162 0.03322 0.10846 0.2422
## 250 18 1 0.153 0.03257 0.10088 0.2323
## 260 17 1 0.144 0.03187 0.09338 0.2223
## 278 16 1 0.135 0.03113 0.08598 0.2122
## 283 15 1 0.126 0.03033 0.07867 0.2020
## 287 14 1 0.117 0.02947 0.07147 0.1917
## 314 13 1 0.108 0.02854 0.06439 0.1813
## 340 12 1 0.099 0.02755 0.05743 0.1708
## 357 11 1 0.090 0.02647 0.05061 0.1602
## 378 10 1 0.081 0.02531 0.04394 0.1495
## 384 9 1 0.072 0.02405 0.03744 0.1386
## 389 8 1 0.063 0.02267 0.03115 0.1275
## 392 7 1 0.054 0.02114 0.02509 0.1163
## 411 6 1 0.045 0.01944 0.01931 0.1049
```

```
## 467  5   1 0.036 0.01751   0.01389    0.0934
## 553  4   1 0.027 0.01528   0.00892    0.0818
## 587  3   1 0.018 0.01256   0.00459    0.0707
## 991  2   1 0.009 0.00894   0.00129    0.0631
## 999  1   1 0.000   NaN       NA        NA
## times to determine the intervals at which we see the survival proportion

summary(km_fit, times = seq(0,999,111))
## Call: survfit(formula = km ~ 1, data = veteran)
##
## time n.risk n.event survival std.err lower 95% CI upper 95% CI
##    0  137     0  1.000 0.0000   1.00000    1.0000
##  111   49    84  0.378 0.0420   0.30395    0.4700
##  222   22    23  0.188 0.0350   0.13075    0.2709
##  333   12     9  0.108 0.0285   0.06439    0.1813
##  444    5     7  0.045 0.0194   0.01931    0.1049
##  555    3     2  0.027 0.0153   0.00892    0.0818
##  666    2     1  0.018 0.0126   0.00459    0.0707
##  777    2     0  0.018 0.0126   0.00459    0.0707
##  888    2     0  0.018 0.0126   0.00459    0.0707
##  999    1     2  0.000   NaN       NA        NA
##To see the 25% and 50%s
summary(km_fit,times=seq(0,100,1))
## Call: survfit(formula = km ~ 1, data = veteran)
##
## time n.risk n.event survival std.err lower 95% CI upper 95% CI
##    0  137     0  1.000 0.0000    1.000    1.000
##    1  137     2  0.985 0.0102    0.966    1.000
##    2  135     1  0.978 0.0125    0.954    1.000
##    3  134     1  0.971 0.0144    0.943    0.999
##    4  133     1  0.964 0.0160    0.933    0.995
##    5  132     0  0.964 0.0160    0.933    0.995
```

```
## 6 132 0 0.964 0.0160 0.933 0.995
## 7 132 3 0.942 0.0200 0.903 0.982
## 8 129 4 0.912 0.0242 0.866 0.961
## 9 125 0 0.912 0.0242 0.866 0.961
## 10 125 2 0.898 0.0259 0.848 0.950
## 11 123 1 0.891 0.0267 0.840 0.944
## 12 122 2 0.876 0.0282 0.822 0.933
## 13 120 2 0.861 0.0295 0.805 0.921
## 14 118 0 0.861 0.0295 0.805 0.921
## 15 118 2 0.847 0.0308 0.788 0.909
## 16 116 1 0.839 0.0314 0.780 0.903
## 17 115 0 0.839 0.0314 0.780 0.903
## 18 115 3 0.818 0.0330 0.755 0.885
## 19 112 2 0.803 0.0340 0.739 0.872
## 20 110 2 0.788 0.0349 0.723 0.860
## 21 108 2 0.774 0.0357 0.707 0.847
## 22 106 1 0.766 0.0361 0.699 0.841
## 23 105 0 0.766 0.0361 0.699 0.841
## 24 105 2 0.752 0.0369 0.683 0.828
## 25 103 3 0.730 0.0379 0.659 0.808
## 26 99 0 0.730 0.0379 0.659 0.808
## 27 99 1 0.723 0.0383 0.651 0.802
## 28 98 0 0.723 0.0383 0.651 0.802
## 29 98 1 0.715 0.0386 0.643 0.795
## 30 97 2 0.700 0.0392 0.628 0.782
## 31 95 2 0.686 0.0397 0.612 0.768
## 32 93 0 0.686 0.0397 0.612 0.768
## 33 93 1 0.678 0.0400 0.604 0.761
## 34 92 0 0.678 0.0400 0.604 0.761
## 35 92 1 0.671 0.0402 0.597 0.755
## 36 91 1 0.664 0.0404 0.589 0.748
```

```
## 37  90  0 0.664 0.0404   0.589   0.748
## 38  90  0 0.664 0.0404   0.589   0.748
## 39  90  0 0.664 0.0404   0.589   0.748
## 40  90  0 0.664 0.0404   0.589   0.748
## 41  90  0 0.664 0.0404   0.589   0.748
## 42  90  1 0.656 0.0406   0.581   0.741
## 43  89  1 0.649 0.0408   0.574   0.734
## 44  88  1 0.641 0.0410   0.566   0.727
## 45  87  1 0.634 0.0412   0.558   0.720
## 46  86  0 0.634 0.0412   0.558   0.720
## 47  86  0 0.634 0.0412   0.558   0.720
## 48  86  1 0.627 0.0414   0.551   0.713
## 49  85  1 0.619 0.0416   0.543   0.706
## 50  84  0 0.619 0.0416   0.543   0.706
## 51  84  3 0.597 0.0420   0.520   0.685
## 52  81  3 0.575 0.0423   0.498   0.664
## 53  78  1 0.568 0.0424   0.490   0.657
## 54  77  2 0.553 0.0426   0.475   0.643
## 55  75  0 0.553 0.0426   0.475   0.643
## 56  75  1 0.546 0.0427   0.468   0.636
## 57  74  0 0.546 0.0427   0.468   0.636
## 58  74  0 0.546 0.0427   0.468   0.636
## 59  74  1 0.538 0.0427   0.461   0.629
## 60  73  0 0.538 0.0427   0.461   0.629
## 61  73  1 0.531 0.0428   0.453   0.622
## 62  72  0 0.531 0.0428   0.453   0.622
## 63  72  1 0.523 0.0428   0.446   0.614
## 64  71  0 0.523 0.0428   0.446   0.614
## 65  71  0 0.523 0.0428   0.446   0.614
## 66  71  0 0.523 0.0428   0.446   0.614
## 67  71  0 0.523 0.0428   0.446   0.614
```

```
## 68 71 0 0.523 0.0428 0.446 0.614
## 69 71 0 0.523 0.0428 0.446 0.614
## 70 71 0 0.523 0.0428 0.446 0.614
## 71 71 0 0.523 0.0428 0.446 0.614
## 72 71 1 0.516 0.0428 0.439 0.607
## 73 70 1 0.509 0.0428 0.431 0.600
## 74 69 0 0.509 0.0428 0.431 0.600
## 75 69 0 0.509 0.0428 0.431 0.600
## 76 69 0 0.509 0.0428 0.431 0.600
## 77 69 0 0.509 0.0428 0.431 0.600
## 78 69 0 0.509 0.0428 0.431 0.600
## 79 69 0 0.509 0.0428 0.431 0.600
## 80 69 2 0.494 0.0429 0.417 0.586
## 81 67 0 0.494 0.0429 0.417 0.586
## 82 67 1 0.487 0.0428 0.409 0.578
## 83 66 0 0.487 0.0428 0.409 0.578
## 84 65 1 0.479 0.0428 0.402 0.571
## 85 64 0 0.479 0.0428 0.402 0.571
## 86 64 0 0.479 0.0428 0.402 0.571
## 87 64 1 0.472 0.0428 0.395 0.563
## 88 62 0 0.472 0.0428 0.395 0.563
## 89 62 0 0.472 0.0428 0.395 0.563
## 90 62 1 0.464 0.0428 0.387 0.556
## 91 61 0 0.464 0.0428 0.387 0.556
## 92 61 1 0.456 0.0428 0.380 0.548
## 93 60 0 0.456 0.0428 0.380 0.548
## 94 60 0 0.456 0.0428 0.380 0.548
## 95 60 2 0.441 0.0427 0.365 0.533
## 96 58 0 0.441 0.0427 0.365 0.533
## 97 58 0 0.441 0.0427 0.365 0.533
## 98 57 0 0.441 0.0427 0.365 0.533
```

```
## 99 57 2 0.426 0.0426 0.350 0.518
## 100 55 1 0.418 0.0425 0.343 0.510
```

- **Interpretation:**

We see from the table that the lowest survival probability (or proportion) is 0.000 or 0% which indicate that all our sample (except the censoring observations) had died at the end of follow up period.

From the third table, we see that 25% is at t=24 where 75% are still alive and 25% has the outcome or died. Also, the 50% (median) is at t=80 where 50% are still alive and 50% has the outcome or died.

It is concluded from this curve that the highest risk of death occurs in the first 100 days of this study where about (1–0.42=0.58 or 58%) of deaths has occurred in this interval.

However, when comparing two groups, IRR does not tell us the risk over time in the different groups as we see from the survival curves.

5.3.3. Plotting Survival Curve for Different Strata

Different survival curves for the two treatments

```
trt_fit<-survfit(km ~trt, data = veteran)
#To see the minimum time and maximum time in our data stratified by treatment
library(dplyr)
veteran %>% group_by(trt) %>% summarise(min=min(time),max = max(time))
## # A tibble: 2 x 3
##   trt min max
##   <dbl> <dbl> <dbl>
## 1 1 3 553
## 2 2 1 999
summary(trt_fit,times = seq(0,999,111))
## Call: survfit(formula = km ~ trt, data = veteran)
##
##        trt=1
## time n.risk n.event survival std.err lower 95% CI upper 95% CI
```

```
## 0 69 0 1.0000 0.0000 1.00000 1.000
## 111 29 37 0.4549 0.0607 0.35018 0.591
## 222 10 17 0.1770 0.0486 0.10340 0.303
## 333 4 6 0.0708 0.0336 0.02793 0.180
## 444 1 3 0.0177 0.0175 0.00256 0.123
##
## trt=2
## time n.risk n.event survival std.err lower 95% CI upper 95% CI
## 0 68 0 1.0000 0.0000 1.00000 1.000
## 111 20 47 0.2994 0.0566 0.20673 0.434
## 222 12 6 0.1996 0.0503 0.12181 0.327
## 333 8 3 0.1464 0.0454 0.07973 0.269
## 444 4 4 0.0732 0.0344 0.02912 0.184
## 555 3 1 0.0549 0.0303 0.01861 0.162
## 666 2 1 0.0366 0.0251 0.00953 0.140
## 777 2 0 0.0366 0.0251 0.00953 0.140
## 888 2 0 0.0366 0.0251 0.00953 0.140
## 999 1 2 0.0000 NaN NA NA
ggsurvplot(trt_fit)
```

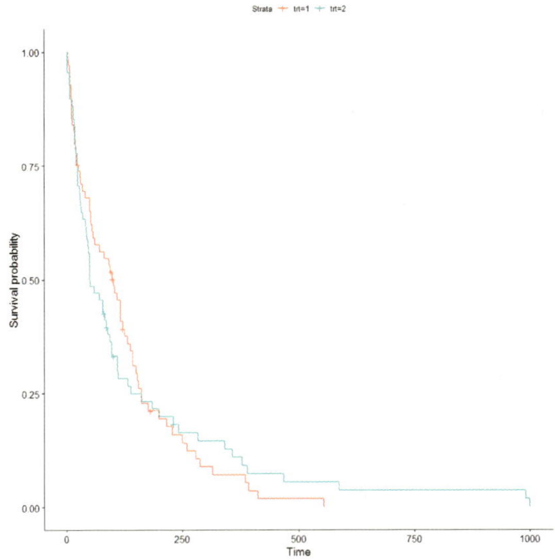

ggsurvplot(trt_fit, surv.median.line = "hv")

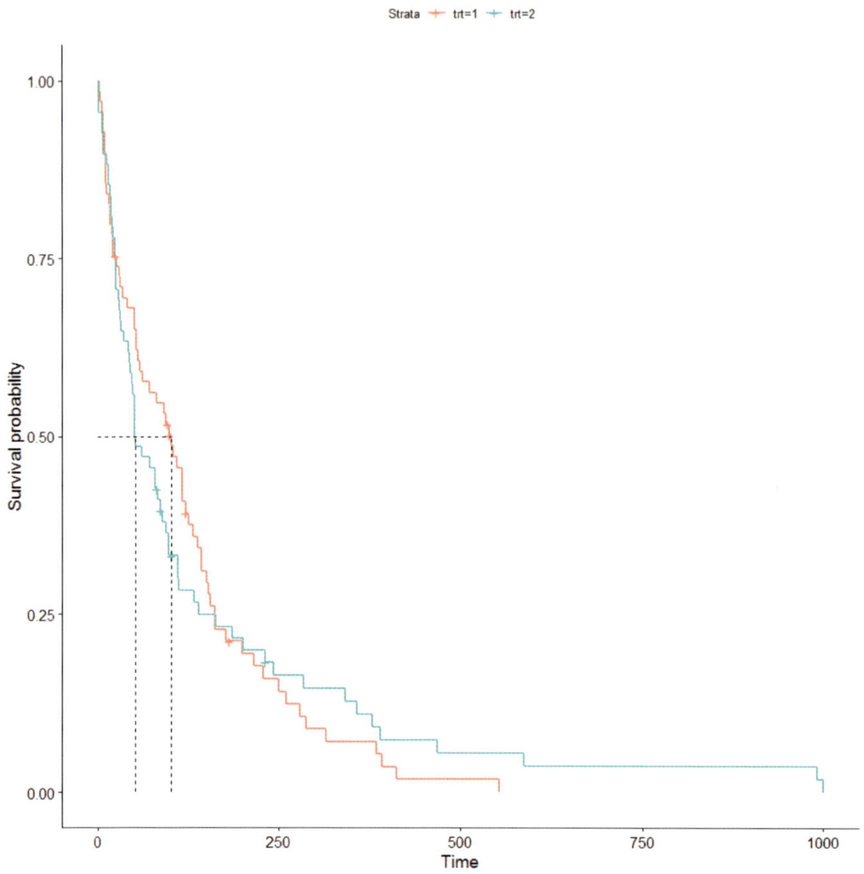

- **Interpretation:** We see that treatment 1 (standard treatment) has higher survival probability or lower death risk in about the first 200 days of life then treatment 2 (test treatment) has higher survival probability afterwards.

The standard treatment has first outcome at day=3 and last outcome (death) at day=553, while the test treatment has first outcome at day=1 and last death event at day=999.

Statistical tests are needed to determine which treatment is better and these tests will be discussed later.

Different survival curves for the different cell types:

cell_fit<-**survfit**(km ~celltype, data = veteran)

```
#To see the minimum time and maximum time in our data stratified by celltype
library(dplyr)
veteran %>% group_by(celltype) %>% summarise(min=min(time),max = max(time))
## # A tibble: 4 x 3
##  celltype  min  max
##  <fct>   <dbl> <dbl>
## 1 squamous    1  999
## 2 smallcell   2  392
## 3 adeno       3  186
## 4 large      12  553
summary(cell_fit,times = seq(0,999,111))
## Call: survfit(formula = km ~ celltype, data = veteran)
##
##         celltype=squamous
## time n.risk n.event survival std.err lower 95% CI upper 95% CI
##    0    35      0  1.000 0.0000    1.0000    1.000
##  111    18     15  0.557 0.0861    0.4112    0.754
##  222    12      5  0.393 0.0865    0.2553    0.605
##  333     7      4  0.252 0.0793    0.1362    0.467
##  444     4      3  0.144 0.0654    0.0592    0.351
##  555     3      1  0.108 0.0581    0.0377    0.310
##  666     2      1  0.072 0.0487    0.0192    0.271
##  777     2      0  0.072 0.0487    0.0192    0.271
##  888     2      0  0.072 0.0487    0.0192    0.271
##  999     1      2  0.000    NaN        NA       NA
##
##         celltype=smallcell
## time n.risk n.event survival std.err lower 95% CI upper 95% CI
##    0    48      0  1.0000 0.0000    1.0000    1.000
##  111     9     37  0.2257 0.0609    0.1330    0.383
```

```
## 222   3    5 0.0878 0.0457    0.0316    0.244
## 333   2    1 0.0585 0.0387    0.0160    0.214
##
##         celltype=adeno
## time n.risk n.event survival std.err lower 95% CI upper 95% CI
##   0   27    0  1.000 0.0000    1.0000     1.000
## 111    5   21  0.206 0.0802    0.0959     0.442
##
##         celltype=large
## time n.risk n.event survival std.err lower 95% CI upper 95% CI
##   0   27    0 1.0000 0.0000  1.00000     1.000
## 111   17   11 0.5926 0.0946  0.43344     0.810
## 222    7    8 0.2881 0.0887  0.15759     0.527
## 333    3    4 0.1235 0.0659  0.04335     0.352
## 444    1    2 0.0412 0.0401  0.00608     0.279
ggsurvplot(cell_fit)
```

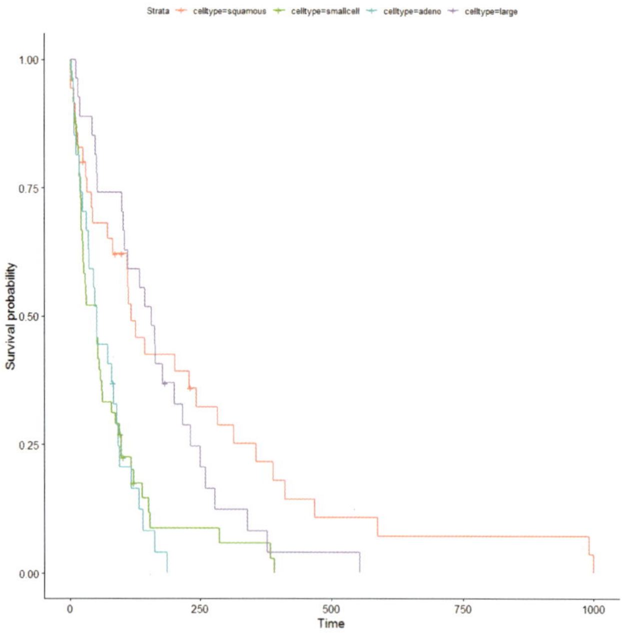


```
ggsurvplot(cell_fit, surv.median.line = "hv")
```

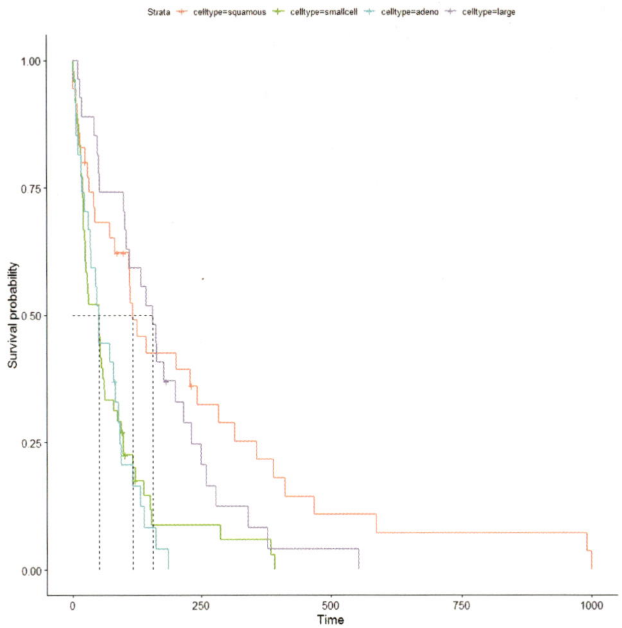

ggsurvplot(cell_fit, surv.median.line = “hv,” pval = TRUE)

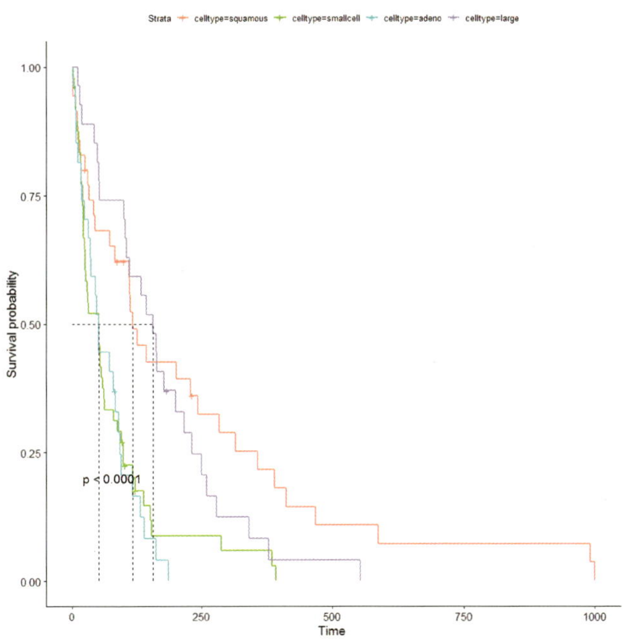

ggsurvplot(cell_fit, surv.median.line = “hv,” pval = TRUE, pval.method = TRUE)

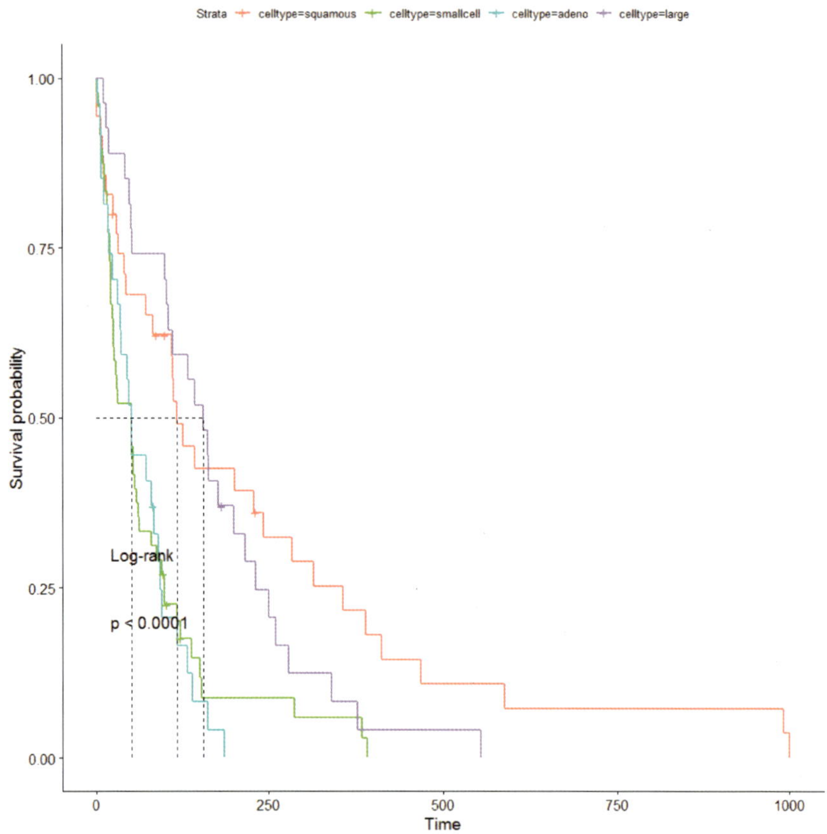

- **Interpretation:** We see that the large celltype has the highest survival in about the first 200 days of the study. After that, the squamous celltype has the highest survival.

Different survival curves for the different age groups.

We can cut the age variable (continuous variables) to four quartiles.

```
#To determine the break points
summary(veteran$age)
## Min. 1st Qu. Median Mean 3rd Qu. Max.
## 34.00 51.00 62.00 58.31 66.00 81.00
#We use 33 not 34 and 82 not 81 to include all the values
age_quart<-cut(veteran$age,breaks = c(33,51,62,66,82),labels = c("Q1,""Q2,""Q3,""Q4"))
table(age_quart, veteran$age)
```

```
##
## age_quart 34 35 36 37 38 39 40 41 42 43 44 45 46 47 48 49 50 51 52 53
54 55 56
##   Q1 1 2 1 2 3 2 1 1 3 4 1 2 1 1 3 3 3 1 0 0 0 0 0
##   Q2 0 0 0 0 0 0 0 0 0 0 0 0 0 0 0 0 0 0 4 3 2 3 1
##   Q3 0 0 0 0 0 0 0 0 0 0 0 0 0 0 0 0 0 0 0 0 0 0 0
##   Q4 0 0 0 0 0 0 0 0 0 0 0 0 0 0 0 0 0 0 0 0 0 0 0
##
## age_quart 57 58 59 60 61 62 63 64 65 66 67 68 69 70 71 72 81
##   Q1 0 0 0 0 0 0 0 0 0 0 0 0 0 0 0 0 0
##   Q2 1 3 1 7 4 10 0 0 0 0 0 0 0 0 0 0 0
##   Q3 0 0 0 0 0 0 12 7 7 7 0 0 0 0 0 0 0
##   Q4 0 0 0 0 0 0 0 0 0 0 6 6 6 5 3 2 2
##Combine the age_quart with our original dataframe
veteran<-data.frame(veteran,age_quart)
table(veteran$age_quart)
##
## Q1 Q2 Q3 Q4
## 35 39 33 30
age_fit<-survfit(km ~ age_quart, data = veteran)
#To see the minimum time and maximum time in our data stratified by age
quartiles
veteran%>%group_by(age_quart)%>%summarise(min=min(time),max
= max(time))

## # A tibble: 4 x 3
##   age_quart  min  max
##   <fct>    <dbl> <dbl>
## 1 Q1         1  991
## 2 Q2         7  999
## 3 Q3         1  467
## 4 Q4         7  392
```

```
summary(age_fit,times = seq(0,999,111))
## Call: survfit(formula = km ~ age_quart, data = veteran)
##
##       age_quart=Q1
## time n.risk n.event survival std.err lower 95% CI upper 95% CI
##  0   35    0  1.0000 0.0000   1.00000    1.000
## 111   11   22  0.3643 0.0823   0.23391    0.567
## 222   7    4  0.2318 0.0744   0.12357    0.435
## 333   3    4  0.0994 0.0538   0.03436    0.287
## 444   2    1  0.0662 0.0449   0.01752    0.250
## 555   1    1  0.0331 0.0325   0.00485    0.226
## 666   1    0  0.0331 0.0325   0.00485    0.226
## 777   1    0  0.0331 0.0325   0.00485    0.226

## 888   1    0  0.0331 0.0325   0.00485    0.226
##
##       age_quart=Q2
## time n.risk n.event survival std.err lower 95% CI upper 95% CI
##  0   39    0  1.0000 0.0000   1.00000    1.000
## 111   16   22  0.4168 0.0809   0.28487    0.610
## 222   6    7  0.2007 0.0696   0.10173    0.396
## 333   4    1  0.1606 0.0662   0.07153    0.360
## 444   2    2  0.0803 0.0520   0.02253    0.286
## 555   2    0  0.0803 0.0520   0.02253    0.286
## 666   1    1  0.0401 0.0385   0.00612    0.263
## 777   1    0  0.0401 0.0385   0.00612    0.263
## 888   1    0  0.0401 0.0385   0.00612    0.263
## 999   1    1  0.0000   NaN      NA      NA
##
##       age_quart=Q3
## time n.risk n.event survival std.err lower 95% CI upper 95% CI
```

```
## 0 33 0 1.0000 0.0000 1.0000 1.000
## 111 17 17 0.4848 0.0870 0.3411 0.689
## 222 6 10 0.1818 0.0671 0.0882 0.375
## 333 4 2 0.1212 0.0568 0.0484 0.304
## 444 1 3 0.0303 0.0298 0.0044 0.209
##
## age_quart=Q4
## time n.risk n.event survival std.err lower 95% CI upper 95% CI
## 0 30 0 1.0000 0.0000 1.00000 1.000
## 111 5 23 0.2188 0.0786 0.10821 0.442
## 222 3 2 0.1313 0.0672 0.04810 0.358
## 333 1 2 0.0438 0.0422 0.00662 0.289
ggsurvplot(age_fit)
```

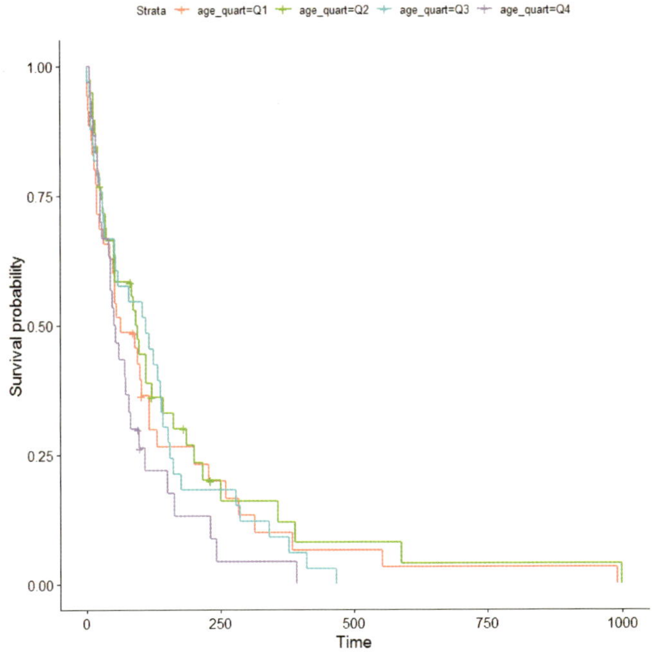


```
ggsurvplot(age_fit, surv.median.line = "hv")
```

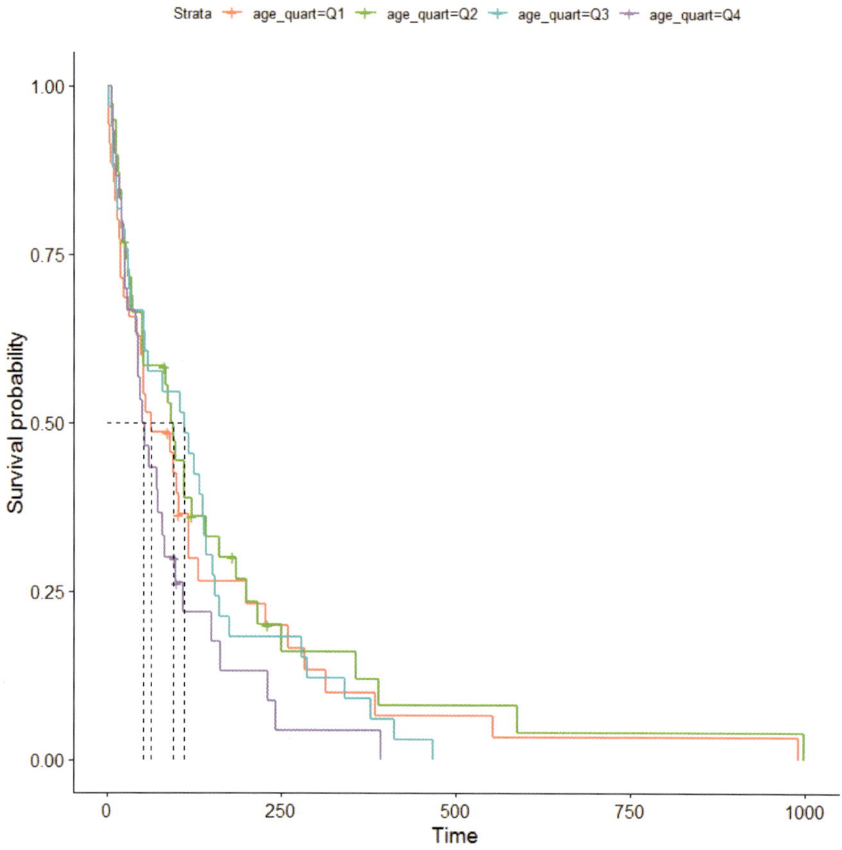

- **Interpretation:** We see that the quartile with the lowest survival probability (highest death risk) is the fourth quartile or age group from 66 to 82 years

5.3.4. More Options

We will examine the more options of ggsurvplot in addition to other different functions for plotting:

1. Standard plotting: the survival plot with survival probability on y_axis

```
ggsurvplot(trt_fit)
```

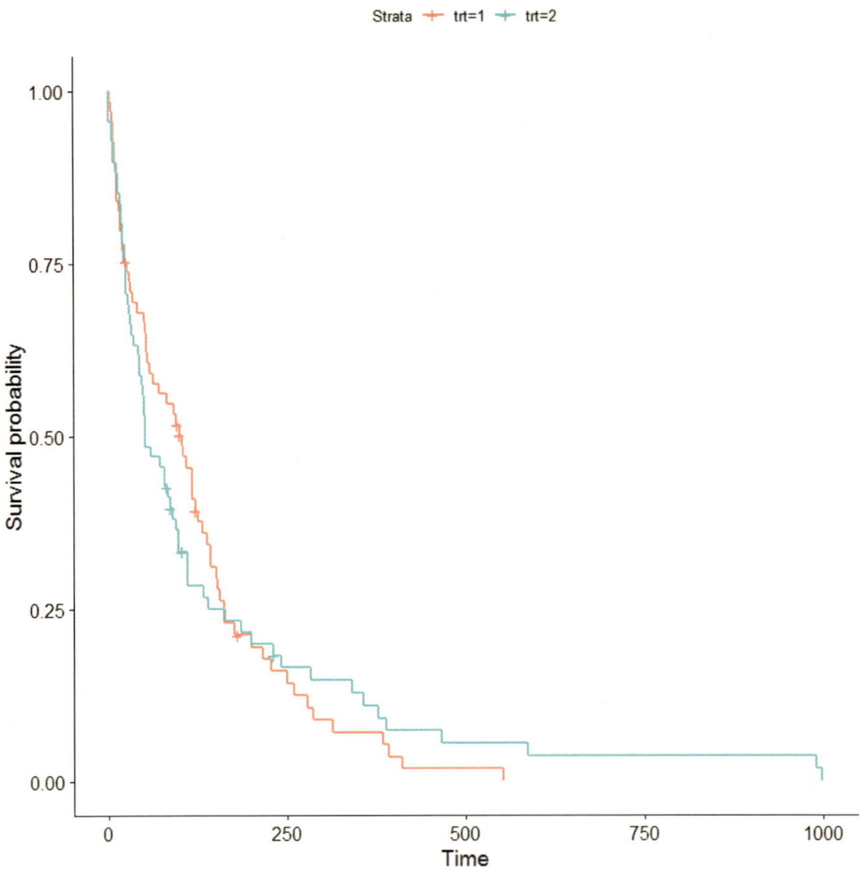

You see that treatment 2 has lower survival probability in the first 200 days (more dangerous).

2. The cumulative event plot with cumulative events on y_axis. Cumulative event = 1-survival probability and it is the mirror image of the survival curve

```
ggsurvplot(trt_fit, fun = "event")
```

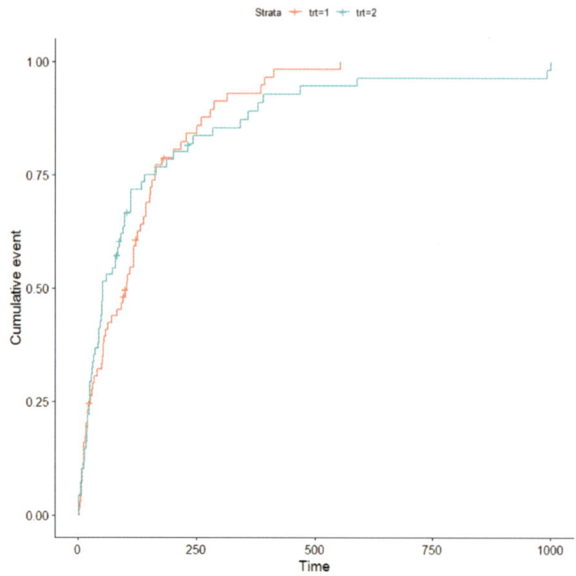

You see now that treatment 2 has higher cumulative death in the first 200 days (more dangerous).

3. The cumulative hazard plot with cumulative hazard on y_axis. Cumulative hazard = -log(survival probability).

```
ggsurvplot(trt_fit, fun = "cumhaz")
```

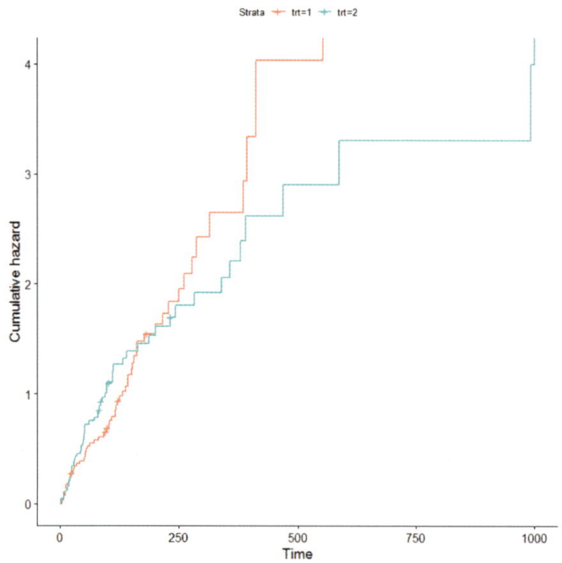

4. The survival plot with the survival probability as percentage.

```
ggsurvplot(trt_fit,data = veteran, fun = "pct")
```

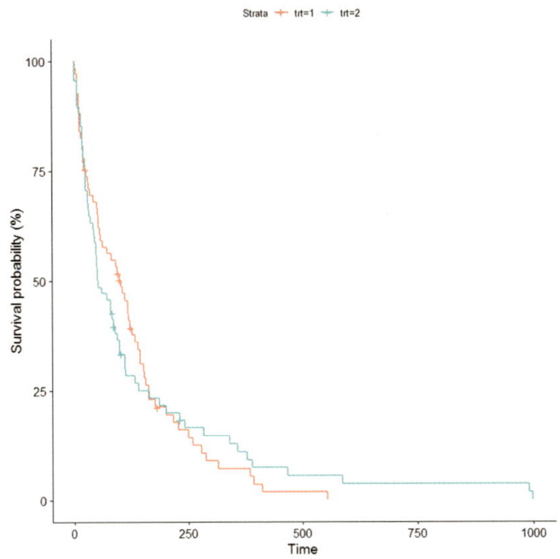

5. Add confidence intervals (CI) to survival curves.

```
##Add confidence bound
ggsurvplot(trt_fit, conf.int = TRUE)
```

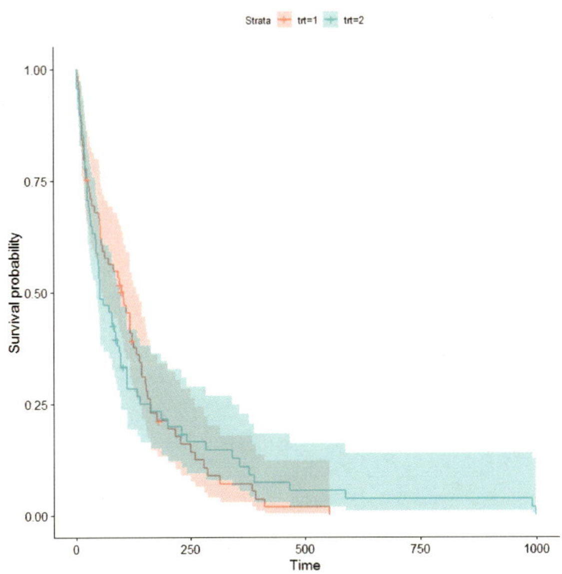

6. Add p-value for the statistical test used. This will be discussed later.

```
ggsurvplot(trt_fit,pval = TRUE, pval.method = TRUE)
```

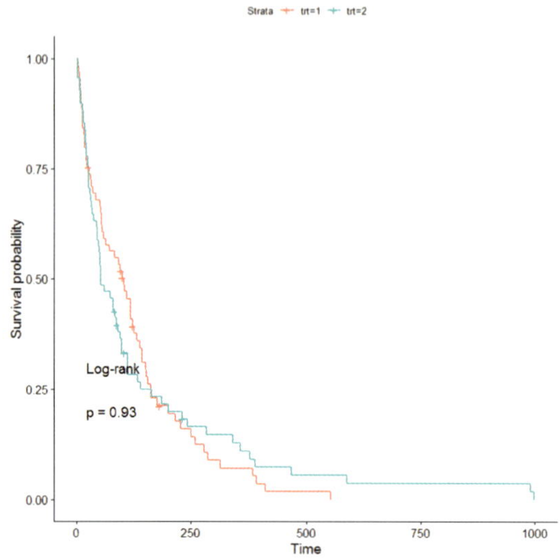

Non-significant difference between the two treatments.

7. Draw a horizontal and vertical line at the median survival.\

```
ggsurvplot(trt_fit, surv.median.line = "hv")
```

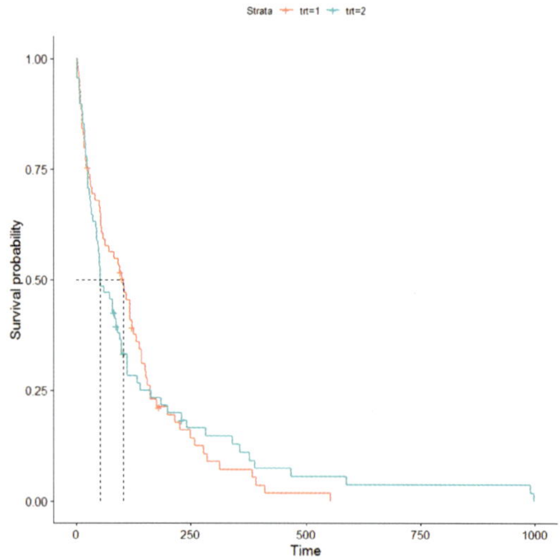

You see that the median survival of treatment 2 occurs earlier in time so it may be more dangerous.

8. Add a risk table (number having the risk of event over time) as numbers and percentage.

```
ggsurvplot(trt_fit, risk.table = "abs_pct")
```

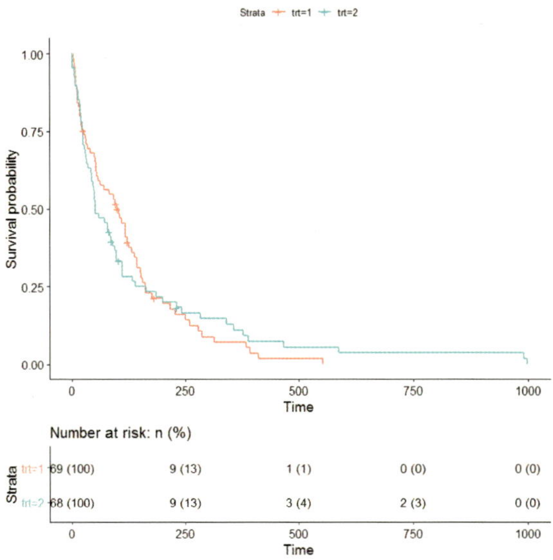

9. Add an event table (cumulative number of events over time).

```
ggsurvplot(trt_fit,cumevents = TRUE)
```

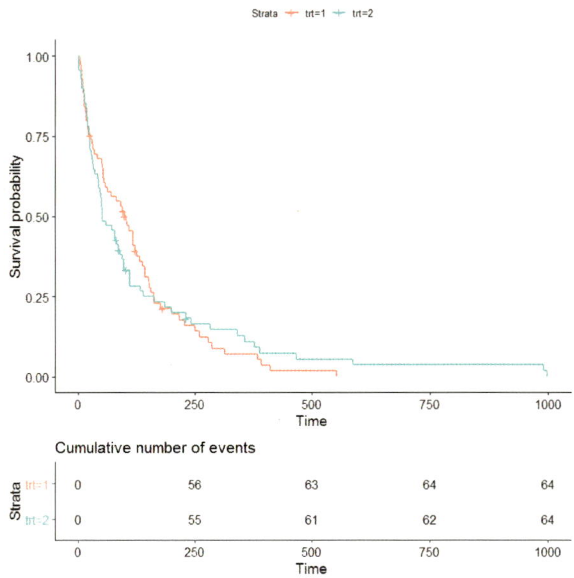

10. Add a table of the cumulative number of censoring subjects over time.

```
ggsurvplot(trt_fit, cumcensor = TRUE)
```

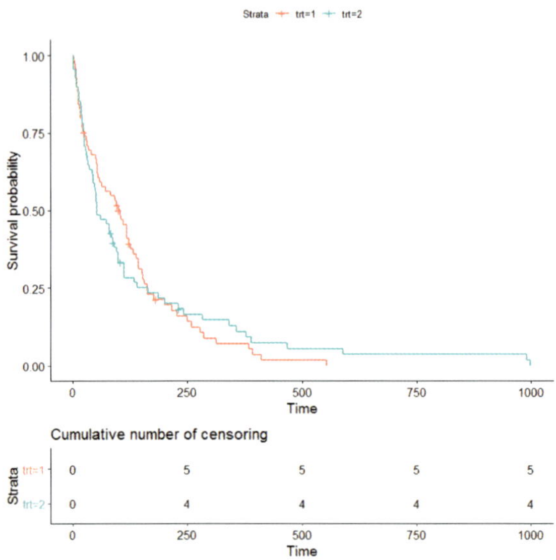

11. Enlarge part of the plot, For example, the first 200 days as they have more death risk.

```
ggsurvplot(trt_fit, break.time.by=20,xlim=c(0,200), surv.median.line = "hv")
```

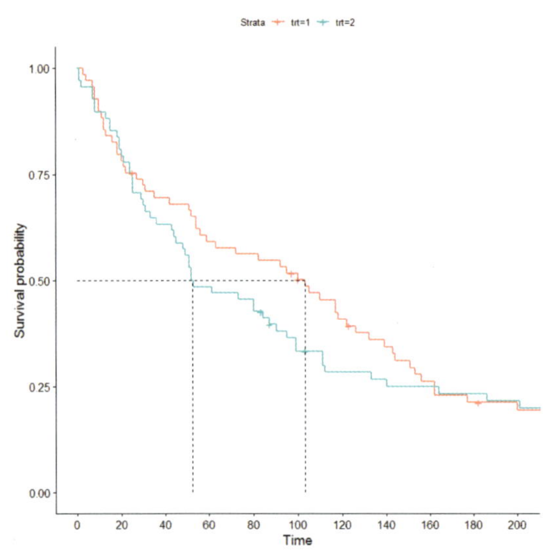

You can see clearly that median survival for treatment 2 occurs at about 50 days while that of treatment 1 occurs at about 105 days; so, treatment 2 (test treatment) may be more dangerous.

5.4. USING RATIOS FOR STATISTICAL TESTS

When the numerator is less than the denominator in the relative risk (RR), odds ratio (OR), and incidence rate ratio (IRR), these ratios will range from 0 to 1.On the other hand, when the numerator is larger than the denominator, these ratios will range from 1 to positive infinity.

For example, in a study of the effect of AZT (azidothymidine, anti-HIV drug) treatment on Maternal to Infant HIV transmission. Comparing AZT to placebo, RR = 0.32 and OR = 0.27. It is concluded that AZT is associated with 68% relative decreased risk in HIV transmission and 73% decreased odds of HIV transmission compared to placebo.

If we compared placebo to AZT, RR = 3.1 and OR = 3.7. It is concluded that placebo is associated with 210% relative increased risk in HIV transmission and 270% increased odds of HIV transmission compared to AZT.

The percent decrease when comparing AZT to placebo is not the same as the percent increase when comparing placebo to AZT because the scaling of ratios is not symmetric around 1 so we use ln or log ratios as they are symmetric around zero.

```
##Comparing AZT to placebo
#relative risk =0.32
#odds ratio=0.27
##Comparing placebo to AZT
#relative risk =1/0.32=3.125
#odds ratio=1/0.27=3.704
##Using the log ratios
##Comparing AZT to placebo
##Relative risk
log(0.32)
## [1] –1.139434
##Odds ratio
log(0.27)
```

```
## [1] –1.309333
##Comparing placebo to AZT
#relative risk
log(3.125)
## [1] 1.139434
#odds ratio
log(3.704)
## [1] 1.309413
```

We see that the log(ratios) are symmetric around zero. Note that log() in R computes natural logarithms, log10 computes common (i.e., base 10) logarithms, and log2 computes binary (i.e., base 2) logarithms.

We use log ratios for:

1. Showing group comparisons as some factors may be negatively or positively associated with some outcome. By using the log ratios, all ratios (larger or smaller than 1) will be comparable numerically.
2. Performing statistical tests on ratios.

CHAPTER 6

SAMPLING DISTRIBUTION

CONTENTS

6.1. INTRODUCTION

Any sample we get randomly from a population is one of many possible samples that we may get by chance. The sample means (for continuous variables), proportions (for categorical variables), and incidence rates (IRs) (for time to event variables) based on certain size vary across different samples of the same size.

Sampling variability is the uncertainty in sample based estimates (means, proportions, IRs,.....).

For example, for describing the distribution of heights in 1 year old children in the US, there are 3 group of researchers:

1. Group 1 take a sample of 100 children and get a mean =
2. Group 2 take a sample of other 100 children and get a mean =
3. Group 3 take a sample of other 100 children and get a mean =

We note that the sample means reported by the 3 groups are not identical. This variability in sample means will decrease by increasing the sample size; so, if these groups has taken samples of 1000 children, the variability observed between the 1000-sample means will be less than 100-sample means.

The cause for decreasing the variability with increasing sample size as the sample estimates (means, proportions, or IRs) will be less affected by sample data (individual observations) with increasing sample size.

The Sampling Distribution of a sample statistic or estimate is a theoretical distribution, that we cannot actually observe, that describes all the possible values of a sample statistic (mean, proportion, IR) from random samples of the same size that are taken from the same population. In real life research, only one sample is taken with a certain size from a specific population.

Understanding the sampling distribution, across samples of the same size, will give us a framework for coupling the sample estimate with some measure of uncertainty to make a statement about the unknown truth (population mean, proportion, or IR) that we cannot observe.

6.2. THE SAMPLING DISTRIBUTION OF THE SAMPLE MEANS

We will use regicor dataset and assume that this dataset is our population. We will take 1000 random samples from this dataset, each of size 20, 50, or

150 rows, calculate the sample mean for each sample, and plot the samples means as histograms and boxplots to see their distribution.

6.2.1. Continuous Normal Distribution

Use the age column (normal continuous data) and generate 1000 random samples, each of size=20.

```
library(tidyverse)
library(compareGroups)
data("regicor")
# population mean for age
mean(regicor$age)
## [1] 54.73627
##For reproducibility, we use the set.seed
set.seed(123)
samples_20<-replicate(1000,sample(regicor$age,20))
str(samples_20)
## int [1:20, 1:1000] 47 62 65 46 71 74 56 48 36 64 ...
##So it consists of 1000 columns (samples) and 20 rows (sample size)
##Take the means of these columns or samples
means_20<-colMeans(samples_20)
str(means_20)
## num [1:1000] 55 54.7 55.3 51.1 54 ...
summary(means_20)
##  Min. 1st Qu. Median  Mean 3rd Qu.  Max.
##  47.50  52.95  54.70  54.67  56.25  62.60
##The means_20 contain the means of these 1000 samples
##To plot a histogram of these means
hist_20<-ggplot()+geom_histogram(aes(x = means_20))
hist_20
```

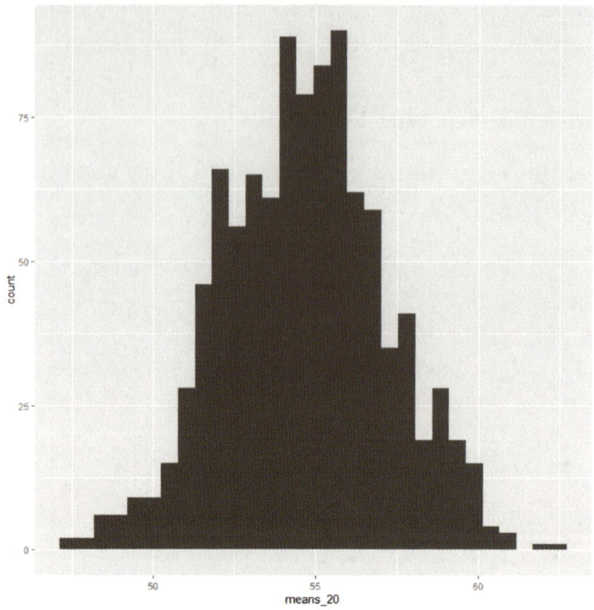

##To plot a boxplot of these means

```
box_20<-ggplot()+geom_boxplot(aes(y=means_20))
box_20
```

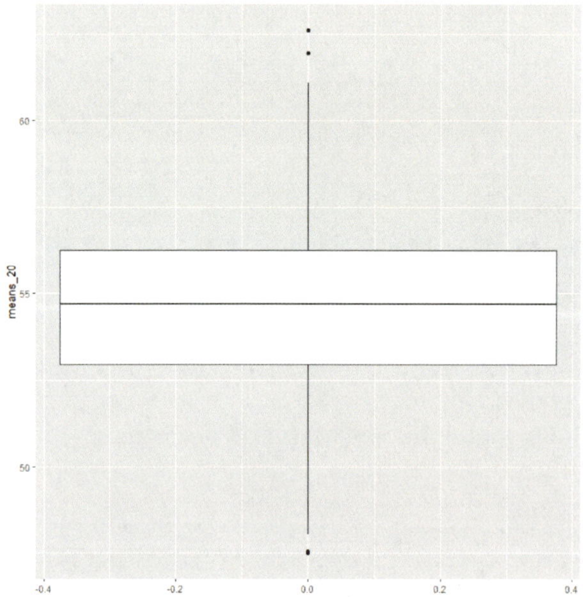

Generate 1000 random samples, each of size=50.

```
set.seed(123)
samples_50<-replicate(1000,sample(regicor$age,50))
str(samples_50)
## int [1:50, 1:1000] 47 62 65 46 71 74 56 48 36 64 ...
means_50<-colMeans(samples_50)
str(means_50)
## num [1:1000] 54.5 53.5 56.3 53.6 54 ...
summary(means_50)
##  Min. 1st Qu. Median  Mean 3rd Qu.  Max.
##  49.10  53.67  54.71  54.74  55.86  59.40
hist_50<-ggplot()+geom_histogram(aes(x = means_50))
hist_50
```

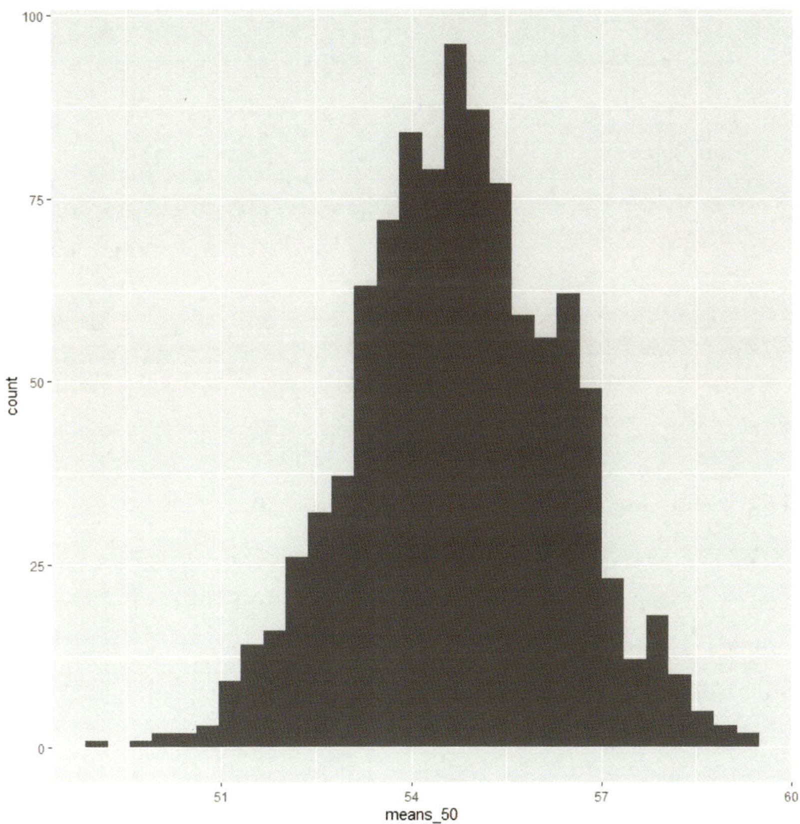

```
##To plot a boxplot of these means
box_50<-ggplot()+geom_boxplot(aes(y=means_50))
box_50
```

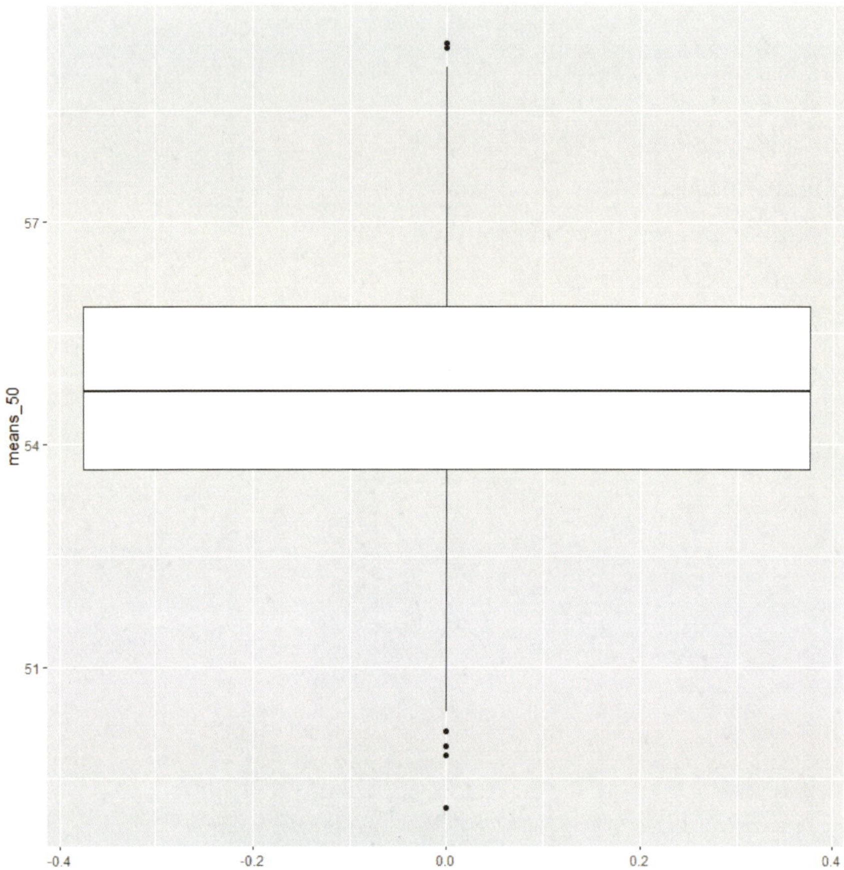

Generate 1000 random samples, each of size=150.

```
set.seed(123)
samples_150<-replicate(1000,sample(regicor$age,150))
str(samples_150)
## int [1:150, 1:1000] 47 62 65 46 71 74 56 48 36 64 ...
means_150<-colMeans(samples_150)
str(means_150)
```

```
## num [1:1000] 55.5 53.9 53.6 54.4 54.6 ...
summary(means_150)
##  Min. 1st Qu. Median  Mean 3rd Qu.  Max.
##  51.70  54.17  54.75  54.74  55.36  57.68
hist_150<-ggplot()+geom_histogram(aes(x = means_150))
hist_150
```

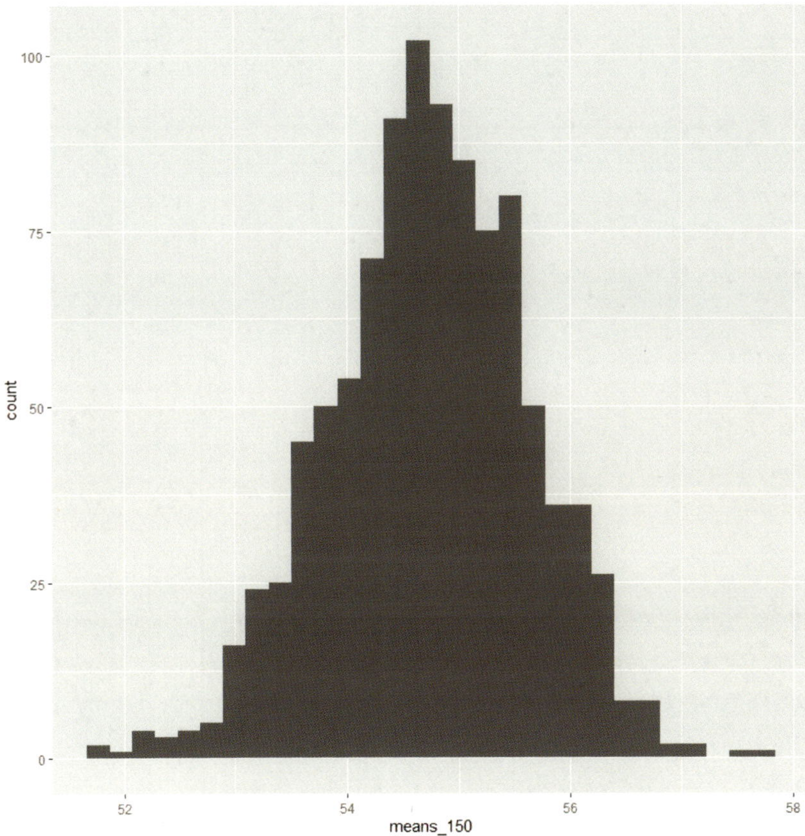


```
##To plot a boxplot of these mean
box_150<-ggplot()+geom_boxplot(aes(y=means_150))
box_150
```

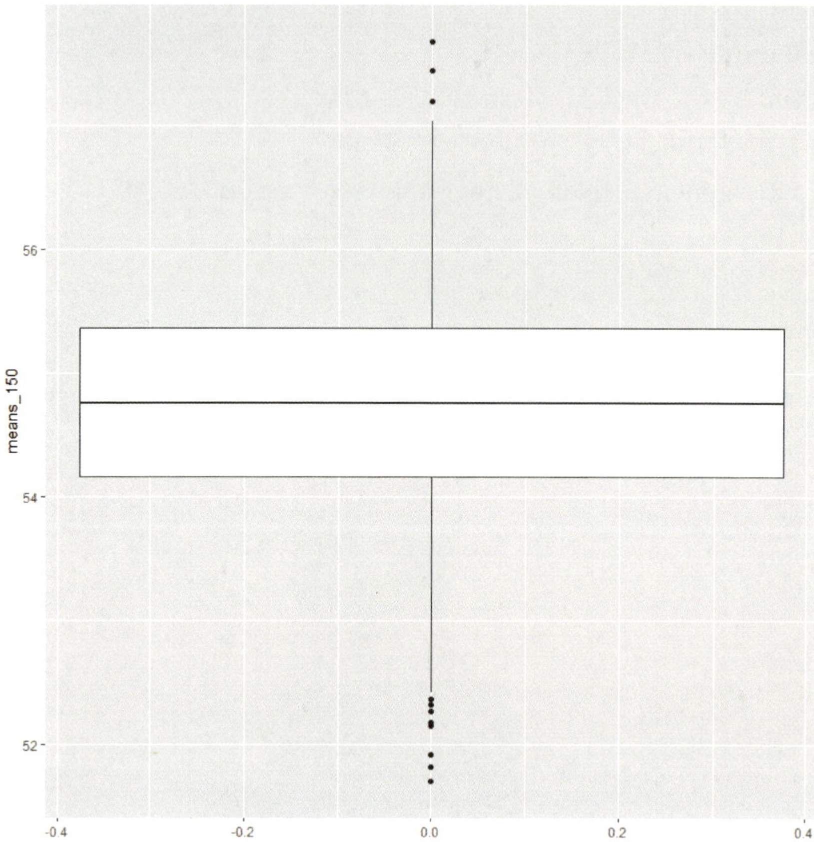

Plot all graphs side by side.

```
##Combine the 3 means in one dataframe
df<-data.frame(means_20=means_20,means_50=means_50,means_150=
means_150)
head(df)
##  means_20 means_50 means_150
## 1  55.00  54.54 55.49333
## 2  54.70  53.52 53.89333
## 3  55.30  56.30 53.60000
## 4  51.10  53.56 54.38667
## 5  54.05  54.00 54.58000
## 6  61.95  57.02 53.69333
# gather the columns
```

```
df<- df %>% gather(key = "mean," value = "value")
table(df$mean)
##
## means_150 means_20 means_50
##   1000    1000    1000
# arrange factor levels
df$mean<- factor(df$mean, levels = c("means_20," "means_50,""means_150"))
ggplot(data = df, aes(x = value, fill = mean)) + geom_histogram()+facet_wrap(~mean)
```

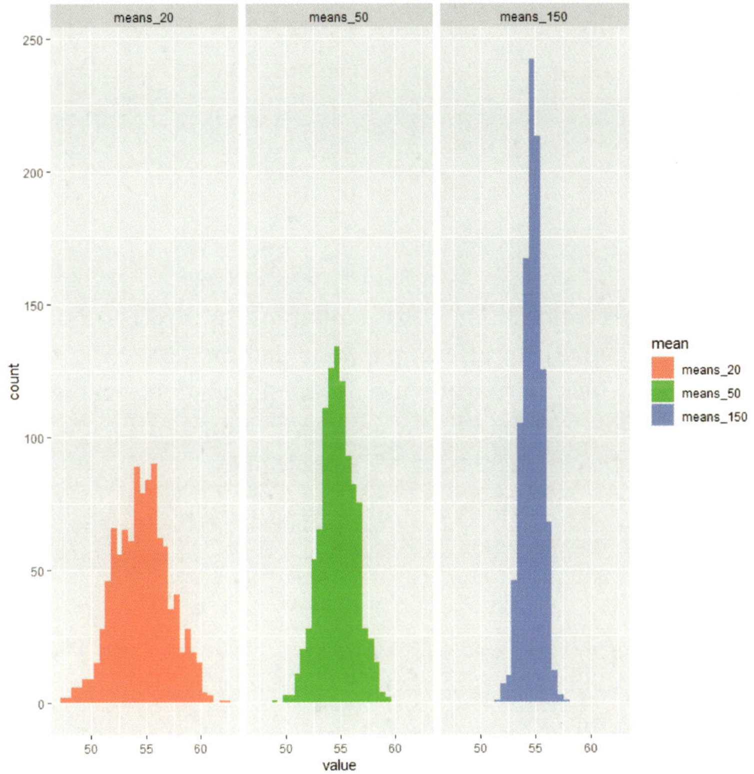


```
ggplot(data = df, aes(x = mean, y = value, col = mean))+geom_boxplot()
```

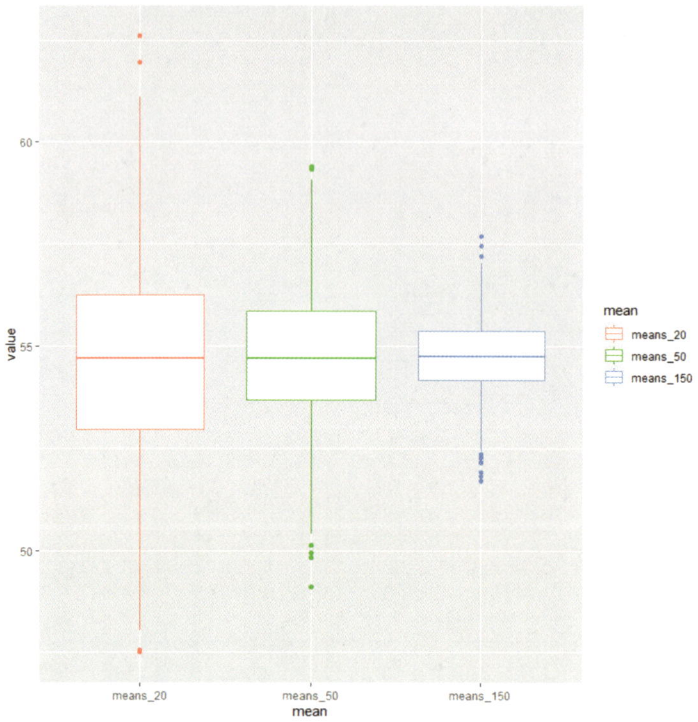

- **Interpretation:** The distribution of means is nearly normal (bell-shaped) for all sample sizes and they have the same mean value which is nearly equal to the population mean (54.7). The variability decreased (the histogram range or box width in the boxplot) with increasing the sample size.

The decrease in variability can also be shown by calculating the standard deviation for each means type.

```
head(df)
##    mean value
## 1 means_20 55.00
## 2 means_20 54.70
## 3 means_20 55.30
## 4 means_20 51.10
## 5 means_20 54.05
## 6 means_20 61.95
df %>% group_by(mean) %>% summarise(sd = sd(value))
```

```
## # A tibble: 3 x 2
## mean sd
## <fct> <dbl>
## 1 means_20 2.47
## 2 means_50 1.61
## 3 means_150 0.891
```

The standard deviation of the sample means decreases as we increase the sample size. It is 2.47 for sample size 20, 1.61 for sample size 50, and 0.891 for sample size 150.

6.2.2. Continuous Skewed Distribution

If you examine the distribution of age in your population (regicor dataset), you will found that it is nearly normal (Bell-shaped). However, if we apply the same methods to another continuous variable that is highly skewed (phyact or physical activity), we will also get the same results.

```
##To see the distribution of age in our population
ggplot(data = regicor,aes(x = age))+geom_histogram(bins = 20)
```

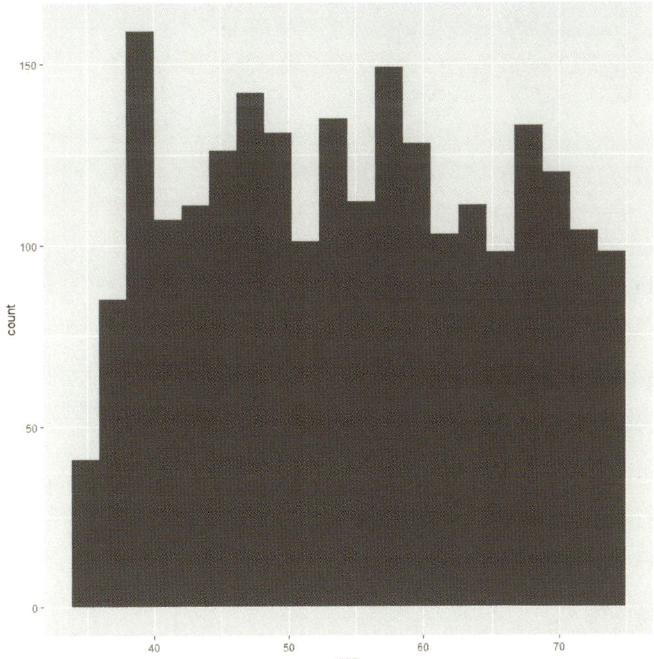

```
##To see the distribution of physical activity in our population
ggplot(data = regicor,aes(x = phyact))+geom_histogram()
```

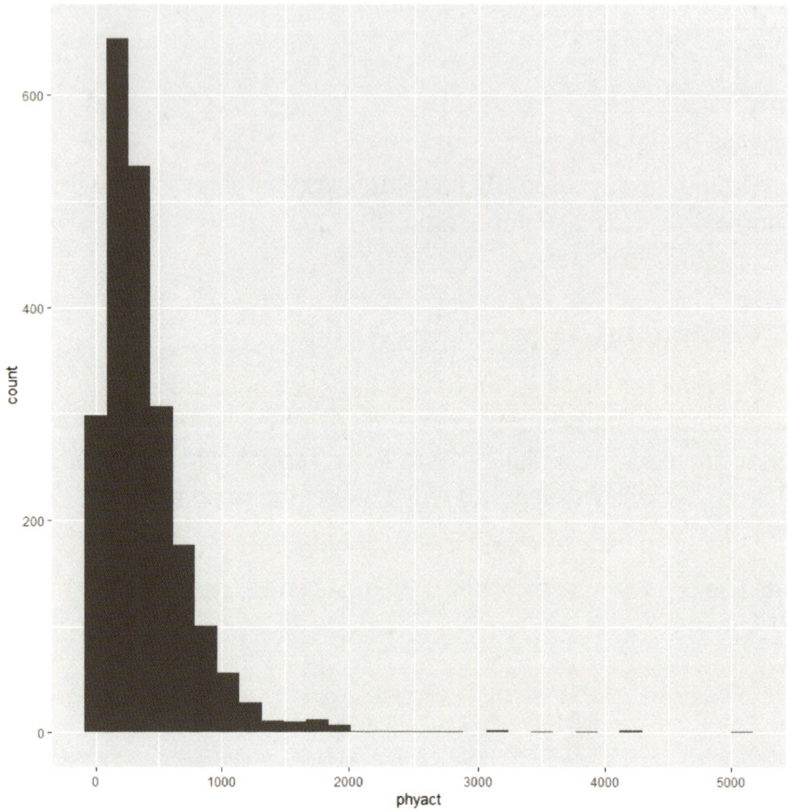


```
# to see population mean
mean(regicor$phyact, na.rm = TRUE)
## [1] 398.8314
```

The mean physical activity = 398.8.

Generate 1000 random samples, each of size = 20.

```
##For reproducibility, we use the set.seed
set.seed(123)
samples_20<-replicate(1000,sample(regicor$phyact,20))
str(samples_20)
## num [1:20, 1:1000] 1028 421 185 430 317 ...
##Take the means of these columns or samples
```

```
means_20<-colMeans(samples_20, na.rm = TRUE)
str(means_20)
## num [1:1000] 418 411 305 331 399 ...
summary(means_20)
##  Min. 1st Qu. Median  Mean 3rd Qu.  Max.
##  208.3  333.8  385.0  394.6  444.3  883.2
##The means_20 contain the means of these 1000 samples
##To plot a histogram of these means
hist_20<-ggplot()+geom_histogram(aes(x = means_20))
hist_20
```

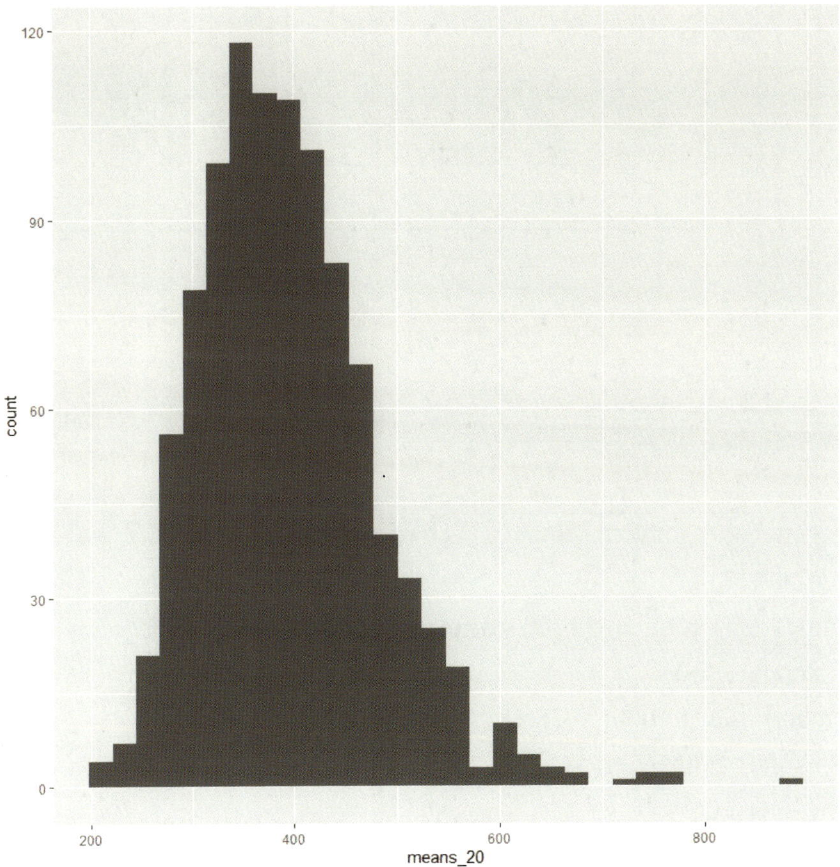


```
##To plot a boxplot of these means
```

```
box_20<-ggplot()+geom_boxplot(aes(y=means_20))
box_20
```

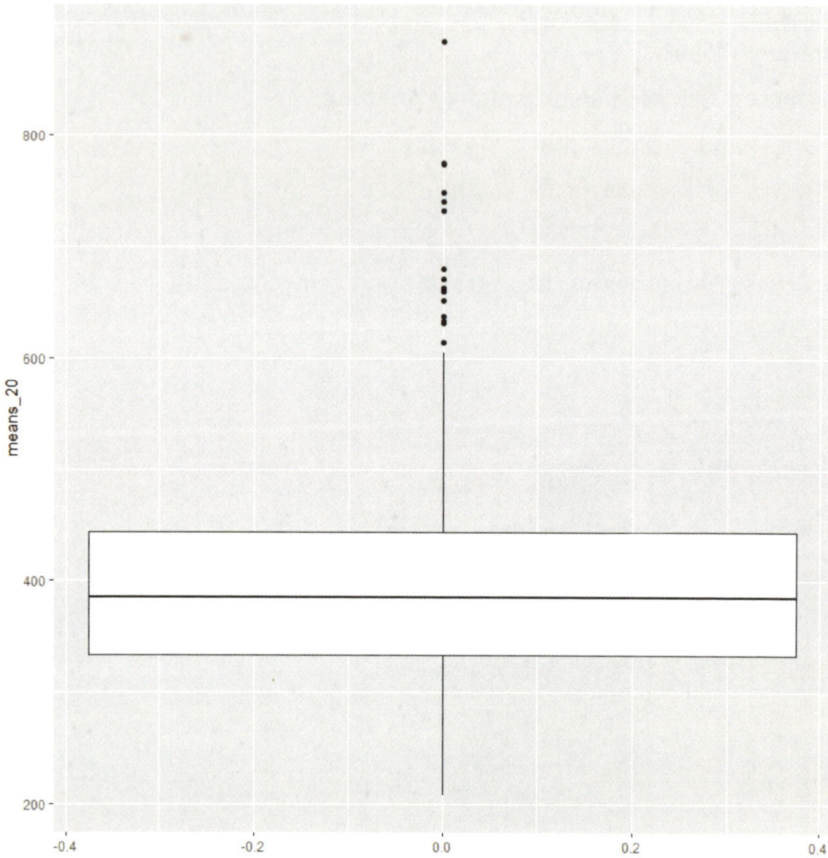

Generate 1000 random samples, each of size=50.

```
set.seed(123)
samples_50<-replicate(1000,sample(regicor$phyact,50))
str(samples_50)
## num [1:50, 1:1000] 1028 421 185 430 317 ...
means_50<-colMeans(samples_50, na.rm = TRUE)
str(means_50)
## num [1:1000] 395 351 378 380 380 ...
```

```
summary(means_50)
##  Min. 1st Qu. Median  Mean 3rd Qu.  Max.
## 248.8  358.4  391.2  396.8  430.5  653.7
hist_50<-ggplot()+geom_histogram(aes(x = means_50))
hist_50
```

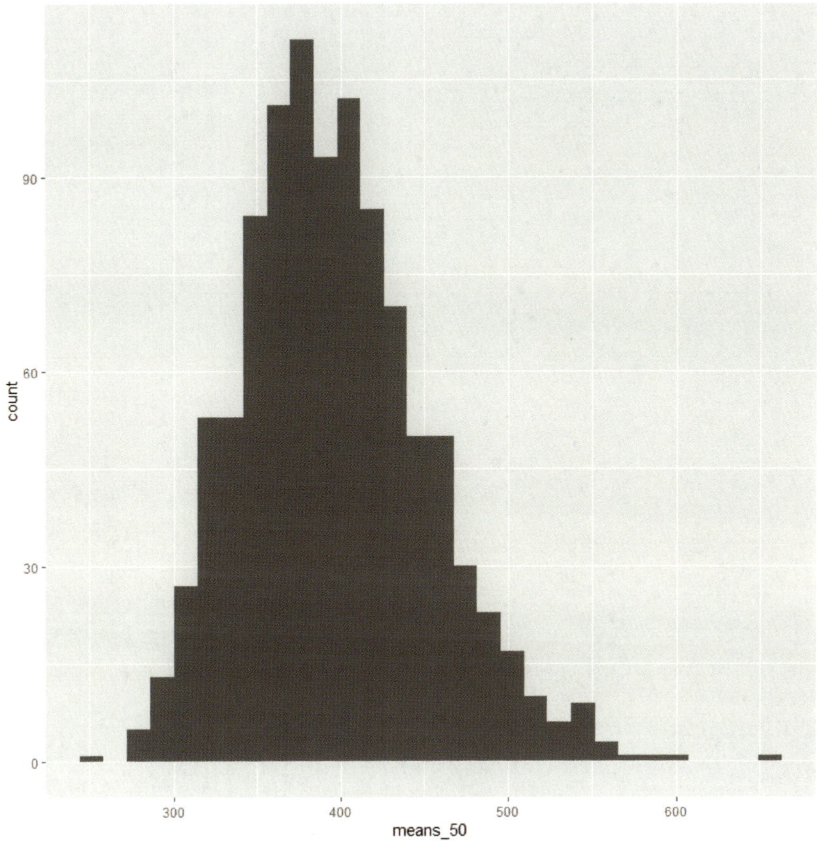


```
##To plot a boxplot of these means
box_50<-ggplot()+geom_boxplot(aes(y=means_50))
box_50
```

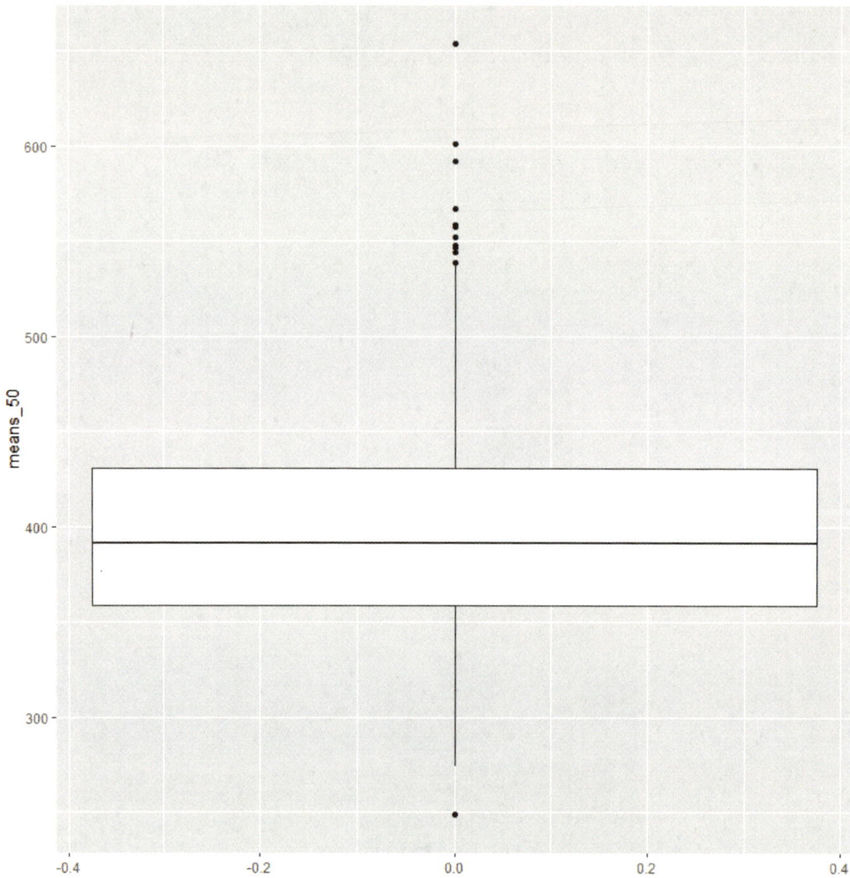

Generate 1000 random samples, each of size=150.

```
set.seed(123)
samples_150<-replicate(1000,sample(regicor$phyact,150))
str(samples_150)
## num [1:150, 1:1000] 1028 421 185 430 317 ...
means_150<-colMeans(samples_150, na.rm = TRUE)
str(means_150)
## num [1:1000] 359 440 419 384 443 ...
summary(means_150)
```

```
## Min. 1st Qu. Median Mean 3rd Qu. Max.
## 314.1 376.5 395.8 398.7 420.1 502.8
hist_150<-ggplot()+geom_histogram(aes(x = means_150))
hist_150
```

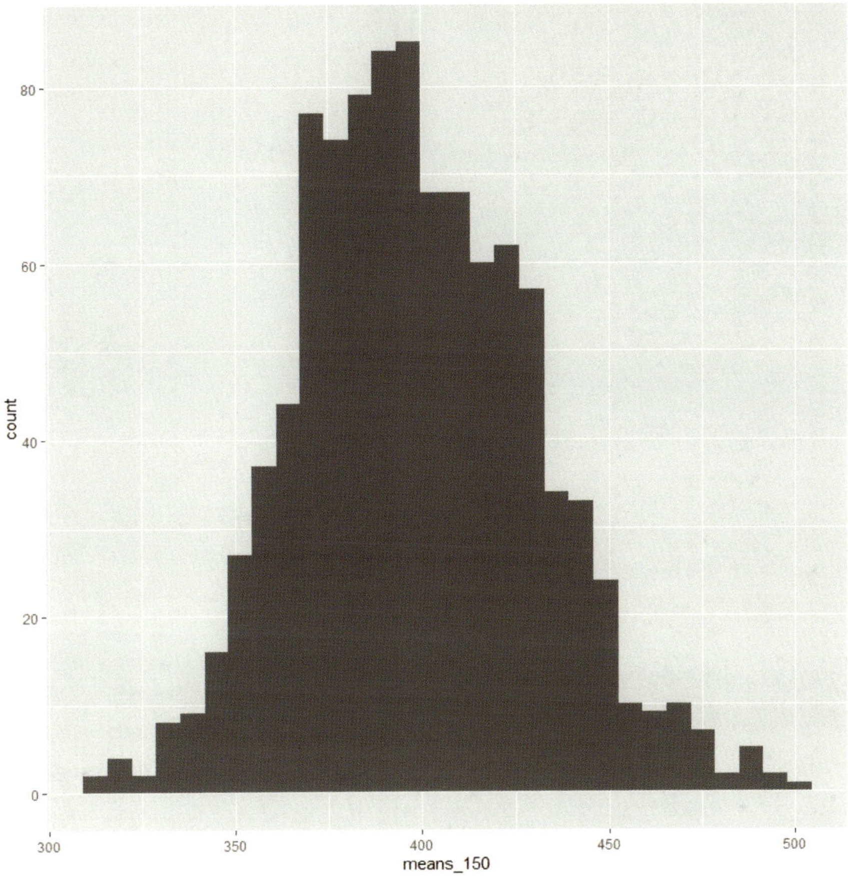


```
##To plot a boxplot of these means
box_150<-ggplot()+geom_boxplot(aes(y=means_150))
box_150
```

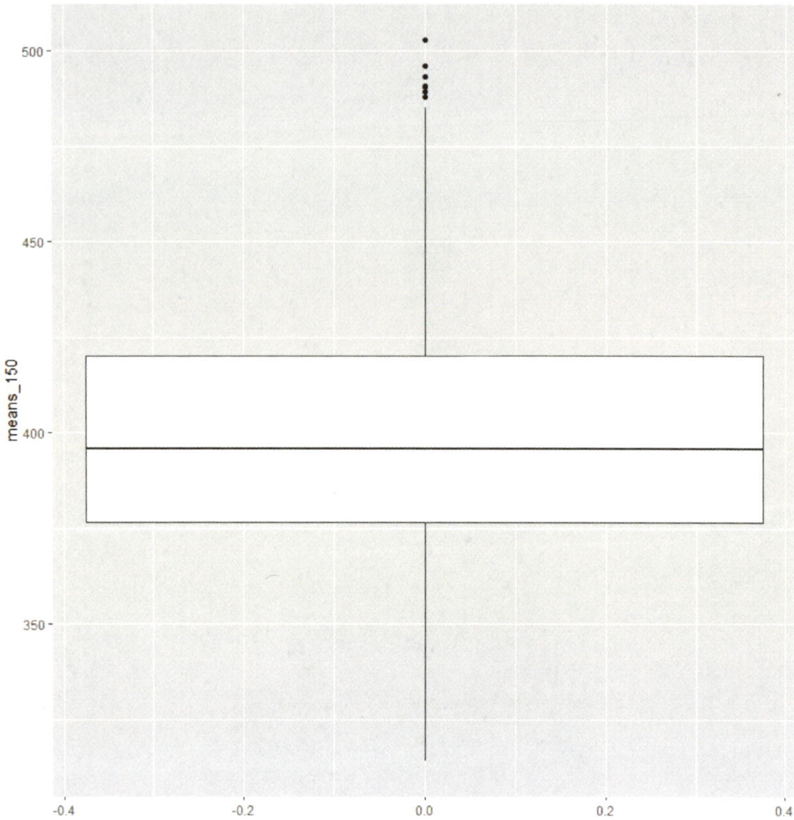

Plot all graphs side by side.

```
##Combine the 3 means in one dataframe
df<-data.frame(means_20=means_20, means_50=means_50, means_150=means_150)
head(df)
## means_20 means_50 means_150
## 1 417.6647 395.2189 359.4413
## 2 411.4778 350.7541 439.5872
## 3 304.6736 378.2334 419.0556
## 4 330.8837 380.3807 384.4075
## 5 398.7416 380.0445 442.9107
## 6 325.9614 517.4790 373.3660
# gather the columns
```

```
df<- df %>% gather(key = "mean," value = "value")
table(df$mean)
##
## means_150 means_20 means_50
##    1000    1000    1000
# arrange factor levels
df$mean<- factor(df$mean, levels = c("means_20," "means_50,""means_150"))
ggplot(data = df, aes(x = value, fill = mean)) + geom_histogram()+facet_wrap(~mean)
```

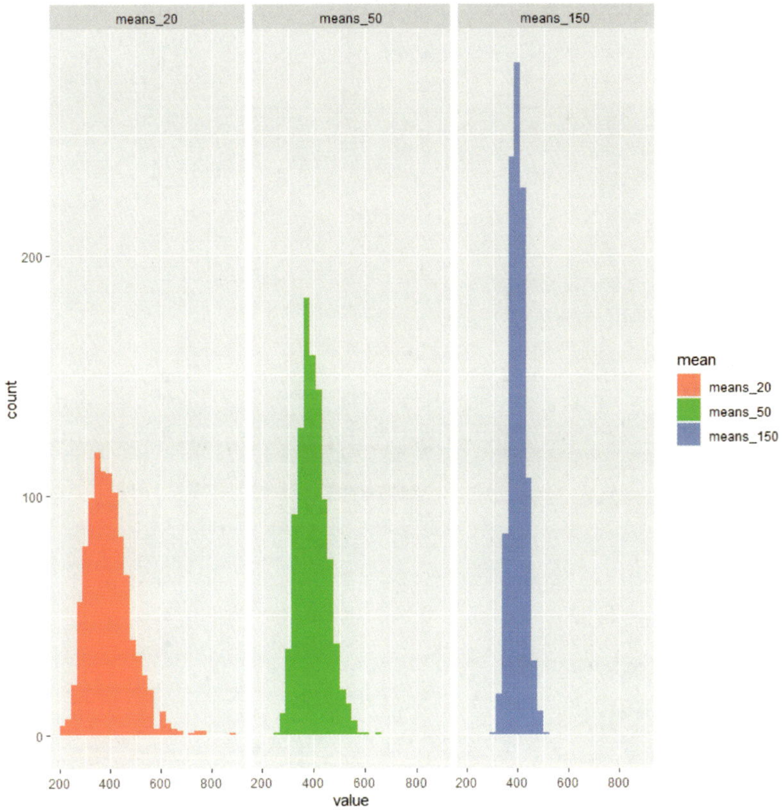


```
ggplot(data = df, aes(x = mean, y = value, col = mean))+geom_boxplot()
```

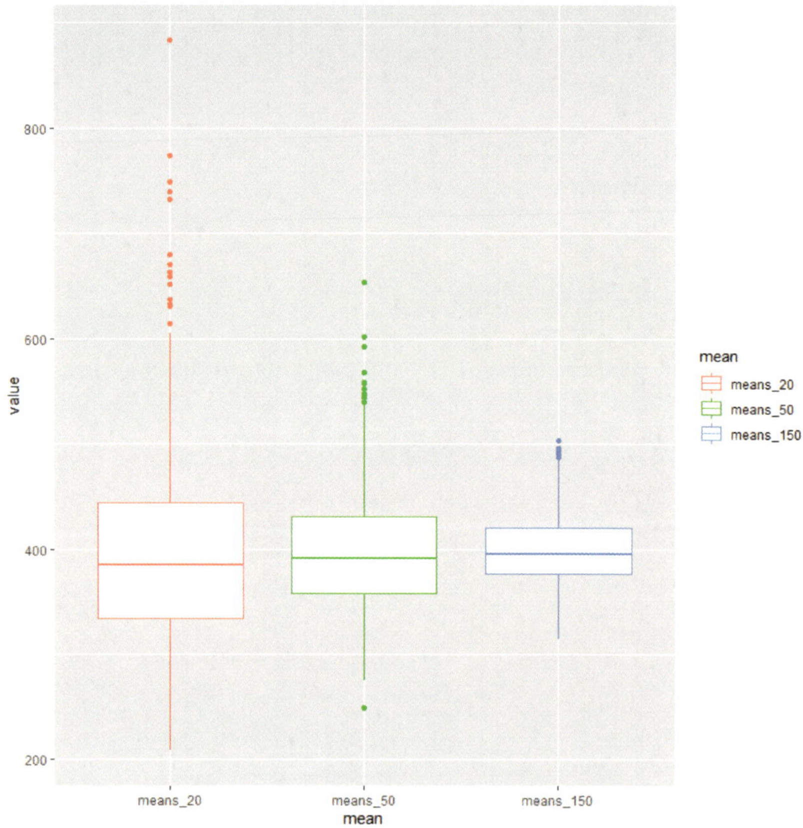


```
# to calculate the standard deviation for each means type
df %>% group_by(mean) %>% summarise(sd = sd(value))
## # A tibble: 3 x 2
##  mean      sd
##  <fct>     <dbl>
## 1 means_20  85.7
## 2 means_50  55.1
## 3 means_150 31.3
```

The distribution of sample means is normal (Bell-shaped) regardless of the sample size and regardless of the distribution of individual values in the population. Although the physical activity has a right skewed distribution in our population (regicor data), the distribution of sample means was nearly normal.

Therefore, we can estimate the sampling distribution of the sample means using only one random sample and the properties of the normal distribution discussed in Chapter 3.

The conclusions are:

1. The variation in the sample means decreases with increasing the sample size.
2. The distribution shape of the sample means becomes more normal with increasing the sample size.
3. The overall mean of the sample means is very close to the true population mean.

6.3. THE SAMPLING DISTRIBUTION OF SAMPLE PROPORTIONS

We will use the regicor data to determine the sampling distribution of the cardiovascular events in this population (categorical variable).

We will replace "No" with 0 and "Yes" with 1 then taking the mean of this variable. This will give us the proportion with ones or cardiovascular events.

The true population proportion:

```
table(regicor$cv)
##
## No Yes
## 2071 92
prop.table(table(regicor$cv))
##
##     No     Yes
## 0.95746648 0.04253352
# replacing No with 0 and Yes with 1
regicor$cv<- ifelse(regicor$cv == "No,"0,1)
mean(regicor$cv, na.rm = TRUE)
## [1] 0.04253352
```

So the proportion of cardiovascular events is 0.043 or 4.3%.

Generate 1000 random samples, each of size=20

```
##For reproducibility, we use the set.seed
set.seed(123)
samples_20<-replicate(1000,sample(regicor$cv,20))
str(samples_20)
## num [1:20, 1:1000] 0 0 0 0 0 0 0 NA 0 0 ...
##Take the means of these columns or samples
means_20<-colMeans(samples_20, na.rm = TRUE)
str(means_20)
## num [1:1000] 0 0.0526 0 0.1111 0.05 ...
summary(means_20)
##   Min. 1st Qu. Median  Mean 3rd Qu.  Max.
## 0.00000 0.00000 0.05000 0.04352 0.05556 0.22222
##The means_20 contain the means of these 1000 samples
##To plot a histogram of these means
hist_20<-ggplot()+geom_histogram(aes(x = means_20))
hist_20
```

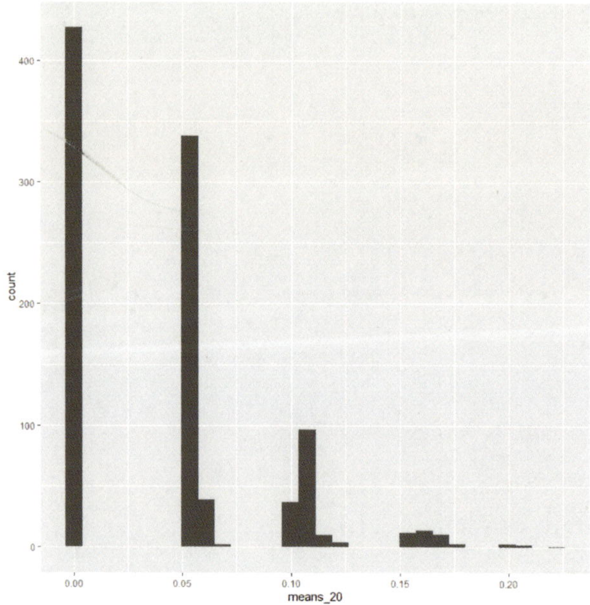


```
##To plot a boxplot of these means
```

```
box_20<-ggplot()+geom_boxplot(aes(y=means_20))
box_20
```

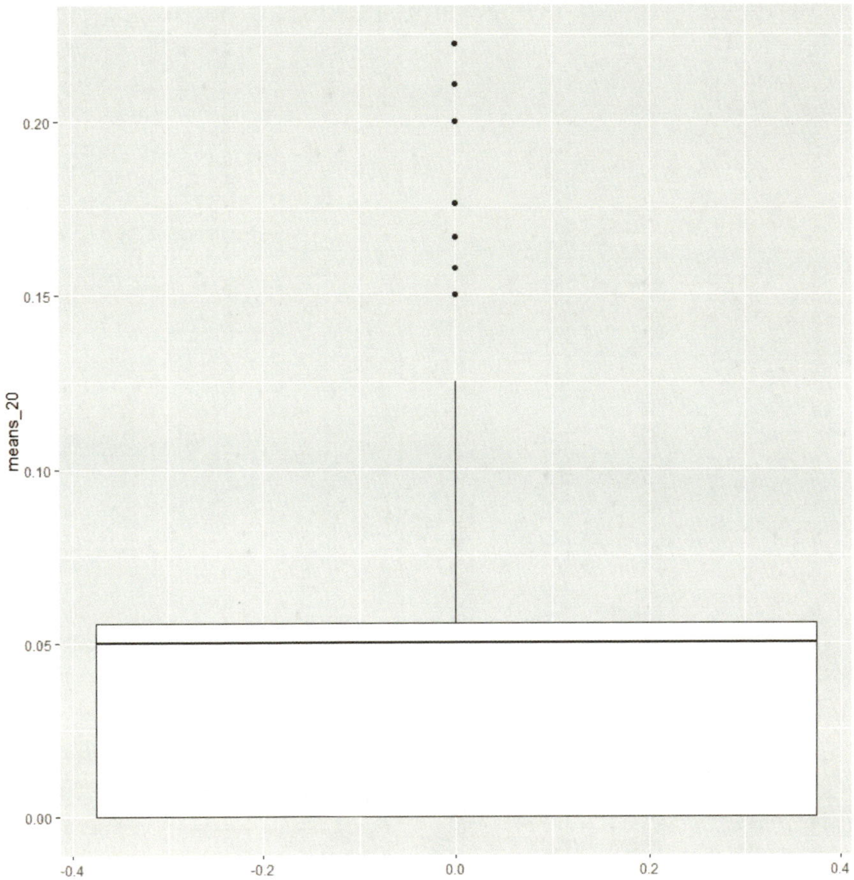

Generate 1000 random samples, each of size=50.

```
set.seed(123)
samples_50<-replicate(1000,sample(regicor$cv,50))
str(samples_50)
## num [1:50, 1:1000] 0 0 0 0 0 0 0 NA 0 0 ...
means_50<-colMeans(samples_50, na.rm = TRUE)
str(means_50)
## num [1:1000] 0.0217 0.0625 0.0612 0.0612 0.0444 ...
summary(means_50)
```

```
## Min. 1st Qu. Median  Mean 3rd Qu.  Max.
## 0.00000 0.02128 0.04167 0.04359 0.06250 0.15217
hist_50<-ggplot()+geom_histogram(aes(x = means_50))
hist_50
```

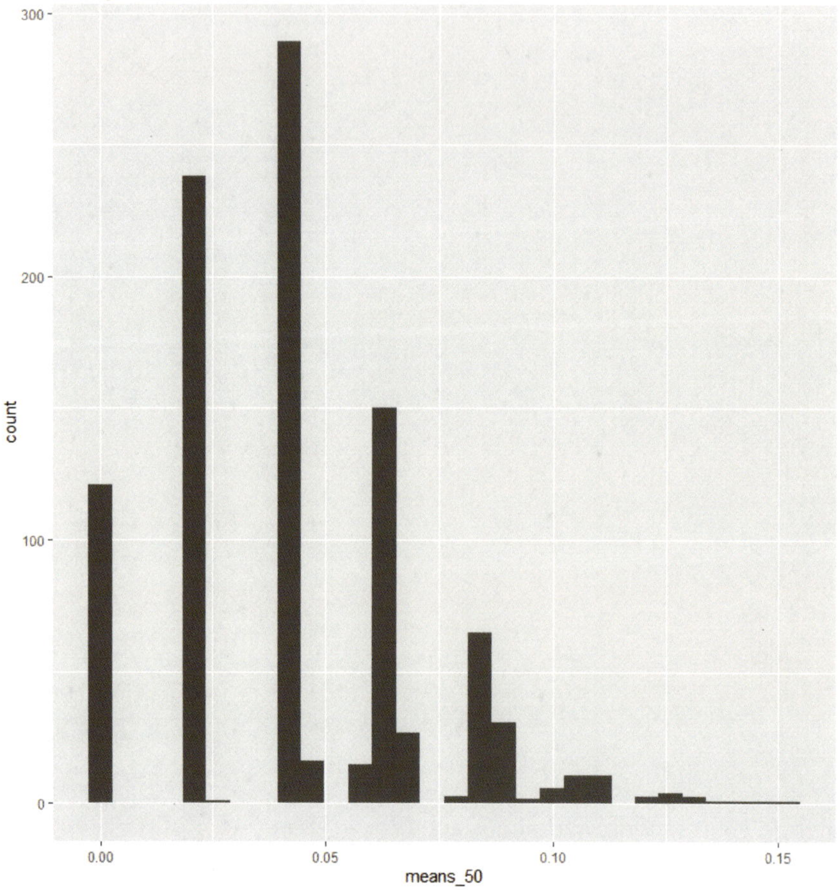


```
##To plot a boxplot of these means
box_50<-ggplot()+geom_boxplot(aes(y=means_50))
box_50
```

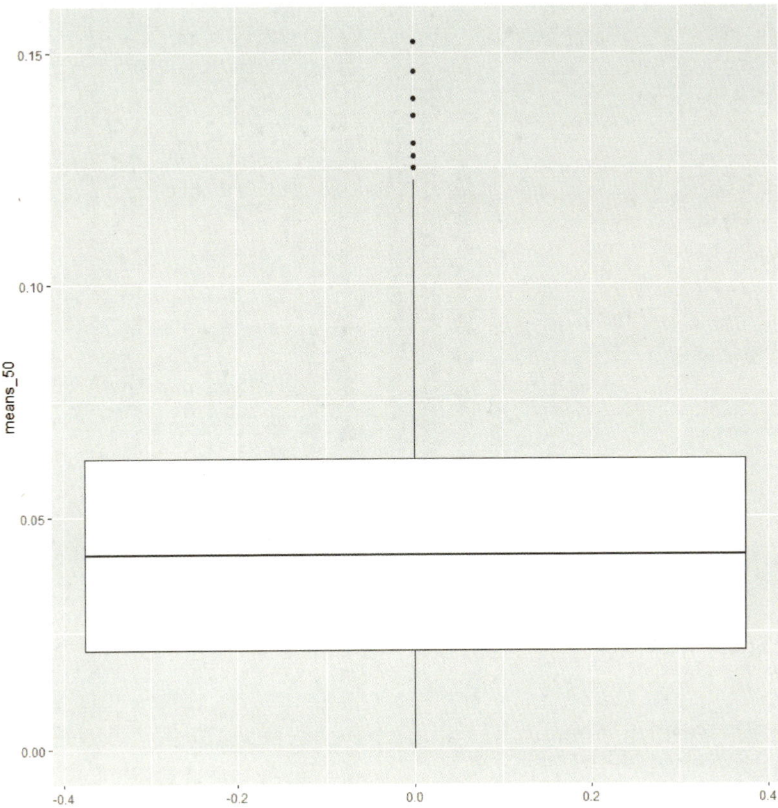

Generate 1000 random samples, each of size=150.

```
set.seed(123)
samples_150<-replicate(1000,sample(regicor$cv,150))
str(samples_150)
## num [1:150, 1:1000] 0 0 0 0 0 0 0 NA 0 0 ...
means_150<-colMeans(samples_150, na.rm = TRUE)
str(means_150)
## num [1:1000] 0.0559 0.0423 0.0567 0.0634 0.047 ...
summary(means_150)
##  Min. 1st Qu. Median  Mean 3rd Qu.  Max.
## 0.00000 0.02920 0.04196 0.04249 0.05147 0.09220
hist_150<-ggplot()+geom_histogram(aes(x = means_150))
hist_150
```

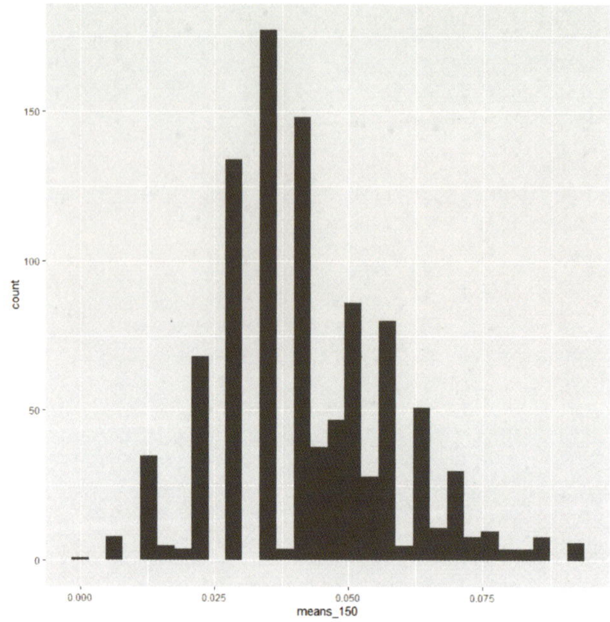


```
##To plot a boxplot of these means
box_150<-ggplot()+geom_boxplot(aes(y=means_150))
box_150
```

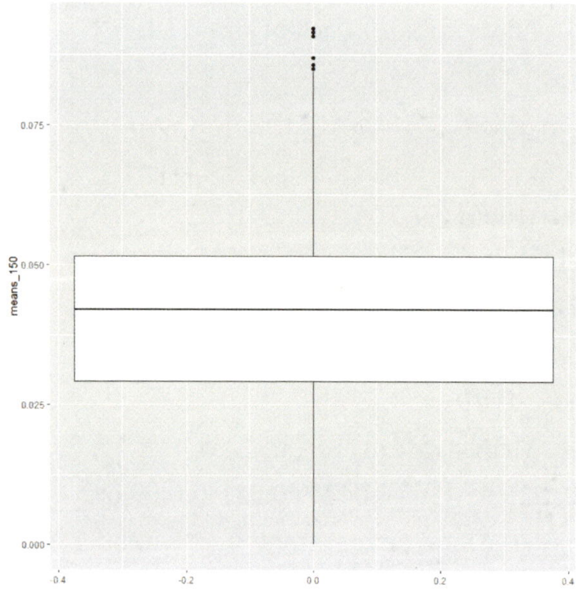

Plot all graphs side by side.

```
##Combine the 3 means in one dataframe
df<-data.frame(means_20=means_20, means_50=means_50,
means_150=means_150)
head(df)
##   means_20  means_50 means_150
## 1 0.00000000 0.02173913 0.05594406
## 2 0.05263158 0.06250000 0.04225352
## 3 0.00000000 0.06122449 0.05673759
## 4 0.11111111 0.06122449 0.06338028
## 5 0.05000000 0.04444444 0.04697987
## 6 0.15000000 0.04166667 0.02877698
# gather the columns
df<- df %>% gather(key = "mean," value = "value")
table(df$mean)
##
## means_150 means_20 means_50
##   1000   1000   1000
# arrange factor levels
df$mean<- factor(df$mean, levels = c("means_20,"
"means_50,""means_150"))
ggplot(data = df, aes(x = value, fill = mean)) + geom_histogram()+facet_
wrap(~mean)
```

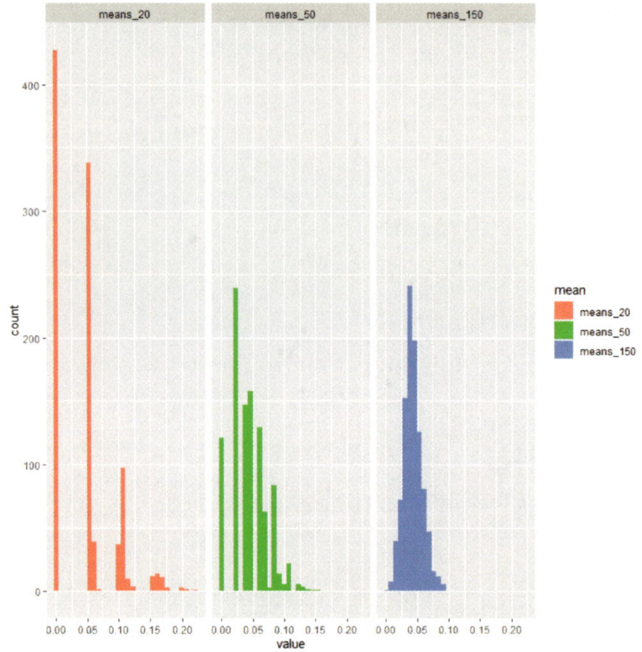


```
ggplot(data = df, aes(x = mean, y = value, col = mean))+geom_boxplot()
```

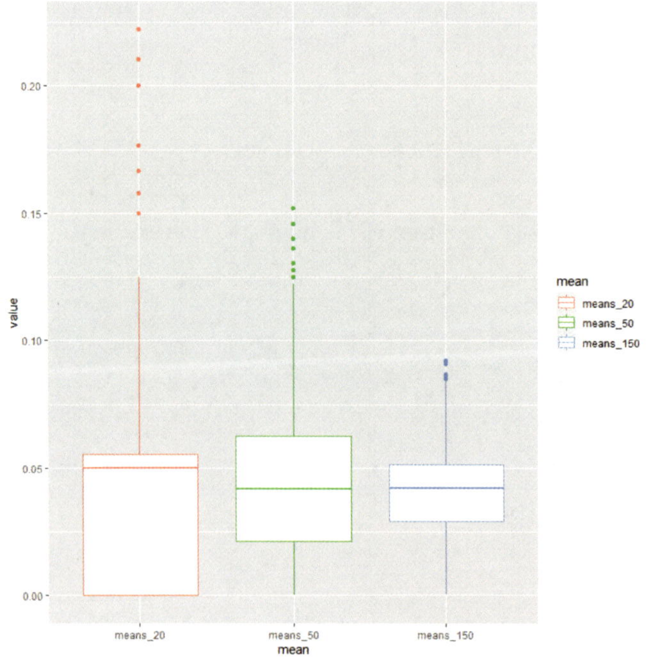

```
# to calculate the standard deviation for each means type
df %>% group_by(mean) %>% summarise(sd = sd(value))
## # A tibble: 3 x 2
##   mean       sd
##   <fct>      <dbl>
## 1 means_20  0.0463
## 2 means_50  0.0286
## 3 means_150 0.0159
```

The same rules apply:

1. The sampling distribution for sample proportions across random samples of the same size is approximately normal.
2. The variation in sample proportions decreases with increasing the sample size and the distribution shape becomes more normal. It decreases from 0.0463 for 20 sample size to 0.0159 for 150 sample size
3. The mean of sample proportions is very close to the true population proportion= 0.043.

6.4. THE SAMPLING DISTRIBUTION OF SAMPLE INCIDENCE RATES (IRS)

We will use the regicor data to determine the sampling distribution of the IR of cardiovascular events in this population (time-to-event variable).

Incidence rate= IR = Total number of cardiovascular events (outcomes) in a sample/Total follow up time. As we have converted cv variable to 0,1 variable so summing the cv variable will give us the total count of cardiovascular events in our data or assumed population.

The true population IR:

```
true_rate <-sum(regicor$cv, na.rm = TRUE)/sum(regicor$tocv, na.rm = TRUE)
true_rate
## [1] 2.424021e–05
```

The true rate is 2.4×10^{-5} or 0.000024.

Generate 1000 random samples, each of size=20, from the two columns cv and tocv as time to event data is a two dimensional data.

```
set.seed(123)
head(regicor[,22:23])
##   cv   tocv
## 6101 0 1024.882
## 5762 0 2756.849
## 2992 0 1905.969
## 2611 0 1055.380
## 2762 NA   NA
## 1516 0 3239.241
samples_20 <- replicate(1000,sample_n(regicor[,22:23],20))
str(samples_20)
## List of 2000
## $ : num [1:20] 0 0 0 0 0 0 0 NA 0 0 ...
## $ : num [1:20] 3626 1703 247 1313 2619 ...
##  ..- attr(*, "label")= chr "Days to cardiovascular event or end of follow-up"
## $ : num [1:20] 0 0 0 NA 0 0 0 0 0 0 ...
## $ : num [1:20] 916 1805 752 NA 3117 ...
##  ..- attr(*, "label")= chr "Days to cardiovascular event or end of follow-up"
## $ : num [1:20] NA 0 0 0 0 0 0 0 NA 0 ...
## $ : num [1:20] NA 2457 184 3549 2718 ...
##  ..- attr(*, "label")= chr "Days to cardiovascular event or end of follow-up"
## $ : num [1:20] 0 0 0 NA 0 0 0 0 0 0 ...
## $ : num [1:20] 1560 2591 269 NA 3293 ...
##  ..- attr(*, "label")= chr "Days to cardiovascular event or end of follow-up"
## $ : num [1:20] 0 0 0 0 0 0 0 0 0 0 ...
## $ : num [1:20] 3474 1479 653 624 2479 ...
##  ..- attr(*, "label")= chr "Days to cardiovascular event or end of follow-
```

```
up"
## $ : num [1:20] 0 0 0 0 0 0 0 1 0 0 ...
## $ : num [1:20] 1958 517 519 3571 3312 ...
##  ..- attr(*, "label")= chr "Days to cardiovascular event or end of follow-up"
## $ : num [1:20] 0 0 0 0 0 0 0 0 0 0 ...
## $ : num [1:20] 3368.8 819.5 242.7 17.6 3463.7 ...
##  ..- attr(*, "label")= chr "Days to cardiovascular event or end of follow-up"
## $ : num [1:20] 0 0 0 0 0 0 0 0 0 0 ...
## $ : num [1:20] 2225 1867 431 952 2498 ...
##  ..- attr(*, "label")= chr "Days to cardiovascular event or end of follow-up"
## $ : num [1:20] 0 0 0 0 0 0 0 0 0 0 ...
## $ : num [1:20] 2603 1181 1646 3123 2302 ...
##  ..- attr(*, "label")= chr "Days to cardiovascular event or end of follow-up"
## $ : num [1:20] 0 0 1 0 0 0 0 0 0 0 ...
## $ : num [1:20] 2696 3057 479 2097 1931 ...
##  ..- attr(*, "label")= chr "Days to cardiovascular event or end of follow-up"
## $ : num [1:20] 0 1 0 0 NA NA 0 0 0 0 ...
## $ : num [1:20] 722.8 7.22 596.15 1355.3 NA ...
##  ..- attr(*, "label")= chr "Days to cardiovascular event or end of follow-up"
## $ : num [1:20] 0 0 0 0 0 0 0 0 0 0 ...
## $ : num [1:20] 670 3074 469 1558 110 ...
##  ..- attr(*, "label")= chr "Days to cardiovascular event or end of follow-up"
## $ : num [1:20] 0 0 0 0 NA 0 0 0 0 0 ...
## $ : num [1:20] 3207 2572 2474 1206 NA ...
##  ..- attr(*, "label")= chr "Days to cardiovascular event or end of follow-up"
```

```
## $ : num [1:20] 1 0 0 0 0 0 0 0 NA 0 ...
## $ : num [1:20] 2003 1458 3389 497 2658 ...
## ..- attr(*, “label”)= chr “Days to cardiovascular event or end of follow-up”
## $ : num [1:20] 0 0 0 0 0 0 0 0 0 0 ...
## $ : num [1:20] 1385 245 2804 1034 1877 ...
## ..- attr(*, “label”)= chr “Days to cardiovascular event or end of follow-up”
## $ : num [1:20] 0 0 0 0 0 0 0 0 0 0 ...
## $ : num [1:20] 332.8 1997 113.9 3072.5 55.8 ...
## ..- attr(*, “label”)= chr “Days to cardiovascular event or end of follow-up”
## $ : num [1:20] 0 0 NA 0 0 0 1 0 0 0 ...
## $ : num [1:20] 546.9 1731.7 NA 15.6 2413.1 ...
## ..- attr(*, “label”)= chr “Days to cardiovascular event or end of follow-up”
## $ : num [1:20] 0 NA 0 0 0 0 0 0 0 0 ...
## $ : num [1:20] 456 NA 149 2505 3420 ...
## ..- attr(*, “label”)= chr “Days to cardiovascular event or end of follow-up”
## $ : num [1:20] 0 NA 0 1 0 0 0 0 0 0 ...
## $ : num [1:20] 2289 NA 2414 102 2712 ...
## ..- attr(*, “label”)= chr “Days to cardiovascular event or end of follow-up”
## $ : num [1:20] 0 1 0 0 0 0 0 0 0 0 ...
## $ : num [1:20] 523.05 7.27 3115.08 1882.53 2885.15 ...
## ..- attr(*, “label”)= chr “Days to cardiovascular event or end of follow-up”
## $ : num [1:20] 0 0 0 1 0 0 0 0 0 0 ...
## $ : num [1:20] 2471 1317 808 70 552 ...
## ..- attr(*, “label”)= chr “Days to cardiovascular event or end of follow-up”
## $ : num [1:20] 0 0 0 0 0 0 0 0 0 0 ...
```

```
## $ : num [1:20] 1821 651 1189 1338 3589 ...
## ..- attr(*, "label")= chr "Days to cardiovascular event or end of follow-up"
## $ : num [1:20] 0 0 0 0 0 0 0 0 0 NA ...
## $ : num [1:20] 1198 1024 2498 1402 2989 ...
## ..- attr(*, "label")= chr "Days to cardiovascular event or end of follow-up"
## $ : num [1:20] 0 0 NA 0 1 0 0 0 1 0 ...
## $ : num [1:20] 3307.8 96.9 NA 293.1 70 ...
## ..- attr(*, "label")= chr "Days to cardiovascular event or end of follow-up"
## $ : num [1:20] 0 0 0 0 0 0 0 0 0 0 ...
## $ : num [1:20] 868 2597 680 1982 464 ...
## ..- attr(*, "label")= chr "Days to cardiovascular event or end of follow-up"
## $ : num [1:20] 0 0 0 0 0 0 0 0 0 NA ...
## $ : num [1:20] 1386 2906 182 3570 2143 ...
## ..- attr(*, "label")= chr "Days to cardiovascular event or end of follow-up"
## $ : num [1:20] 0 0 0 0 0 NA NA 0 0 0 ...
## $ : num [1:20] 2957 3507 332 1746 3048 ...
## ..- attr(*, "label")= chr "Days to cardiovascular event or end of follow-up"
## $ : num [1:20] NA 0 0 0 0 1 0 0 0 0 ...
## $ : num [1:20] NA 1488 3384 943 292 ...
## ..- attr(*, "label")= chr "Days to cardiovascular event or end of follow-up"
## $ : num [1:20] 0 0 0 0 NA 0 0 0 0 0 ...
## $ : num [1:20] 3202 1668 2257 2440 NA ...
## ..- attr(*, "label")= chr "Days to cardiovascular event or end of follow-up"
## $ : num [1:20] 0 0 0 0 NA 1 0 0 0 0 ...
## $ : num [1:20] 3293 1958 1029 1714 NA ...
```

..- attr(*, "label")= chr "Days to cardiovascular event or end of follow-up"
$: num [1:20] 0 0 0 0 0 0 0 0 0 0 ...
$: num [1:20] 1097 2460 1590 1659 3609 ...
..- attr(*, "label")= chr "Days to cardiovascular event or end of follow-up"
$: num [1:20] 0 0 0 0 0 0 0 0 0 0 ...
$: num [1:20] 636 2451 933 3435 3133 ...
..- attr(*, "label")= chr "Days to cardiovascular event or end of follow-up"
$: num [1:20] 0 0 0 0 0 0 0 0 0 0 ...
$: num [1:20] 1806 1598 1707 583 1322 ...
..- attr(*, "label")= chr "Days to cardiovascular event or end of follow-up"
$: num [1:20] 1 0 0 0 0 0 0 0 0 0 ...
$: num [1:20] 1136 1644.3 487.2 3410.2 26.4 ...
..- attr(*, "label")= chr "Days to cardiovascular event or end of follow-up"
$: num [1:20] 0 0 0 0 0 0 0 0 0 1 ...
$: num [1:20] 2389 742 2385 1638 1491 ...
..- attr(*, "label")= chr "Days to cardiovascular event or end of follow-up"
$: num [1:20] 0 0 0 0 0 0 0 0 0 NA ...
$: num [1:20] 1696.8 64.5 708.2 1804.9 1710.7 ...
..- attr(*, "label")= chr "Days to cardiovascular event or end of follow-up"
$: num [1:20] 0 0 0 0 0 0 0 0 0 0 ...
$: num [1:20] 2455.6 72.9 2756.8 2024.2 1840.2 ...
..- attr(*, "label")= chr "Days to cardiovascular event or end of follow-up"
$: num [1:20] 0 0 0 0 0 0 0 0 0 0 ...
$: num [1:20] 1432 581 1551 928 1122 ...
..- attr(*, "label")= chr "Days to cardiovascular event or end of follow-

```
up"
## $ : num [1:20] 0 0 0 0 0 0 0 0 0 0 ...
## $ : num [1:20] 1355 245 348 3044 2156 ...
##  ..- attr(*, "label")= chr "Days to cardiovascular event or end of follow-
up"
## $ : num [1:20] 0 0 0 0 0 0 0 0 0 0 ...
## $ : num [1:20] 2366 1616 2793 3287 1143 ...
##  ..- attr(*, "label")= chr "Days to cardiovascular event or end of follow-
up"
## $ : num [1:20] 0 0 0 0 0 0 0 0 0 0 ...
## $ : num [1:20] 2788 182 162 2879 2215 ...
##  ..- attr(*, "label")= chr "Days to cardiovascular event or end of follow-
up"
## $ : num [1:20] NA 1 0 0 0 0 0 0 0 0 ...
## $ : num [1:20] NA 881 1579 1647 1656 ...
##  ..- attr(*, "label")= chr "Days to cardiovascular event or end of follow-
up"
## $ : num [1:20] 0 0 0 0 0 0 0 0 0 0 ...
## $ : num [1:20] 1198 1958 673 363 1955 ...
##  ..- attr(*, "label")= chr "Days to cardiovascular event or end of follow-
up"
## $ : num [1:20] 0 0 0 0 0 0 NA 0 0 0 ...
## $ : num [1:20] 1686.3 1974.5 34.1 2414 839.8 ...
##  ..- attr(*, "label")= chr "Days to cardiovascular event or end of follow-
up"
## $ : num [1:20] 0 0 0 NA 0 0 0 0 0 0 ...
## $ : num [1:20] 2613 924 3119 NA 1711 ...
##  ..- attr(*, "label")= chr "Days to cardiovascular event or end of follow-
up"
## $ : num [1:20] 0 0 0 0 0 0 0 0 0 0 ...
## $ : num [1:20] 2458 3216 295 268 543 ...
##  ..- attr(*, "label")= chr "Days to cardiovascular event or end of follow-
up"
```

```
## $ : num [1:20] NA 0 0 0 0 0 0 0 0 0 ...
## $ : num [1:20] NA 1058 2287 1644 467 ...
##  ..- attr(*, "label")= chr "Days to cardiovascular event or end of follow-up"
## $ : num [1:20] 0 0 0 0 0 0 0 0 0 0 ...
## $ : num [1:20] 2934 2101 1698 620 2983 ...
##  ..- attr(*, "label")= chr "Days to cardiovascular event or end of follow-up"
## $ : num [1:20] NA 0 0 0 0 0 0 NA 0 0 ...
## $ : num [1:20] NA 2793 521 188 3293 ...
##  ..- attr(*, "label")= chr "Days to cardiovascular event or end of follow-up"
## $ : num [1:20] 0 0 0 0 0 0 NA 0 0 0 ...
##  [list output truncated]
## – attr(*, "dim")= int [1:2] 2 1000
## – attr(*, "dimnames")=List of 2
##  ..$ : chr [1:2] "cv" "tocv"
##  ..$ : NULL
dim(samples_20)
## [1]  2 1000
##It is a list of 2000 elements, each element has 2 rows one for cardiovascular event and the other for time and 1000 columns or 1000 samples. Each row has 20 numbers
##To view the elements of cardiovascular events or 1st row only
str(samples_20[1,])
## List of 1000
## $ : num [1:20] 0 0 0 0 0 0 0 NA 0 0 ...
## $ : num [1:20] 0 0 0 NA 0 0 0 0 0 0 ...
## $ : num [1:20] NA 0 0 0 0 0 0 0 NA 0 ...
## $ : num [1:20] 0 0 0 NA 0 0 0 0 0 0 ...
## $ : num [1:20] 0 0 0 0 0 0 0 0 0 0 ...
## $ : num [1:20] 0 0 0 0 0 0 0 1 0 0 ...
```

```
## $ : num [1:20] 0 0 0 0 0 0 0 0 0 0 ...
## $ : num [1:20] 0 0 0 0 0 0 0 0 0 0 ...
## $ : num [1:20] 0 0 0 0 0 0 0 0 0 0 ...
## $ : num [1:20] 0 0 1 0 0 0 0 0 0 0 ...
## $ : num [1:20] 0 1 0 0 NA NA 0 0 0 0 ...
## $ : num [1:20] 0 0 0 0 0 0 0 0 0 0 ...
## $ : num [1:20] 0 0 0 0 NA 0 0 0 0 0 ...
## $ : num [1:20] 1 0 0 0 0 0 0 0 NA 0 ...
## $ : num [1:20] 0 0 0 0 0 0 0 0 0 0 ...
## $ : num [1:20] 0 0 0 0 0 0 0 0 0 0 ...
## $ : num [1:20] 0 0 NA 0 0 0 1 0 0 0 ...
## $ : num [1:20] 0 NA 0 0 0 0 0 0 0 0 ...
## $ : num [1:20] 0 NA 0 1 0 0 0 0 0 0 ...
## $ : num [1:20] 0 1 0 0 0 0 0 0 0 0 ...
## $ : num [1:20] 0 0 0 1 0 0 0 0 0 0 ...
## $ : num [1:20] 0 0 0 0 0 0 0 0 0 0 ...
## $ : num [1:20] 0 0 0 0 0 0 0 0 0 NA ...
## $ : num [1:20] 0 0 NA 0 1 0 0 0 1 0 ...
## $ : num [1:20] 0 0 0 0 0 0 0 0 0 0 ...
## $ : num [1:20] 0 0 0 0 0 0 0 0 0 NA ...
## $ : num [1:20] 0 0 0 0 0 NA NA 0 0 0 ...
## $ : num [1:20] NA 0 0 0 0 1 0 0 0 0 ...
## $ : num [1:20] 0 0 0 0 NA 0 0 0 0 0 ...
## $ : num [1:20] 0 0 0 0 NA 1 0 0 0 0 ...
## $ : num [1:20] 0 0 0 0 0 0 0 0 0 0 ...
## $ : num [1:20] 0 0 0 0 0 0 0 0 0 0 ...
## $ : num [1:20] 0 0 0 0 0 0 0 0 0 0 ...
## $ : num [1:20] 1 0 0 0 0 0 0 0 0 0 ...
## $ : num [1:20] 0 0 0 0 0 0 0 0 0 1 ...
## $ : num [1:20] 0 0 0 0 0 0 0 0 0 NA ...
## $ : num [1:20] 0 0 0 0 0 0 0 0 0 0 ...
```

```
## $ : num [1:20] 0 0 0 0 0 0 0 0 0 0 ...
## $ : num [1:20] 0 0 0 0 0 0 0 0 0 0 ...
## $ : num [1:20] 0 0 0 0 0 0 0 0 0 0 ...
## $ : num [1:20] 0 0 0 0 0 0 0 0 0 0 ...
## $ : num [1:20] NA 1 0 0 0 0 0 0 0 0 ...
## $ : num [1:20] 0 0 0 0 0 0 0 0 0 0 ...
## $ : num [1:20] 0 0 0 0 0 0 NA 0 0 0 ...
## $ : num [1:20] 0 0 0 NA 0 0 0 0 0 0 ...
## $ : num [1:20] 0 0 0 0 0 0 0 0 0 0 ...
## $ : num [1:20] NA 0 0 0 0 0 0 0 0 0 ...
## $ : num [1:20] 0 0 0 0 0 0 0 0 0 0 ...
## $ : num [1:20] NA 0 0 0 0 0 0 NA 0 0 ...
## $ : num [1:20] 0 0 0 0 0 0 NA 0 0 0 ...
## $ : num [1:20] 0 0 0 NA 0 0 0 0 0 0 ...
## $ : num [1:20] 0 0 1 0 0 0 0 0 0 0 ...
## $ : num [1:20] 0 0 0 0 0 0 0 0 0 0 ...
## $ : num [1:20] 0 0 0 0 0 0 0 0 0 0 ...
## $ : num [1:20] 0 0 0 0 0 0 0 0 0 0 ...
## $ : num [1:20] 0 0 0 NA 0 0 0 0 0 0 ...
## $ : num [1:20] 0 0 0 0 0 0 0 0 0 0 ...
## $ : num [1:20] 0 0 0 1 0 0 0 0 0 0 ...
## $ : num [1:20] 0 0 0 0 0 1 0 0 0 0 ...
## $ : num [1:20] 0 0 0 NA NA NA 0 0 0 1 ...
## $ : num [1:20] 0 0 0 0 NA 0 0 0 0 0 ...
## $ : num [1:20] 0 0 0 0 0 NA 0 0 NA 0 ...
## $ : num [1:20] 0 0 0 0 0 1 0 0 0 0 ...
## $ : num [1:20] 0 0 0 0 0 NA 0 0 0 0 ...
## $ : num [1:20] 0 0 0 0 0 0 0 0 1 0 ...
## $ : num [1:20] 0 0 0 0 0 0 0 0 0 NA ...
## $ : num [1:20] 0 0 0 0 NA 1 0 0 0 0 ...
## $ : num [1:20] 0 0 0 1 0 0 0 0 0 0 ...
```

```
## $ : num [1:20] 0 0 0 0 0 0 NA 0 0 NA ...
## $ : num [1:20] 0 NA 0 NA 0 0 0 NA 0 0 ...
## $ : num [1:20] 0 0 0 0 0 0 0 0 0 0 ...
## $ : num [1:20] 0 0 1 0 0 0 NA 0 0 0 ...
## $ : num [1:20] 0 0 0 0 0 0 0 0 0 0 ...
## $ : num [1:20] 0 0 0 0 0 NA NA NA 0 0 ...
## $ : num [1:20] NA 0 0 0 0 0 0 NA 0 1 ...
## $ : num [1:20] NA 0 0 0 0 0 0 0 0 0 ...
## $ : num [1:20] 0 0 0 NA 0 NA 0 0 0 0 ...
## $ : num [1:20] 0 1 0 0 0 0 0 0 0 0 ...
## $ : num [1:20] 0 0 0 0 0 1 0 0 0 0 ...
## $ : num [1:20] 0 0 0 0 NA 0 0 0 0 0 ...
## $ : num [1:20] 0 0 0 0 0 0 0 0 0 0 ...
## $ : num [1:20] 0 0 0 0 0 1 0 0 0 0 ...
## $ : num [1:20] 0 0 0 0 0 0 0 0 0 NA ...
## $ : num [1:20] NA 0 1 NA 0 0 0 0 0 0 ...
## $ : num [1:20] 0 0 0 1 0 0 0 1 NA 0 ...
## $ : num [1:20] 0 0 0 0 0 0 0 0 0 0 ...
## $ : num [1:20] 0 0 0 0 0 0 0 0 0 0 ...
## $ : num [1:20] 0 0 0 1 0 0 0 0 0 0 ...
## $ : num [1:20] 0 0 0 0 0 0 0 0 NA 0 ...
## $ : num [1:20] NA 0 0 0 0 1 0 0 NA NA ...
## $ : num [1:20] 0 0 0 0 0 0 0 0 0 0 ...
## $ : num [1:20] 0 0 0 0 0 NA 0 0 0 0 ...
## $ : num [1:20] 0 0 0 0 0 0 0 0 0 0 ...
## $ : num [1:20] 0 0 0 0 0 0 0 0 0 0 ...
## $ : num [1:20] 0 0 0 NA 0 0 0 0 0 0 ...
## $ : num [1:20] 0 0 0 0 0 0 0 0 1 0 ...
## $ : num [1:20] 0 0 0 0 0 0 0 0 0 0 ...
## $ : num [1:20] 0 0 0 NA 0 0 0 0 0 0 ...
## $ : num [1:20] 0 0 0 0 0 0 0 0 0 0 ...
```

```
## [list output truncated]
##To view the cardiovascular events of the 1st element
str(samples_20[1,][1])
## List of 1
## $ : num [1:20] 0 0 0 0 0 0 0 NA 0 0 ...
##To view the elements of time only
str(samples_20[2,])
## List of 1000
## $ : num [1:20] 3626 1703 247 1313 2619 ...
## ..- attr(*, "label")= chr "Days to cardiovascular event or end of follow-up"
## $ : num [1:20] 916 1805 752 NA 3117 ...
## ..- attr(*, "label")= chr "Days to cardiovascular event or end of follow-up"
## $ : num [1:20] NA 2457 184 3549 2718 ...
## ..- attr(*, "label")= chr "Days to cardiovascular event or end of follow-up"
## $ : num [1:20] 1560 2591 269 NA 3293 ...
## ..- attr(*, "label")= chr "Days to cardiovascular event or end of follow-up"
## $ : num [1:20] 3474 1479 653 624 2479 ...
## ..- attr(*, "label")= chr "Days to cardiovascular event or end of follow-up"
## $ : num [1:20] 1958 517 519 3571 3312 ...
## ..- attr(*, "label")= chr "Days to cardiovascular event or end of follow-up"
## $ : num [1:20] 3368.8 819.5 242.7 17.6 3463.7 ...
## ..- attr(*, "label")= chr "Days to cardiovascular event or end of follow-up"
## $ : num [1:20] 2225 1867 431 952 2498 ...
## ..- attr(*, "label")= chr "Days to cardiovascular event or end of follow-up"
## $ : num [1:20] 2603 1181 1646 3123 2302 ...
```

```
## ..- attr(*, "label")= chr "Days to cardiovascular event or end of follow-up"
## $ : num [1:20] 2696 3057 479 2097 1931 ...
## ..- attr(*, "label")= chr "Days to cardiovascular event or end of follow-up"
## $ : num [1:20] 722.8 7.22 596.15 1355.3 NA ...
## ..- attr(*, "label")= chr "Days to cardiovascular event or end of follow-up"
## $ : num [1:20] 670 3074 469 1558 110 ...
## ..- attr(*, "label")= chr "Days to cardiovascular event or end of follow-up"
## $ : num [1:20] 3207 2572 2474 1206 NA ...
## ..- attr(*, "label")= chr "Days to cardiovascular event or end of follow-up"
## $ : num [1:20] 2003 1458 3389 497 2658 ...
## ..- attr(*, "label")= chr "Days to cardiovascular event or end of follow-up"
## $ : num [1:20] 1385 245 2804 1034 1877 ...
## ..- attr(*, "label")= chr "Days to cardiovascular event or end of follow-up"
## $ : num [1:20] 332.8 1997 113.9 3072.5 55.8 ...
## ..- attr(*, "label")= chr "Days to cardiovascular event or end of follow-up"
## $ : num [1:20] 546.9 1731.7 NA 15.6 2413.1 ...
## ..- attr(*, "label")= chr "Days to cardiovascular event or end of follow-up"
## $ : num [1:20] 456 NA 149 2505 3420 ...
## ..- attr(*, "label")= chr "Days to cardiovascular event or end of follow-up"
## $ : num [1:20] 2289 NA 2414 102 2712 ...
## ..- attr(*, "label")= chr "Days to cardiovascular event or end of follow-up"
## $ : num [1:20] 523.05 7.27 3115.08 1882.53 2885.15 ...
## ..- attr(*, "label")= chr "Days to cardiovascular event or end of follow-
```

up"

```
## $ : num [1:20] 2471 1317 808 70 552 ...
##  ..- attr(*, "label")= chr "Days to cardiovascular event or end of follow-up"
## $ : num [1:20] 1821 651 1189 1338 3589 ...
##  ..- attr(*, "label")= chr "Days to cardiovascular event or end of follow-up"
## $ : num [1:20] 1198 1024 2498 1402 2989 ...
##  ..- attr(*, "label")= chr "Days to cardiovascular event or end of follow-up"
## $ : num [1:20] 3307.8 96.9 NA 293.1 70 ...
##  ..- attr(*, "label")= chr "Days to cardiovascular event or end of follow-up"
## $ : num [1:20] 868 2597 680 1982 464 ...
##  ..- attr(*, "label")= chr "Days to cardiovascular event or end of follow-up"
## $ : num [1:20] 1386 2906 182 3570 2143 ...
##  ..- attr(*, "label")= chr "Days to cardiovascular event or end of follow-up"
## $ : num [1:20] 2957 3507 332 1746 3048 ...
##  ..- attr(*, "label")= chr "Days to cardiovascular event or end of follow-up"
## $ : num [1:20] NA 1488 3384 943 292 ...
##  ..- attr(*, "label")= chr "Days to cardiovascular event or end of follow-up"
## $ : num [1:20] 3202 1668 2257 2440 NA ...
##  ..- attr(*, "label")= chr "Days to cardiovascular event or end of follow-up"
## $ : num [1:20] 3293 1958 1029 1714 NA ...
##  ..- attr(*, "label")= chr "Days to cardiovascular event or end of follow-up"
## $ : num [1:20] 1097 2460 1590 1659 3609 ...
##  ..- attr(*, "label")= chr "Days to cardiovascular event or end of follow-
```

```
up"
## $ : num [1:20] 636 2451 933 3435 3133 ...
## ..- attr(*, "label")= chr "Days to cardiovascular event or end of follow-up"
## $ : num [1:20] 1806 1598 1707 583 1322 ...
## ..- attr(*, "label")= chr "Days to cardiovascular event or end of follow-up"
## $ : num [1:20] 1136 1644.3 487.2 3410.2 26.4 ...
## ..- attr(*, "label")= chr "Days to cardiovascular event or end of follow-up"
## $ : num [1:20] 2389 742 2385 1638 1491 ...
## ..- attr(*, "label")= chr "Days to cardiovascular event or end of follow-up"
## $ : num [1:20] 1696.8 64.5 708.2 1804.9 1710.7 ...
## ..- attr(*, "label")= chr "Days to cardiovascular event or end of follow-up"
## $ : num [1:20] 2455.6 72.9 2756.8 2024.2 1840.2 ...
## ..- attr(*, "label")= chr "Days to cardiovascular event or end of follow-up"
## $ : num [1:20] 1432 581 1551 928 1122 ...
## ..- attr(*, "label")= chr "Days to cardiovascular event or end of follow-up"
## $ : num [1:20] 1355 245 348 3044 2156 ...
## ..- attr(*, "label")= chr "Days to cardiovascular event or end of follow-up"
## $ : num [1:20] 2366 1616 2793 3287 1143 ...
## ..- attr(*, "label")= chr "Days to cardiovascular event or end of follow-up"
## $ : num [1:20] 2788 182 162 2879 2215 ...
## ..- attr(*, "label")= chr "Days to cardiovascular event or end of follow-up"
## $ : num [1:20] NA 881 1579 1647 1656 ...
## ..- attr(*, "label")= chr "Days to cardiovascular event or end of follow-
```

up"
$: num [1:20] 1198 1958 673 363 1955 ...
..- attr(*, "label")= chr "Days to cardiovascular event or end of follow-up"
$: num [1:20] 1686.3 1974.5 34.1 2414 839.8 ...
..- attr(*, "label")= chr "Days to cardiovascular event or end of follow-up"
$: num [1:20] 2613 924 3119 NA 1711 ...
..- attr(*, "label")= chr "Days to cardiovascular event or end of follow-up"
$: num [1:20] 2458 3216 295 268 543 ...
..- attr(*, "label")= chr "Days to cardiovascular event or end of follow-up"
$: num [1:20] NA 1058 2287 1644 467 ...
..- attr(*, "label")= chr "Days to cardiovascular event or end of follow-up"
$: num [1:20] 2934 2101 1698 620 2983 ...
..- attr(*, "label")= chr "Days to cardiovascular event or end of follow-up"
$: num [1:20] NA 2793 521 188 3293 ...
..- attr(*, "label")= chr "Days to cardiovascular event or end of follow-up"
$: num [1:20] 1009.2 94.8 1997.7 468.8 1697.7 ...
..- attr(*, "label")= chr "Days to cardiovascular event or end of follow-up"
$: num [1:20] 179 2031 1366 NA 1628 ...
..- attr(*, "label")= chr "Days to cardiovascular event or end of follow-up"
$: num [1:20] 2988.81 438.46 9.94 1547.39 506.65 ...
..- attr(*, "label")= chr "Days to cardiovascular event or end of follow-up"
$: num [1:20] 510 2426 796 1994 3115 ...
..- attr(*, "label")= chr "Days to cardiovascular event or end of follow-

```
up"
## $ : num [1:20] 604 3403 2041 489 2839 ...
## ..- attr(*, "label")= chr "Days to cardiovascular event or end of follow-
up"
## $ : num [1:20] 3638 3097 1616 519 3003 ...
## ..- attr(*, "label")= chr "Days to cardiovascular event or end of follow-
up"
## $ : num [1:20] 293 2724 2742 NA 3570 ...
## ..- attr(*, "label")= chr "Days to cardiovascular event or end of follow-
up"
## $ : num [1:20] 2949 2471 636 3561 1067 ...
## ..- attr(*, "label")= chr "Days to cardiovascular event or end of follow-
up"
## $ : num [1:20] 3611.1 1450.4 3100 94.1 1111 ...
## ..- attr(*, "label")= chr "Days to cardiovascular event or end of follow-
up"
## $ : num [1:20] 1152 409 310 2031 1909 ...
## ..- attr(*, "label")= chr "Days to cardiovascular event or end of follow-
up"
## $ : num [1:20] 3170 3050 628 NA NA ...
## ..- attr(*, "label")= chr "Days to cardiovascular event or end of follow-
up"
## $ : num [1:20] 1427 1437 2964 515 NA ...
## ..- attr(*, "label")= chr "Days to cardiovascular event or end of follow-
up"
## $ : num [1:20] 2815 3067 1385 2634 138 ...
## ..- attr(*, "label")= chr "Days to cardiovascular event or end of follow-
up"
## $ : num [1:20] 729 3338 2086 1684 1573 ...
## ..- attr(*, "label")= chr "Days to cardiovascular event or end of follow-
up"
## $ : num [1:20] 1663 639 3163 1718 1199 ...
## ..- attr(*, "label")= chr "Days to cardiovascular event or end of follow-
```

up"

$: num [1:20] 113.9 66.9 3548.1 1057.2 1578.1 ...

..- attr(*, "label")= chr "Days to cardiovascular event or end of follow-up"

$: num [1:20] 2287 1424 484 1981 2192 ...

..- attr(*, "label")= chr "Days to cardiovascular event or end of follow-up"

$: num [1:20] 1763 1399 3171 3445 NA ...

..- attr(*, "label")= chr "Days to cardiovascular event or end of follow-up"

$: num [1:20] 3320 2232 736 2063 1991 ...

..- attr(*, "label")= chr "Days to cardiovascular event or end of follow-up"

$: num [1:20] 2804 520 812 2910 1957 ...

..- attr(*, "label")= chr "Days to cardiovascular event or end of follow-up"

$: num [1:20] 777 NA 3394 NA 3481 ...

..- attr(*, "label")= chr "Days to cardiovascular event or end of follow-up"

$: num [1:20] 2896 1548 3414 2793 3122 ...

..- attr(*, "label")= chr "Days to cardiovascular event or end of follow-up"

$: num [1:20] 3481 2598 1216 1385 2531 ...

..- attr(*, "label")= chr "Days to cardiovascular event or end of follow-up"

$: num [1:20] 1408.7 1662.1 37.6 1334.6 2111 ...

..- attr(*, "label")= chr "Days to cardiovascular event or end of follow-up"

$: num [1:20] 561 1623 791 901 3555 ...

..- attr(*, "label")= chr "Days to cardiovascular event or end of follow-up"

$: num [1:20] NA 2429 2742 3357 1231 ...

..- attr(*, "label")= chr "Days to cardiovascular event or end of follow-

```
up"
## $ : num [1:20] NA 792 3617 2636 2571 ...
## ..- attr(*, "label")= chr "Days to cardiovascular event or end of follow-up"
## $ : num [1:20] 598 1043 1763 NA 674 ...
## ..- attr(*, "label")= chr "Days to cardiovascular event or end of follow-up"
## $ : num [1:20] 479 894 2757 1213 849 ...
## ..- attr(*, "label")= chr "Days to cardiovascular event or end of follow-up"
## $ : num [1:20] 3009 1419 1918 1366 1931 ...
## ..- attr(*, "label")= chr "Days to cardiovascular event or end of follow-up"
## $ : num [1:20] 1724 536 1475 2919 NA ...
## ..- attr(*, "label")= chr "Days to cardiovascular event or end of follow-up"
## $ : num [1:20] 1158 1141 433 1023 239 ...
## ..- attr(*, "label")= chr "Days to cardiovascular event or end of follow-up"
## $ : num [1:20] 3320 2165 3438 1491 2500 ...
## ..- attr(*, "label")= chr "Days to cardiovascular event or end of follow-up"
## $ : num [1:20] 2692 1874 1782 3624 618 ...
## ..- attr(*, "label")= chr "Days to cardiovascular event or end of follow-up"
## $ : num [1:20] NA 467 700 NA 2989 ...
## ..- attr(*, "label")= chr "Days to cardiovascular event or end of follow-up"
## $ : num [1:20] 817 1409 3205 417 3022 ...
## ..- attr(*, "label")= chr "Days to cardiovascular event or end of follow-up"
## $ : num [1:20] 3152 2766 790 3552 2044 ...
## ..- attr(*, "label")= chr "Days to cardiovascular event or end of follow-
```

```
up"
## $ : num [1:20] 746 1100 804 377 2754 ...
## ..- attr(*, "label")= chr "Days to cardiovascular event or end of follow-
up"
## $ : num [1:20] 520 543 3057 198 409 ...
## ..- attr(*, "label")= chr "Days to cardiovascular event or end of follow-
up"
## $ : num [1:20] 1500 2797 3518 2396 1967 ...
## ..- attr(*, "label")= chr "Days to cardiovascular event or end of follow-
up"
## $ : num [1:20] NA 3423 1129 661 2726 ...
## ..- attr(*, "label")= chr "Days to cardiovascular event or end of follow-
up"
## $ : num [1:20] 943 2787 2304 339 2785 ...
## ..- attr(*, "label")= chr "Days to cardiovascular event or end of follow-
up"
## $ : num [1:20] 849 3153 2687 103 124 ...
## ..- attr(*, "label")= chr "Days to cardiovascular event or end of follow-
up"
## $ : num [1:20] 3456 2385 973 1355 848 ...
## ..- attr(*, "label")= chr "Days to cardiovascular event or end of follow-
up"
## $ : num [1:20] 3006 2268 1352 3230 843 ...
## ..- attr(*, "label")= chr "Days to cardiovascular event or end of follow-
up"
## $ : num [1:20] 3398 2513 2666 NA 3216 ...
## ..- attr(*, "label")= chr "Days to cardiovascular event or end of follow-
up"
## $ : num [1:20] 1607 2111 3363 1113 410 ...
## ..- attr(*, "label")= chr "Days to cardiovascular event or end of follow-
up"
## $ : num [1:20] 422 2312 3071 2879 3586 ...
## ..- attr(*, "label")= chr "Days to cardiovascular event or end of follow-
```

```
up"
## $ : num [1:20] 3152 3379 112 NA 1411 ...
## ..- attr(*, "label")= chr "Days to cardiovascular event or end of follow-up"
## $ : num [1:20] 1452 2914 2531 1907 715 ...
## ..- attr(*, "label")= chr "Days to cardiovascular event or end of follow-up"
## [list output truncated]
##To view the follow-up times of the 1st element
str(samples_20[2,][1])
## List of 1
## $ : num [1:20] 3626 1703 247 1313 2619 ...
## ..- attr(*, "label")= chr "Days to cardiovascular event or end of follow-up"
##To obtain the sums of cardiovascular events only, use sapply where it sums every element of the list
sums_cv<-sapply(samples_20[1,],sum, na.rm = TRUE)
summary(sums_cv)
## Min. 1st Qu. Median Mean 3rd Qu. Max.
## 0.000 0.000 1.000 0.819 1.000 4.000
##To obtain the sums of time only, use sapply where it sums every element of the list
sums_time<-sapply(samples_20[2,],sum, na.rm = TRUE)
summary(sums_time)
## Min. 1st Qu. Median Mean 3rd Qu. Max.
## 17246 29486 32831 32847 36108 48114
##To obtain the incidence rate for each sample
rates_20<-sums_cv/sums_time
head(rates_20)
## [1] 0.000000e+00 3.341490e-05 0.000000e+00 6.451571e-05 2.427611e-05
## [6] 9.313591e-05
```

```
summary(rates_20)
## Min. 1st Qu. Median Mean 3rd Qu. Max.
## 0.000e+00 0.000e+00 2.719e-05 2.637e-05 3.628e-05 2.280e-04
##To view a histogram of incidence rates based on samples of size=20
hist_20<-ggplot()+geom_histogram(aes(x = rates_20))
hist_20
```

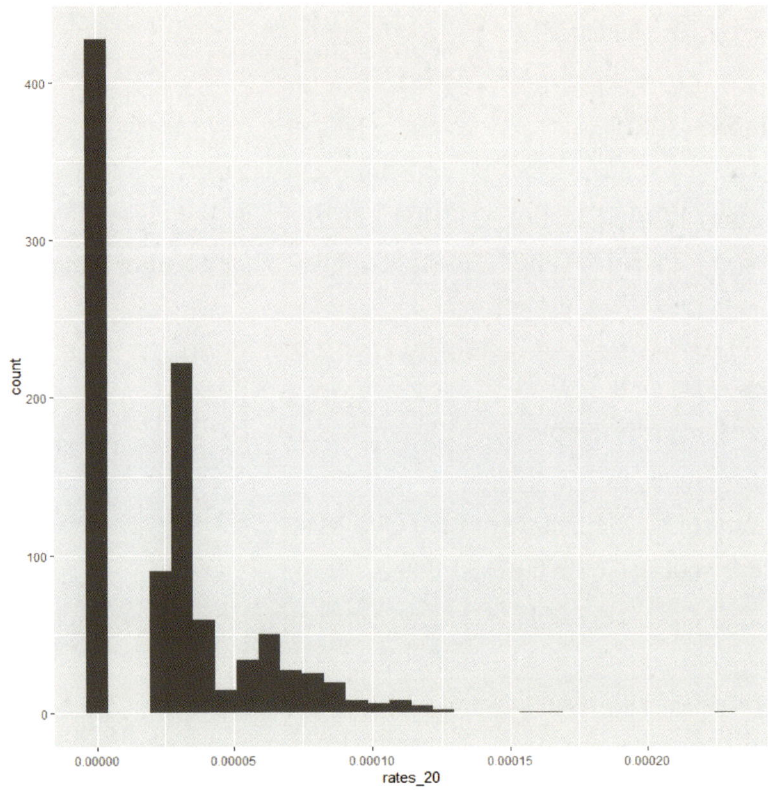


```
##To plot a boxplot of these incidence rates
box_20<-ggplot()+geom_boxplot(aes(y=rates_20))
box_20
```

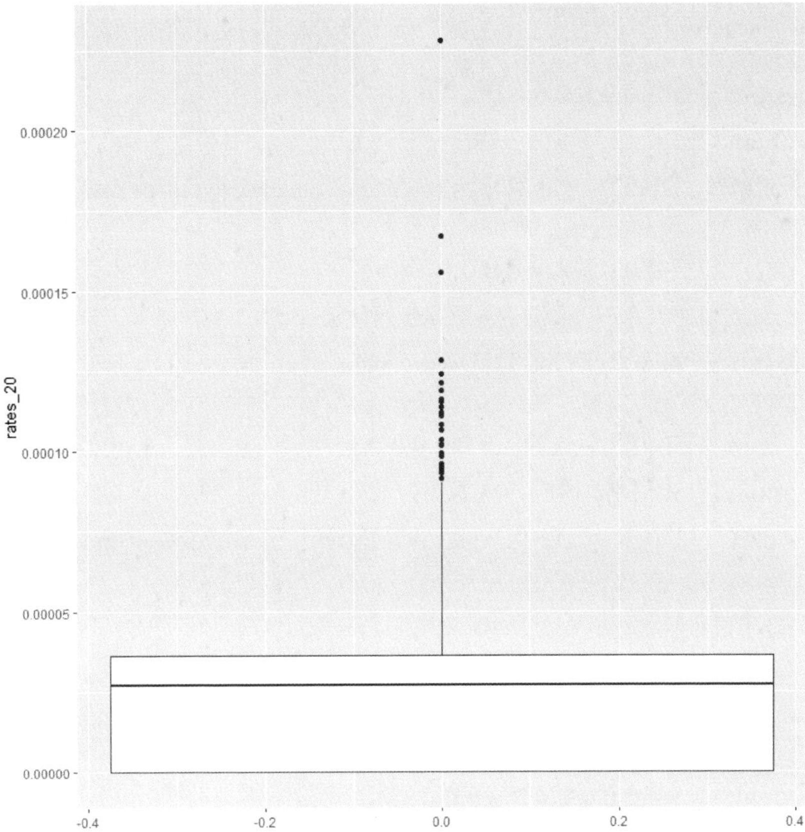

Generate 1000 random samples, each of size=50, from the two columns cv and tocv as time to event data is a two dimensional data.

```
set.seed(123)
samples_50<-replicate(1000,sample_n(regicor[,22:23],50))
str(samples_50)
## List of 2000
## $ : num [1:50] 0 0 0 0 0 0 0 NA 0 0 ...
## $ : num [1:50] 3626 1703 247 1313 2619 ...
##  ..- attr(*, "label")= chr "Days to cardiovascular event or end of follow-up"
## $ : num [1:50] 0 0 0 0 0 0 0 0 0 0 ...
## $ : num [1:50] 3540 1313 2427 1782 2426 ...
```

```
## ..- attr(*, "label")= chr "Days to cardiovascular event or end of follow-up"
## $ : num [1:50] 0 0 0 0 0 0 0 1 0 0 ...
## $ : num [1:50] 1958 517 519 3571 3312 ...
## ..- attr(*, "label")= chr "Days to cardiovascular event or end of follow-up"
## $ : num [1:50] 0 0 NA 0 0 0 0 0 1 0 ...
## $ : num [1:50] 173 1752 NA 438 3356 ...
## ..- attr(*, "label")= chr "Days to cardiovascular event or end of follow-up"
## $ : num [1:50] 0 0 NA NA 0 0 0 0 0 0 ...
## $ : num [1:50] 596 1355 NA NA 2572 ...
## ..- attr(*, "label")= chr "Days to cardiovascular event or end of follow-up"
## $ : num [1:50] 0 0 0 0 0 0 1 0 1 0 ...
## $ : num [1:50] 171 2490 362 1200 2144 ...
## ..- attr(*, "label")= chr "Days to cardiovascular event or end of follow-up"
## $ : num [1:50] 0 0 0 0 0 0 0 0 0 0 ...
## $ : num [1:50] 113.9 3072.5 55.8 1437.2 541.9 ...
## ..- attr(*, "label")= chr "Days to cardiovascular event or end of follow-up"
## $ : num [1:50] 0 0 0 0 0 0 0 0 0 NA ...
## $ : num [1:50] 2804 2785 1967 1039 768 ...
## ..- attr(*, "label")= chr "Days to cardiovascular event or end of follow-up"
## $ : num [1:50] 1 0 0 0 0 0 0 1 0 NA ...
## $ : num [1:50] 70 552 2010 3071 1931 ...
## ..- attr(*, "label")= chr "Days to cardiovascular event or end of follow-up"
## $ : num [1:50] 0 0 1 0 1 0 0 0 0 0 ...
## $ : num [1:50] 1982 2106 427 3357 106 ...
```

```
## ..- attr(*, "label")= chr "Days to cardiovascular event or end of follow-up"
## $ : num [1:50] 0 0 0 0 0 0 0 NA 0 0 ...
## $ : num [1:50] 182 3570 2143 2932 1698 ...
## ..- attr(*, "label")= chr "Days to cardiovascular event or end of follow-up"
## $ : num [1:50] 0 0 0 0 NA 0 0 0 0 0 ...
## $ : num [1:50] 246.6 2930.2 120.9 12.8 NA ...
## ..- attr(*, "label")= chr "Days to cardiovascular event or end of follow-up"
## $ : num [1:50] 0 0 0 0 0 0 0 0 0 0 ...
## $ : num [1:50] 1590 1659 3609 342 2196 ...
## ..- attr(*, "label")= chr "Days to cardiovascular event or end of follow-up"
## $ : num [1:50] 0 0 0 0 0 0 0 0 0 1 ...
## $ : num [1:50] 3374 3344 3105 1409 1452 ...
## ..- attr(*, "label")= chr "Days to cardiovascular event or end of follow-up"
## $ : num [1:50] 0 0 0 0 0 0 0 0 NA 0 ...
## $ : num [1:50] 64.5 708.2 1804.9 1710.7 1815.1 ...
## ..- attr(*, "label")= chr "Days to cardiovascular event or end of follow-up"
## $ : num [1:50] 0 0 0 0 0 0 0 0 0 0 ...
## $ : num [1:50] 2490 1998 2408 2392 348 ...
## ..- attr(*, "label")= chr "Days to cardiovascular event or end of follow-up"
## $ : num [1:50] 0 0 0 0 0 0 0 0 0 0 ...
## $ : num [1:50] 182 162 2879 2215 2413 ...
## ..- attr(*, "label")= chr "Days to cardiovascular event or end of follow-up"
## $ : num [1:50] NA 0 0 0 0 0 0 0 0 0 ...
## $ : num [1:50] NA 1025 1343 835 1352 ...
```

```
## ..- attr(*, "label")= chr "Days to cardiovascular event or end of follow-up"
## $ : num [1:50] 0 0 0 0 0 0 0 0 0 0 ...
## $ : num [1:50] 3216 295 268 543 598 ...
## ..- attr(*, "label")= chr "Days to cardiovascular event or end of follow-up"
## $ : num [1:50] 0 0 NA 0 0 1 0 0 NA 0 ...
## $ : num [1:50] 203 2628 NA 3171 3153 ...
## ..- attr(*, "label")= chr "Days to cardiovascular event or end of follow-up"
## $ : num [1:50] NA 0 0 0 0 0 0 0 NA 0 ...
## $ : num [1:50] NA 1628 2557 2696 2110 ...
## ..- attr(*, "label")= chr "Days to cardiovascular event or end of follow-up"
## $ : num [1:50] 0 0 0 0 1 0 0 0 0 0 ...
## $ : num [1:50] 178.92 2177.83 3401.55 112.6 7.27 ...
## ..- attr(*, "label")= chr "Days to cardiovascular event or end of follow-up"
## $ : num [1:50] 0 0 0 0 0 0 0 0 0 0 ...
## $ : num [1:50] 3570 1638 3210 3529 3295 ...
## ..- attr(*, "label")= chr "Days to cardiovascular event or end of follow-up"
## $ : num [1:50] 0 0 0 1 0 0 0 0 0 0 ...
## $ : num [1:50] 1632 2084 2392 106 2930 ...
## ..- attr(*, "label")= chr "Days to cardiovascular event or end of follow-up"
## $ : num [1:50] NA 0 0 0 0 0 0 0 0 NA ...
## $ : num [1:50] NA 972 2724 1698 234 ...
## ..- attr(*, "label")= chr "Days to cardiovascular event or end of follow-up"
## $ : num [1:50] 1 0 0 0 0 0 0 0 0 0 ...
## $ : num [1:50] 894 102 3384 3179 3423 ...
```

```
## ..- attr(*, "label")= chr "Days to cardiovascular event or end of follow-up"
## $ : num [1:50] 0 0 0 0 NA 0 0 0 NA 0 ...
## $ : num [1:50] 3117 849 2513 1723 NA ...
## ..- attr(*, "label")= chr "Days to cardiovascular event or end of follow-up"
## $ : num [1:50] 0 0 0 0 0 0 0 0 0 0 ...
## $ : num [1:50] 542 2459 594 331 1191 ...
## ..- attr(*, "label")= chr "Days to cardiovascular event or end of follow-up"
## $ : num [1:50] 0 0 0 0 0 0 0 0 NA 0 ...
## $ : num [1:50] 2196 3325 3433 3556 293 ...
## ..- attr(*, "label")= chr "Days to cardiovascular event or end of follow-up"
## $ : num [1:50] 0 0 0 0 0 0 0 0 0 0 ...
## $ : num [1:50] 1870 3082 285 3170 1036 ...
## ..- attr(*, "label")= chr "Days to cardiovascular event or end of follow-up"
## $ : num [1:50] 0 0 0 0 0 0 0 0 0 0 ...
## $ : num [1:50] 3486 2531 444 1840 1422 ...
## ..- attr(*, "label")= chr "Days to cardiovascular event or end of follow-up"
## $ : num [1:50] 0 0 0 0 0 0 0 0 1 0 ...
## $ : num [1:50] 2259 1990 82 3009 1419 ...
## ..- attr(*, "label")= chr "Days to cardiovascular event or end of follow-up"
## $ : num [1:50] 0 0 0 1 0 0 0 0 NA 0 ...
## $ : num [1:50] 293 1317 1297 2063 3549 ...
## ..- attr(*, "label")= chr "Days to cardiovascular event or end of follow-up"
## $ : num [1:50] 0 NA 0 NA 0 1 NA 0 0 0 ...
## $ : num [1:50] 1811 NA 2475 NA 467 ...
```

```
## ..- attr(*, "label")= chr "Days to cardiovascular event or end of follow-up"
## $ : num [1:50] 0 0 0 0 0 NA 0 0 0 0 ...
## $ : num [1:50] 463 1235 561 1570 3477 ...
## ..- attr(*, "label")= chr "Days to cardiovascular event or end of follow-up"
## $ : num [1:50] 0 0 0 0 0 0 0 0 0 0 ...
## $ : num [1:50] 2426 2126 2686 1500 2797 ...
## ..- attr(*, "label")= chr "Days to cardiovascular event or end of follow-up"
## $ : num [1:50] 0 0 0 0 0 0 1 NA 0 0 ...
## $ : num [1:50] 2908 3108 1589 1413 928 ...
## ..- attr(*, "label")= chr "Days to cardiovascular event or end of follow-up"
## $ : num [1:50] 0 0 0 0 0 0 0 0 0 0 ...
## $ : num [1:50] 1827 2421 634 3006 2268 ...
## ..- attr(*, "label")= chr "Days to cardiovascular event or end of follow-up"
## $ : num [1:50] 1 0 0 0 0 0 0 0 0 0 ...
## $ : num [1:50] 1319 3540 337 636 117 ...
## ..- attr(*, "label")= chr "Days to cardiovascular event or end of follow-up"
## $ : num [1:50] 0 0 0 0 0 0 0 0 0 0 ...
## $ : num [1:50] 2979 1510 1452 2914 2531 ...
## ..- attr(*, "label")= chr "Days to cardiovascular event or end of follow-up"
## $ : num [1:50] 0 0 0 0 0 0 0 0 0 0 ...
## $ : num [1:50] 1364 397 2414 2675 782 ...
## ..- attr(*, "label")= chr "Days to cardiovascular event or end of follow-up"
## $ : num [1:50] 0 0 0 0 0 0 0 0 1 0 ...
## $ : num [1:50] 806 1697 516 443 2479 ...
```

```
## ..- attr(*, "label")= chr "Days to cardiovascular event or end of follow-up"
## $ : num [1:50] 0 0 0 0 1 1 0 0 0 0 ...
## $ : num [1:50] 494.34 545.03 3164.12 2224.79 3.17 ...
## ..- attr(*, "label")= chr "Days to cardiovascular event or end of follow-up"
## $ : num [1:50] NA 0 0 0 0 0 0 0 0 0 ...
## $ : num [1:50] NA 2132.8 1625.3 1296.5 50.9 ...
## ..- attr(*, "label")= chr "Days to cardiovascular event or end of follow-up"
## $ : num [1:50] 0 0 0 0 0 0 0 0 0 0 ...
## $ : num [1:50] 1600 223 160 3118 1560 ...
## ..- attr(*, "label")= chr "Days to cardiovascular event or end of follow-up"
## $ : num [1:50] NA 0 0 NA 0 0 1 0 0 1 ...
## $ : num [1:50] NA 3216 1578 NA 427 ...
## ..- attr(*, "label")= chr "Days to cardiovascular event or end of follow-up"
## $ : num [1:50] 0 0 0 0 0 0 NA 0 0 0 ...
## $ : num [1:50] 3589 638 2192 2963 2853 ...
## ..- attr(*, "label")= chr "Days to cardiovascular event or end of follow-up"
## $ : num [1:50] 0 0 NA NA 0 0 0 0 0 0 ...
## $ : num [1:50] 3293 1013 NA NA 598 ...
## ..- attr(*, "label")= chr "Days to cardiovascular event or end of follow-up"
## $ : num [1:50] 0 0 0 0 0 0 0 0 0 0 ...
## $ : num [1:50] 2389 513 213 2712 939 ...
## ..- attr(*, "label")= chr "Days to cardiovascular event or end of follow-up"
## $ : num [1:50] 0 0 1 0 0 0 0 0 0 0 ...
## [list output truncated]
```

```
## – attr(*, "dim")= int [1:2] 2 1000
## – attr(*, "dimnames")=List of 2
##  ..$ : chr [1:2] "cv" "tocv"
##  ..$ : NULL
dim(samples_50)
## [1]  2 1000
```

##It is a list of 2000 elements, each element has 2 rows one for cardiovascular event and the other for time and 1000 columns or 1000 samples. Each row has 50 numbers

##To view the elements of cardiovascular events or 1st row only

```
str(samples_50[1,])
## List of 1000
## $ : num [1:50] 0 0 0 0 0 0 0 NA 0 0 ...
## $ : num [1:50] 0 0 0 0 0 0 0 0 0 0 ...
## $ : num [1:50] 0 0 0 0 0 0 0 1 0 0 ...
## $ : num [1:50] 0 0 NA 0 0 0 0 0 1 0 ...
## $ : num [1:50] 0 0 NA NA 0 0 0 0 0 0 ...
## $ : num [1:50] 0 0 0 0 0 0 1 0 1 0 ...
## $ : num [1:50] 0 0 0 0 0 0 0 0 0 0 ...
## $ : num [1:50] 0 0 0 0 0 0 0 0 0 NA ...
## $ : num [1:50] 1 0 0 0 0 0 0 1 0 NA ...
## $ : num [1:50] 0 0 1 0 1 0 0 0 0 0 ...
## $ : num [1:50] 0 0 0 0 0 0 0 NA 0 0 ...
## $ : num [1:50] 0 0 0 0 NA 0 0 0 0 0 ...
## $ : num [1:50] 0 0 0 0 0 0 0 0 0 0 ...
## $ : num [1:50] 0 0 0 0 0 0 0 0 0 1 ...
## $ : num [1:50] 0 0 0 0 0 0 0 0 NA 0 ...
## $ : num [1:50] 0 0 0 0 0 0 0 0 0 0 ...
## $ : num [1:50] 0 0 0 0 0 0 0 0 0 0 ...
## $ : num [1:50] NA 0 0 0 0 0 0 0 0 0 ...
## $ : num [1:50] 0 0 0 0 0 0 0 0 0 0 ...
## $ : num [1:50] 0 0 NA 0 0 1 0 0 NA 0 ...
```

```
## $ : num [1:50] NA 0 0 0 0 0 0 0 NA 0 ...
## $ : num [1:50] 0 0 0 0 1 0 0 0 0 0 ...
## $ : num [1:50] 0 0 0 0 0 0 0 0 0 0 ...
## $ : num [1:50] 0 0 0 1 0 0 0 0 0 0 ...
## $ : num [1:50] NA 0 0 0 0 0 0 0 0 NA ...
## $ : num [1:50] 1 0 0 0 0 0 0 0 0 0 ...
## $ : num [1:50] 0 0 0 0 NA 0 0 0 NA 0 ...
## $ : num [1:50] 0 0 0 0 0 0 0 0 0 0 ...
## $ : num [1:50] 0 0 0 0 0 0 0 0 NA 0 ...
## $ : num [1:50] 0 0 0 0 0 0 0 0 0 0 ...
## $ : num [1:50] 0 0 0 0 0 0 0 0 0 0 ...
## $ : num [1:50] 0 0 0 0 0 0 0 0 1 0 ...
## $ : num [1:50] 0 0 0 1 0 0 0 0 NA 0 ...
## $ : num [1:50] 0 NA 0 NA 0 1 NA 0 0 0 ...
## $ : num [1:50] 0 0 0 0 0 NA 0 0 0 0 ...
## $ : num [1:50] 0 0 0 0 0 0 0 0 0 0 ...
## $ : num [1:50] 0 0 0 0 0 0 1 NA 0 0 ...
## $ : num [1:50] 0 0 0 0 0 0 0 0 0 0 ...
## $ : num [1:50] 1 0 0 0 0 0 0 0 0 0 ...
## $ : num [1:50] 0 0 0 0 0 0 0 0 0 0 ...
## $ : num [1:50] 0 0 0 0 0 0 0 0 0 0 ...
## $ : num [1:50] 0 0 0 0 0 0 0 0 1 0 ...
## $ : num [1:50] 0 0 0 0 1 1 0 0 0 0 ...
## $ : num [1:50] NA 0 0 0 0 0 0 0 0 0 ...
## $ : num [1:50] 0 0 0 0 0 0 0 0 0 0 ...
## $ : num [1:50] NA 0 0 NA 0 0 1 0 0 1 ...
## $ : num [1:50] 0 0 0 0 0 0 NA 0 0 0 ...
## $ : num [1:50] 0 0 NA NA 0 0 0 0 0 0 ...
## $ : num [1:50] 0 0 0 0 0 0 0 0 0 0 ...
## $ : num [1:50] 0 0 1 0 0 0 0 0 0 0 ...
## $ : num [1:50] 0 0 0 0 NA 0 0 0 0 0 ...
```

```
## $ : num [1:50] 0 0 0 0 0 0 0 0 0 0 ...
## $ : num [1:50] 0 0 0 0 0 0 0 0 0 0 ...
## $ : num [1:50] 0 0 NA NA 0 1 0 NA 0 0 ...
## $ : num [1:50] 0 0 0 0 1 0 0 0 0 0 ...
## $ : num [1:50] 0 0 0 0 NA 0 0 0 0 0 ...
## $ : num [1:50] 0 0 0 0 0 0 0 0 1 0 ...
## $ : num [1:50] 0 0 0 0 0 0 0 0 0 NA ...
## $ : num [1:50] 0 0 0 0 1 0 0 0 0 0 ...
## $ : num [1:50] 0 0 0 0 0 0 0 0 0 0 ...
## $ : num [1:50] 0 0 0 0 0 0 0 0 0 0 ...
## $ : num [1:50] NA 0 0 0 0 0 0 0 0 0 ...
## $ : num [1:50] 0 0 NA 0 0 0 0 0 0 0 ...
## $ : num [1:50] 0 0 0 1 0 0 0 0 0 0 ...
## $ : num [1:50] 0 0 0 0 0 0 0 0 0 0 ...
## $ : num [1:50] 0 0 0 0 0 0 0 0 0 0 ...
## $ : num [1:50] 0 0 0 0 0 0 0 0 1 0 ...
## $ : num [1:50] 0 0 0 0 0 0 1 0 0 0 ...
## $ : num [1:50] 0 0 0 0 0 0 0 0 0 0 ...
## $ : num [1:50] 0 0 0 0 0 0 0 0 0 0 ...
## $ : num [1:50] 0 0 0 0 0 0 0 0 0 0 ...
## $ : num [1:50] 0 0 0 0 NA 0 0 0 0 0 ...
## $ : num [1:50] 0 0 0 0 0 0 0 0 NA 0 ...
## $ : num [1:50] 0 0 0 0 0 0 0 0 0 0 ...
## $ : num [1:50] 0 0 0 0 NA 0 NA 0 0 0 ...
## $ : num [1:50] 0 0 0 0 0 1 0 0 0 0 ...
## $ : num [1:50] 0 NA 0 0 0 0 0 NA 0 0 ...
## $ : num [1:50] 0 0 0 0 0 0 0 NA 0 0 ...
## $ : num [1:50] 0 0 0 1 0 0 0 0 1 0 ...
## $ : num [1:50] 0 0 0 0 0 0 0 0 0 0 ...
## $ : num [1:50] 0 0 0 0 0 0 0 0 0 0 ...
## $ : num [1:50] 0 0 0 0 0 0 0 NA 0 NA ...
```

```
## $ : num [1:50] 0 0 NA 0 0 0 0 0 NA 0 ...
## $ : num [1:50] 0 0 0 0 0 0 0 0 0 0 ...
## $ : num [1:50] NA 0 0 0 0 0 0 0 0 0 ...
## $ : num [1:50] 0 NA 0 0 0 0 0 0 0 0 ...
## $ : num [1:50] 0 1 NA 0 0 0 0 0 0 0 ...
## $ : num [1:50] 0 0 0 NA 0 0 0 1 0 0 ...
## $ : num [1:50] 0 0 0 0 0 0 0 0 0 0 ...
## $ : num [1:50] 0 0 0 0 0 0 0 0 0 0 ...
## $ : num [1:50] 0 0 0 NA 0 0 0 0 0 0 ...
## $ : num [1:50] 0 0 0 1 0 0 1 0 0 0 ...
## $ : num [1:50] 0 1 0 0 0 0 0 0 0 NA ...
## $ : num [1:50] 0 0 0 0 0 0 0 NA 0 0 ...
## $ : num [1:50] 0 0 0 0 0 0 0 NA 0 0 ...
## $ : num [1:50] 0 0 0 0 0 0 NA 0 0 0 ...
## $ : num [1:50] 0 0 0 0 0 0 0 0 0 1 ...
## $ : num [1:50] 0 0 0 0 0 0 0 0 0 0 ...
## $ : num [1:50] 0 1 0 0 0 0 0 0 NA 0 ...
##  [list output truncated]
##To view the cardiovascular events of the 1st element
str(samples_50[1,][1])
## List of 1
## $ : num [1:50] 0 0 0 0 0 0 0 NA 0 0 ...
##To view the elements of time only
str(samples_50[2,])
## List of 1000
## $ : num [1:50] 3626 1703 247 1313 2619 ...
##  ..- attr(*, "label")= chr "Days to cardiovascular event or end of follow-up"
## $ : num [1:50] 3540 1313 2427 1782 2426 ...
##  ..- attr(*, "label")= chr "Days to cardiovascular event or end of follow-up"
```

```
## $ : num [1:50] 1958 517 519 3571 3312 ...
## ..- attr(*, "label")= chr "Days to cardiovascular event or end of follow-up"
## $ : num [1:50] 173 1752 NA 438 3356 ...
## ..- attr(*, "label")= chr "Days to cardiovascular event or end of follow-up"
## $ : num [1:50] 596 1355 NA NA 2572 ...
## ..- attr(*, "label")= chr "Days to cardiovascular event or end of follow-up"
## $ : num [1:50] 171 2490 362 1200 2144 ...
## ..- attr(*, "label")= chr "Days to cardiovascular event or end of follow-up"
## $ : num [1:50] 113.9 3072.5 55.8 1437.2 541.9 ...
## ..- attr(*, "label")= chr "Days to cardiovascular event or end of follow-up"
## $ : num [1:50] 2804 2785 1967 1039 768 ...
## ..- attr(*, "label")= chr "Days to cardiovascular event or end of follow-up"
## $ : num [1:50] 70 552 2010 3071 1931 ...
## ..- attr(*, "label")= chr "Days to cardiovascular event or end of follow-up"
## $ : num [1:50] 1982 2106 427 3357 106 ...
## ..- attr(*, "label")= chr "Days to cardiovascular event or end of follow-up"
## $ : num [1:50] 182 3570 2143 2932 1698 ...
## ..- attr(*, "label")= chr "Days to cardiovascular event or end of follow-up"
## $ : num [1:50] 246.6 2930.2 120.9 12.8 NA ...
## ..- attr(*, "label")= chr "Days to cardiovascular event or end of follow-up"
## $ : num [1:50] 1590 1659 3609 342 2196 ...
## ..- attr(*, "label")= chr "Days to cardiovascular event or end of follow-up"
```

```
## $ : num [1:50] 3374 3344 3105 1409 1452 ...
## ..- attr(*, “label”)= chr “Days to cardiovascular event or end of follow-up”
## $ : num [1:50] 64.5 708.2 1804.9 1710.7 1815.1 ...
## ..- attr(*, “label”)= chr “Days to cardiovascular event or end of follow-up”
## $ : num [1:50] 2490 1998 2408 2392 348 ...
## ..- attr(*, “label”)= chr “Days to cardiovascular event or end of follow-up”
## $ : num [1:50] 182 162 2879 2215 2413 ...
## ..- attr(*, “label”)= chr “Days to cardiovascular event or end of follow-up”
## $ : num [1:50] NA 1025 1343 835 1352 ...
## ..- attr(*, “label”)= chr “Days to cardiovascular event or end of follow-up”
## $ : num [1:50] 3216 295 268 543 598 ...
## ..- attr(*, “label”)= chr “Days to cardiovascular event or end of follow-up”
## $ : num [1:50] 203 2628 NA 3171 3153 ...
## ..- attr(*, “label”)= chr “Days to cardiovascular event or end of follow-up”
## $ : num [1:50] NA 1628 2557 2696 2110 ...
## ..- attr(*, “label”)= chr “Days to cardiovascular event or end of follow-up”
## $ : num [1:50] 178.92 2177.83 3401.55 112.6 7.27 ...
## ..- attr(*, “label”)= chr “Days to cardiovascular event or end of follow-up”
## $ : num [1:50] 3570 1638 3210 3529 3295 ...
## ..- attr(*, “label”)= chr “Days to cardiovascular event or end of follow-up”
## $ : num [1:50] 1632 2084 2392 106 2930 ...

## ..- attr(*, “label”)= chr “Days to cardiovascular event or end of follow-up”
```

```
## $ : num [1:50] NA 972 2724 1698 234 ...
## ..- attr(*, "label")= chr "Days to cardiovascular event or end of follow-up"
## $ : num [1:50] 894 102 3384 3179 3423 ...
## ..- attr(*, "label")= chr "Days to cardiovascular event or end of follow-up"
## $ : num [1:50] 3117 849 2513 1723 NA ...
## ..- attr(*, "label")= chr "Days to cardiovascular event or end of follow-up"
## $ : num [1:50] 542 2459 594 331 1191 ...
## ..- attr(*, "label")= chr "Days to cardiovascular event or end of follow-up"
## $ : num [1:50] 2196 3325 3433 3556 293 ...
## ..- attr(*, "label")= chr "Days to cardiovascular event or end of follow-up"
## $ : num [1:50] 1870 3082 285 3170 1036 ...
## ..- attr(*, "label")= chr "Days to cardiovascular event or end of follow-up"
## $ : num [1:50] 3486 2531 444 1840 1422 ...
## ..- attr(*, "label")= chr "Days to cardiovascular event or end of follow-up"
## $ : num [1:50] 2259 1990 82 3009 1419 ...
## ..- attr(*, "label")= chr "Days to cardiovascular event or end of follow-up"
## $ : num [1:50] 293 1317 1297 2063 3549 ...
## ..- attr(*, "label")= chr "Days to cardiovascular event or end of follow-up"
## $ : num [1:50] 1811 NA 2475 NA 467 ...
## ..- attr(*, "label")= chr "Days to cardiovascular event or end of follow-up"
## $ : num [1:50] 463 1235 561 1570 3477 ...
## ..- attr(*, "label")= chr "Days to cardiovascular event or end of follow-up"
## $ : num [1:50] 2426 2126 2686 1500 2797 ...
```

```
## ..- attr(*, "label")= chr "Days to cardiovascular event or end of follow-up"
## $ : num [1:50] 2908 3108 1589 1413 928 ...
## ..- attr(*, "label")= chr "Days to cardiovascular event or end of follow-up"
## $ : num [1:50] 1827 2421 634 3006 2268 ...
## ..- attr(*, "label")= chr "Days to cardiovascular event or end of follow-up"
## $ : num [1:50] 1319 3540 337 636 117 ...
## ..- attr(*, "label")= chr "Days to cardiovascular event or end of follow-up"
## $ : num [1:50] 2979 1510 1452 2914 2531 ...
## ..- attr(*, "label")= chr "Days to cardiovascular event or end of follow-up"
## $ : num [1:50] 1364 397 2414 2675 782 ...
## ..- attr(*, "label")= chr "Days to cardiovascular event or end of follow-up"
## $ : num [1:50] 806 1697 516 443 2479 ...
## ..- attr(*, "label")= chr "Days to cardiovascular event or end of follow-up"
## $ : num [1:50] 494.34 545.03 3164.12 2224.79 3.17 ...
## ..- attr(*, "label")= chr "Days to cardiovascular event or end of follow-up"
## $ : num [1:50] NA 2132.8 1625.3 1296.5 50.9 ...
## ..- attr(*, "label")= chr "Days to cardiovascular event or end of follow-up"
## $ : num [1:50] 1600 223 160 3118 1560 ...
## ..- attr(*, "label")= chr "Days to cardiovascular event or end of follow-up"
## $ : num [1:50] NA 3216 1578 NA 427 ...
## ..- attr(*, "label")= chr "Days to cardiovascular event or end of follow-up"
## $ : num [1:50] 3589 638 2192 2963 2853 ...
## ..- attr(*, "label")= chr "Days to cardiovascular event or end of follow-
```

```
up"
## $ : num [1:50] 3293 1013 NA NA 598 ...
## ..- attr(*, "label")= chr "Days to cardiovascular event or end of follow-up"
## $ : num [1:50] 2389 513 213 2712 939 ...
## ..- attr(*, "label")= chr "Days to cardiovascular event or end of follow-up"
## $ : num [1:50] 2608 3312 1319 782 2330 ...
## ..- attr(*, "label")= chr "Days to cardiovascular event or end of follow-up"
## $ : num [1:50] 459 515 3287 2839 NA ...
## ..- attr(*, "label")= chr "Days to cardiovascular event or end of follow-up"
## $ : num [1:50] 721 2219 1652 3225 883 ...
## ..- attr(*, "label")= chr "Days to cardiovascular event or end of follow-up"
## $ : num [1:50] 3591 2531 2864 2142 1697 ...
## ..- attr(*, "label")= chr "Days to cardiovascular event or end of follow-up"
## $ : num [1:50] 1730 2414 NA NA 2957 ...
## ..- attr(*, "label")= chr "Days to cardiovascular event or end of follow-up"
## $ : num [1:50] 3624.5 3347.6 3637.5 638.4 94.1 ...
## ..- attr(*, "label")= chr "Days to cardiovascular event or end of follow-up"
## $ : num [1:50] 2614 55 2345 3216 NA ...
## ..- attr(*, "label")= chr "Days to cardiovascular event or end of follow-up"
## $ : num [1:50] 2903 1489 464 2136 3131 ...
## ..- attr(*, "label")= chr "Days to cardiovascular event or end of follow-up"
## $ : num [1:50] 606 1423 2712 798 985 ...
## ..- attr(*, "label")= chr "Days to cardiovascular event or end of follow-up"
```

```
## $ : num [1:50] 2974 2337 1236 3054 2552 ...
## ..- attr(*, "label")= chr "Days to cardiovascular event or end of follow-up"
## $ : num [1:50] 3366 1711 2551 1460 2121 ...
## ..- attr(*, "label")= chr "Days to cardiovascular event or end of follow-up"
## $ : num [1:50] 601 592 1243 173 2107 ...
## ..- attr(*, "label")= chr "Days to cardiovascular event or end of follow-up"
## $ : num [1:50] NA 1923 2781 2725 2539 ...
## ..- attr(*, "label")= chr "Days to cardiovascular event or end of follow-up"
## $ : num [1:50] 2983 2966 NA 669 790 ...
## ..- attr(*, "label")= chr "Days to cardiovascular event or end of follow-up"
## $ : num [1:50] 3609 1541 1131 198 2529 ...
## ..- attr(*, "label")= chr "Days to cardiovascular event or end of follow-up"
## $ : num [1:50] 148 521 1036 736 435 ...
## ..- attr(*, "label")= chr "Days to cardiovascular event or end of follow-up"
## $ : num [1:50] 1870 431 205 2104 2982 ...
## ..- attr(*, "label")= chr "Days to cardiovascular event or end of follow-up"
## $ : num [1:50] 3423 3308 1152 3105 2065 ...
## ..- attr(*, "label")= chr "Days to cardiovascular event or end of follow-up"
## $ : num [1:50] 1409 863 3510 2451 3486 ...
## ..- attr(*, "label")= chr "Days to cardiovascular event or end of follow-up"
## $ : num [1:50] 1775 3399 1364 1838 3084 ...
## ..- attr(*, "label")= chr "Days to cardiovascular event or end of follow-up"
## $ : num [1:50] 476 1061 200 1638 3302 ...
```

```
## ..- attr(*, "label")= chr "Days to cardiovascular event or end of follow-up"
## $ : num [1:50] 686 1470 3549 2804 283 ...
## ..- attr(*, "label")= chr "Days to cardiovascular event or end of follow-up"
## $ : num [1:50] 1957 868 1752 432 NA ...
## ..- attr(*, "label")= chr "Days to cardiovascular event or end of follow-up"
## $ : num [1:50] 385 2544 2456 2079 2143 ...
## ..- attr(*, "label")= chr "Days to cardiovascular event or end of follow-up"
## $ : num [1:50] 934 1826 1354 934 2664 ...
## ..- attr(*, "label")= chr "Days to cardiovascular event or end of follow-up"
## $ : num [1:50] 2797 167 1062 2903 NA ...
## ..- attr(*, "label")= chr "Days to cardiovascular event or end of follow-up"
## $ : num [1:50] 396 1746 1543 2400 124 ...
## ..- attr(*, "label")= chr "Days to cardiovascular event or end of follow-up"
## $ : num [1:50] 742 NA 752 557 2287 ...
## ..- attr(*, "label")= chr "Days to cardiovascular event or end of follow-up"
## $ : num [1:50] 36.9 3603.6 1986.2 1715.4 2375.2 ...
## ..- attr(*, "label")= chr "Days to cardiovascular event or end of follow-up"
## $ : num [1:50] 3442.1 3617.5 2771.2 94.1 3239.2 ...
## ..- attr(*, "label")= chr "Days to cardiovascular event or end of follow-up"
## $ : num [1:50] 3343 2839 2268 3030 869 ...
## ..- attr(*, "label")= chr "Days to cardiovascular event or end of follow-up"
## $ : num [1:50] 2661 348 3484 3071 2219 ...
## ..- attr(*, "label")= chr "Days to cardiovascular event or end of follow-
```

```
up"
## $ : num [1:50] 2008 2766 2104 698 1185 ...
##  ..- attr(*, "label")= chr "Days to cardiovascular event or end of follow-
up"
## $ : num [1:50] 265 3607 NA 1909 1113 ...
##  ..- attr(*, "label")= chr "Days to cardiovascular event or end of follow-
up"
## $ : num [1:50] 594 708 2603 906 2470 ...
##  ..- attr(*, "label")= chr "Days to cardiovascular event or end of follow-
up"
## $ : num [1:50] NA 3601 1646 930 2691 ...
##  ..- attr(*, "label")= chr "Days to cardiovascular event or end of follow-
up"
## $ : num [1:50] 3005 NA 985 1834 2842 ...
##  ..- attr(*, "label")= chr "Days to cardiovascular event or end of follow-
up"
## $ : num [1:50] 2671 2621 NA 493 2463 ...
##  ..- attr(*, "label")= chr "Days to cardiovascular event or end of follow-
up"
## $ : num [1:50] 1098 1381 307 NA 1158 ...
##  ..- attr(*, "label")= chr "Days to cardiovascular event or end of follow-
up"
## $ : num [1:50] 2671 247 2862 325 1065 ...
##  ..- attr(*, "label")= chr "Days to cardiovascular event or end of follow-
up"
## $ : num [1:50] 2781 2465 2170 2431 2539 ...
##  ..- attr(*, "label")= chr "Days to cardiovascular event or end of follow-
up"
## $ : num [1:50] 2418 3477 138 NA 1776 ...
##  ..- attr(*, "label")= chr "Days to cardiovascular event or end of follow-
up"
## $ : num [1:50] 3520.17 1763.11 2591.22 7.22 1550.33 ...
##  ..- attr(*, "label")= chr "Days to cardiovascular event or end of follow-
```

```
up"
## $ : num [1:50] 469 132 784 2099 1724 ...
##  ..- attr(*, "label")= chr "Days to cardiovascular event or end of follow-up"
## $ : num [1:50] 1194 3153 1082 1782 3433 ...
##  ..- attr(*, "label")= chr "Days to cardiovascular event or end of follow-up"
## $ : num [1:50] 3131 937 917 2906 1055 ...
##  ..- attr(*, "label")= chr "Days to cardiovascular event or end of follow-up"
## $ : num [1:50] 934 3445 2111 3317 2995 ...
##  ..- attr(*, "label")= chr "Days to cardiovascular event or end of follow-up"
## $ : num [1:50] 1879 2480 144 3428 1508 ...
##  ..- attr(*, "label")= chr "Days to cardiovascular event or end of follow-up"
## $ : num [1:50] 543 2144 886 1800 3650 ...
##  ..- attr(*, "label")= chr "Days to cardiovascular event or end of follow-up"
## $ : num [1:50] 3133.1 10.1 3201.6 3216.1 2465.4 ...
##  ..- attr(*, "label")= chr "Days to cardiovascular event or end of follow-up"
##  [list output truncated]
##To view the follow-up times of the 1st element
str(samples_50[2,][1])
## List of 1
## $ : num [1:50] 3626 1703 247 1313 2619 ...
##  ..- attr(*, "label")= chr "Days to cardiovascular event or end of follow-up"
##To view the follow-up times of the last or 1000th element
str(samples_50[2,][1000])
## List of 1
## $ : num [1:50] 2787 1718 214 1753 2097 ...
```

```
## ..- attr(*, "label")= chr "Days to cardiovascular event or end of follow-up"
##To obtain the sums of cardiovascular events only, use sapply where it sums every element of the list
sums_cv<-sapply(samples_50[1,],sum, na.rm = TRUE)
summary(sums_cv)
## Min. 1st Qu. Median Mean 3rd Qu. Max.
## 0.000 1.000 2.000 2.056 3.000 7.000
##To obtain the sums of time only, use sapply where it sums every element of the list
sums_time<-sapply(samples_50[2,],sum, na.rm = TRUE)
summary(sums_time)
## Min. 1st Qu. Median Mean 3rd Qu. Max.
## 55009 77219 82810 82650 87859 108378
##To obtain the incidence rate for each sample
rates_50<-sums_cv/sums_time
head(rates_50)
## [1] 1.281609e-05 3.147441e-05 3.347042e-05 3.444698e-05 3.020894e-05
## [6] 2.232570e-05
summary(rates_50)
## Min. 1st Qu. Median Mean 3rd Qu. Max.
## 0.000e+00 1.212e-05 2.400e-05 2.541e-05 3.629e-05 1.056e-04
##To view a histogram of incidence rates based on samples of size=50
hist_50<-ggplot()+geom_histogram(aes(x = rates_50))
hist_50
```

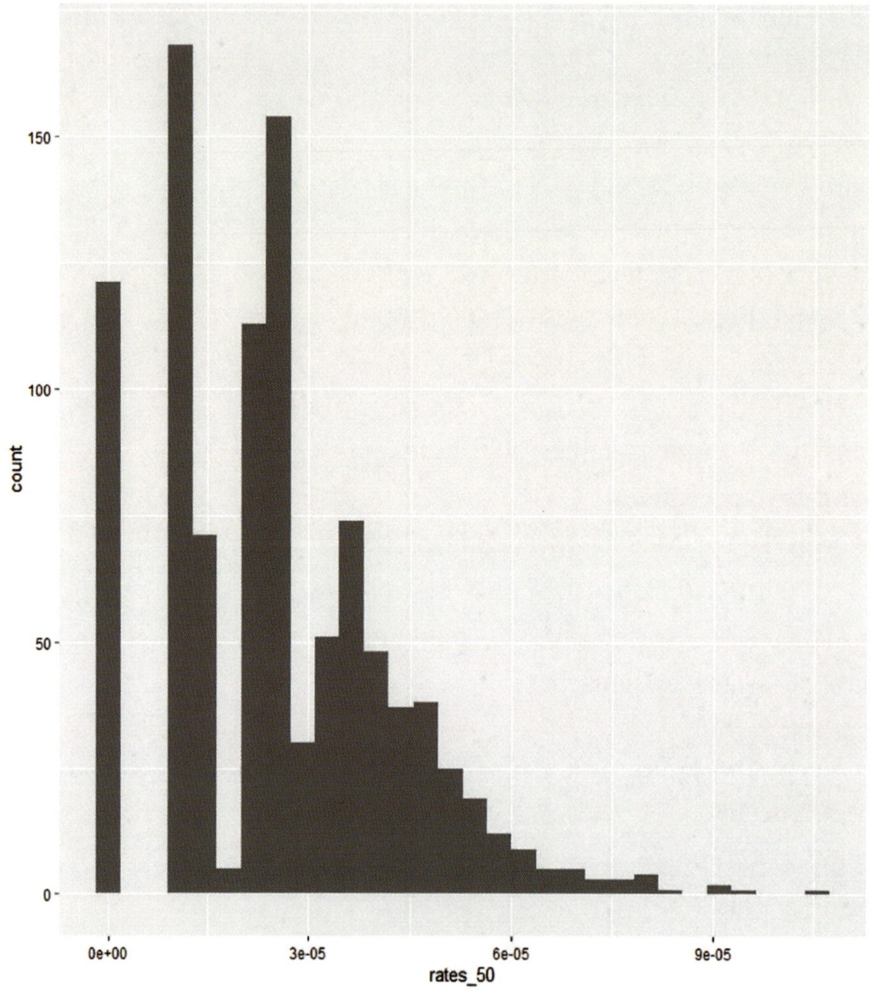


```
##To plot a boxplot of these incidence rates
box_50<-ggplot()+geom_boxplot(aes(y=rates_50))
box_50
```

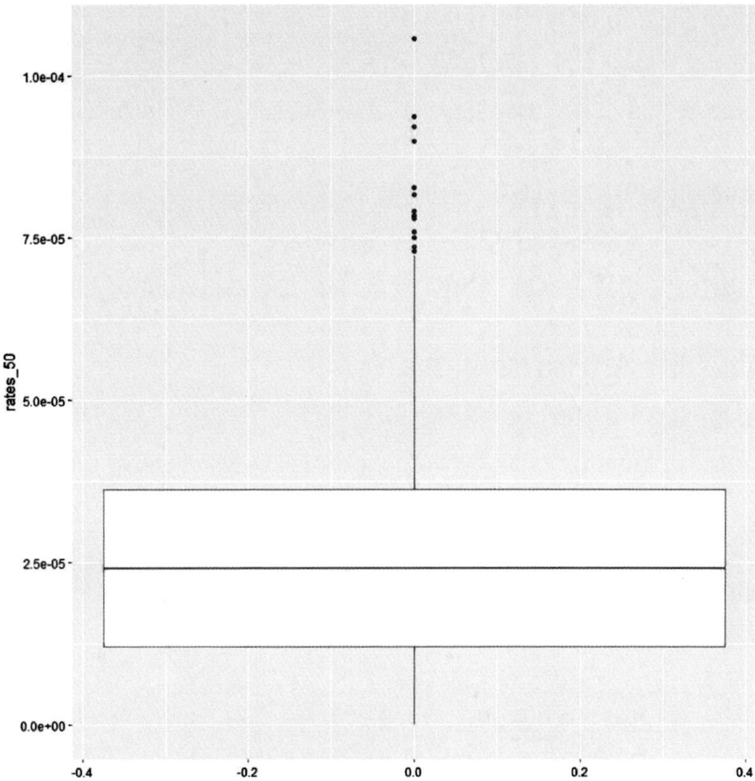

Generate 1000 random samples, each of size= 150, from the two columns cv and tocv as time to event data is a two dimensional data.

```
set.seed(123)
samples_150<-replicate(1000, sample_n(regicor[,22:23],150))
str(samples_150)
## List of 2000
## $ : num [1:150] 0 0 0 0 0 0 0 NA 0 0 ...
## $ : num [1:150] 3626 1703 247 1313 2619 ...
##  ..- attr(*, "label")= chr "Days to cardiovascular event or end of follow-up"
## $ : num [1:150] 1 0 0 0 0 0 0 0 0 0 ...
## $ : num [1:150] 1895 2603 1181 1646 3123 ...
##  ..- attr(*, "label")= chr "Days to cardiovascular event or end of follow-up"
```

$: num [1:150] 0 0 0 0 0 0 0 0 0 0 ...

$: num [1:150] 409 235 2678 1836 1959 ...

..- attr(*, "label")= chr "Days to cardiovascular event or end of follow-up"

$: num [1:150] 1 0 0 0 0 0 0 0 0 0 ...

$: num [1:150] 278 295 3570 1446 681 ...

..- attr(*, "label")= chr "Days to cardiovascular event or end of follow-up"

$: num [1:150] 0 0 0 0 0 0 0 0 0 0 ...

$: num [1:150] 1061 636 2451 933 3435 ...

..- attr(*, "label")= chr "Days to cardiovascular event or end of follow-up"

$: num [1:150] 0 0 NA 0 0 0 0 0 0 0 ...

$: num [1:150] 500 709 NA 3397 3418 ...

..- attr(*, "label")= chr "Days to cardiovascular event or end of follow-up"

$: num [1:150] 0 0 0 0 0 0 0 0 0 0 ...

$: num [1:150] 2692 269 1958 2712 2305 ...

..- attr(*, "label")= chr "Days to cardiovascular event or end of follow-up"

$: num [1:150] 0 0 0 0 0 0 0 0 0 0 ...

$: num [1:150] 624.4 3624.5 2120.7 66.9 111.6 ...

..- attr(*, "label")= chr "Days to cardiovascular event or end of follow-up"

$: num [1:150] NA 0 0 0 0 0 1 0 0 0 ...

$: num [1:150] NA 729 3338 2086 1684 ...

..- attr(*, "label")= chr "Days to cardiovascular event or end of follow-up"

$: num [1:150] 0 0 0 NA 0 0 0 0 0 0 ...

$: num [1:150] 1531 181 3153 NA 3162 ...

..- attr(*, "label")= chr "Days to cardiovascular event or end of follow-up"

$: num [1:150] 0 0 0 0 0 0 NA 0 0 0 ...

```
## $ : num [1:150] 2983 3507 703 978 2712 ...
##  ..- attr(*, "label")= chr "Days to cardiovascular event or end of follow-up"
## $ : num [1:150] 0 0 0 0 0 0 0 0 0 0 ...
## $ : num [1:150] 3152 2766 790 3552 2044 ...
##  ..- attr(*, "label")= chr "Days to cardiovascular event or end of follow-up"
## $ : num [1:150] 0 NA 0 0 0 0 0 0 0 0 ...
## $ : num [1:150] 510 NA 3394 1827 2421 ...
##  ..- attr(*, "label")= chr "Days to cardiovascular event or end of follow-up"
## $ : num [1:150] 0 0 0 0 0 0 0 0 0 0 ...
## $ : num [1:150] 550 2064 1364 397 2414 ...
##  ..- attr(*, "label")= chr "Days to cardiovascular event or end of follow-up"
## $ : num [1:150] 0 0 0 0 0 0 0 0 0 0 ...
## $ : num [1:150] 2132.8 1625.3 1296.5 50.9 3069.8 ...
##  ..- attr(*, "label")= chr "Days to cardiovascular event or end of follow-up"
## $ : num [1:150] 0 0 NA 0 0 0 0 0 0 0 ...
## $ : num [1:150] 2853 3201 NA 1695 1087 ...
##  ..- attr(*, "label")= chr "Days to cardiovascular event or end of follow-up"
## $ : num [1:150] 0 0 0 0 0 0 0 0 0 0 ...
## $ : num [1:150] 2330 3296 197 3443 2781 ...
##  ..- attr(*, "label")= chr "Days to cardiovascular event or end of follow-up"
## $ : num [1:150] 0 0 0 0 0 0 0 NA 0 0 ...
## $ : num [1:150] 668 3344 1694 353 245 ...
##  ..- attr(*, "label")= chr "Days to cardiovascular event or end of follow-up"
## $ : num [1:150] 0 0 0 0 0 0 0 0 0 NA ...
## $ : num [1:150] 2097 3096 849 507 1842 ...
```

```
## ..- attr(*, "label")= chr "Days to cardiovascular event or end of follow-up"
## $ : num [1:150] NA 1 0 0 0 0 NA 0 0 0 ...
## $ : num [1:150] NA 110.2 1379 29.8 449.3 ...
## ..- attr(*, "label")= chr "Days to cardiovascular event or end of follow-up"
## $ : num [1:150] 0 1 0 0 NA 0 NA 0 0 0 ...
## $ : num [1:150] 1739 510 2291 774 NA ...
## ..- attr(*, "label")= chr "Days to cardiovascular event or end of follow-up"
## $ : num [1:150] 0 0 0 NA NA 0 0 0 0 0 ...
## $ : num [1:150] 284 1865 179 NA NA ...
## ..- attr(*, "label")= chr "Days to cardiovascular event or end of follow-up"
## $ : num [1:150] 0 0 0 1 0 0 0 0 0 0 ...
## $ : num [1:150] 1724 2107 2640 645 110 ...
## ..- attr(*, "label")= chr "Days to cardiovascular event or end of follow-up"
## $ : num [1:150] 0 0 NA 0 0 0 0 0 0 0 ...
## $ : num [1:150] 1285 2687 NA 2658 1897 ...
## ..- attr(*, "label")= chr "Days to cardiovascular event or end of follow-up"
## $ : num [1:150] 0 0 0 0 1 0 0 0 NA 0 ...
## $ : num [1:150] 736 2332 2537 2973 196 ...
## ..- attr(*, "label")= chr "Days to cardiovascular event or end of follow-up"
## $ : num [1:150] 0 0 0 0 0 0 0 0 0 0 ...
## $ : num [1:150] 1668 1105 1550 2899 1774 ...
## ..- attr(*, "label")= chr "Days to cardiovascular event or end of follow-up"
## $ : num [1:150] 0 0 0 0 1 0 0 0 0 0 ...
## $ : num [1:150] 359 2119 1751 3397 1136 ...
## ..- attr(*, "label")= chr "Days to cardiovascular event or end of follow-
```

```
up"
## $ : num [1:150] 0 NA 1 0 NA 0 0 0 0 0 ...
## $ : num [1:150] 2110 NA 198 312 NA ...
## ..- attr(*, "label")= chr "Days to cardiovascular event or end of follow-
up"
## $ : num [1:150] 0 0 0 0 0 0 0 0 1 0 ...
## $ : num [1:150] 3066 3608 1271 814 985 ...
## ..- attr(*, "label")= chr "Days to cardiovascular event or end of follow-
up"
## $ : num [1:150] 0 0 0 0 0 0 0 0 0 0 ...
## $ : num [1:150] 2572 1718 1747 3307 817 ...
## ..- attr(*, "label")= chr "Days to cardiovascular event or end of follow-
up"
## $ : num [1:150] 1 0 0 0 1 0 0 0 0 0 ...
## $ : num [1:150] 1356 3030 3484 469 132 ...
## ..- attr(*, "label")= chr "Days to cardiovascular event or end of follow-
up"
## $ : num [1:150] 0 0 0 0 NA 0 0 0 0 0 ...
## $ : num [1:150] 2111 3317 2995 484 NA ...
## ..- attr(*, "label")= chr "Days to cardiovascular event or end of follow-
up"
## $ : num [1:150] 0 0 0 NA 0 0 0 0 0 0 ...
## $ : num [1:150] 1752 410 1055 NA 737 ...
## ..- attr(*, "label")= chr "Days to cardiovascular event or end of follow-
up"
## $ : num [1:150] 1 0 0 0 NA 0 0 0 0 0 ...
## $ : num [1:150] 1214 704 1720 2340 NA ...
## ..- attr(*, "label")= chr "Days to cardiovascular event or end of follow-
up"
## $ : num [1:150] 0 0 0 0 0 1 0 0 0 0 ...
## $ : num [1:150] 1597 2914 1974 2024 493 ...
## ..- attr(*, "label")= chr "Days to cardiovascular event or end of follow-
up"
```

```
## $ : num [1:150] 0 0 0 0 0 0 0 0 0 0 ...
## $ : num [1:150] 3618 2218 2100 32 1965 ...
## ..- attr(*, "label")= chr "Days to cardiovascular event or end of follow-up"
## $ : num [1:150] 0 0 0 NA 0 0 0 0 0 0 ...
## $ : num [1:150] 3153 363 2422 NA 1013 ...
## ..- attr(*, "label")= chr "Days to cardiovascular event or end of follow-up"
## $ : num [1:150] NA 0 0 0 0 0 0 0 0 0 ...
## $ : num [1:150] NA 2748 460 469 1022 ...
## ..- attr(*, "label")= chr "Days to cardiovascular event or end of follow-up"
## $ : num [1:150] 0 0 0 0 0 NA 0 0 0 0 ...
## $ : num [1:150] 3110 2812 828 2184 337 ...
## ..- attr(*, "label")= chr "Days to cardiovascular event or end of follow-up"
## $ : num [1:150] 0 0 0 0 0 0 NA 0 0 0 ...
## $ : num [1:150] 1518 932 2664 484 2498 ...
## ..- attr(*, "label")= chr "Days to cardiovascular event or end of follow-up"
## $ : num [1:150] NA 0 0 1 0 0 0 0 0 0 ...
## $ : num [1:150] NA 3339.7 3225.1 37.1 2837.1 ...
## ..- attr(*, "label")= chr "Days to cardiovascular event or end of follow-up"
## $ : num [1:150] 0 0 0 0 0 0 0 0 1 0 ...
## $ : num [1:150] 2041 3445 1838 358 868 ...
## ..- attr(*, "label")= chr "Days to cardiovascular event or end of follow-up"
## $ : num [1:150] 0 0 0 0 0 0 0 0 0 0 ...
## $ : num [1:150] 521 636 2571 2621 1087 ...

## ..- attr(*, "label")= chr "Days to cardiovascular event or end of follow-up"
```

```
## $ : num [1:150] 0 0 0 0 0 1 0 0 0 0 ...
## $ : num [1:150] 952 2974 886 3271 3291 ...
##  ..- attr(*, "label")= chr "Days to cardiovascular event or end of follow-up"
## $ : num [1:150] 0 0 NA 0 0 0 0 0 1 0 ...
## $ : num [1:150] 3527 2142 NA 2543 3625 ...
##  ..- attr(*, "label")= chr "Days to cardiovascular event or end of follow-up"
## $ : num [1:150] 0 1 0 1 0 0 0 0 0 0 ...
## $ : num [1:150] 1409 1356 1752 2538 215 ...
##  ..- attr(*, "label")= chr "Days to cardiovascular event or end of follow-up"
## $ : num [1:150] 0 0 0 0 0 0 0 1 NA 0 ...
## $ : num [1:150] 484 877 1429 3061 1334 ...
##  ..- attr(*, "label")= chr "Days to cardiovascular event or end of follow-up"
## $ : num [1:150] 0 0 0 0 0 0 0 0 0 0 ...
## $ : num [1:150] 3057 2212 2272 3651 1368 ...
##  ..- attr(*, "label")= chr "Days to cardiovascular event or end of follow-up"
## $ : num [1:150] 0 0 0 0 0 0 0 0 0 0 ...
## $ : num [1:150] 66.9 2287.3 2342.9 1960.9 3021.9 ...
##  ..- attr(*, "label")= chr "Days to cardiovascular event or end of follow-up"
## $ : num [1:150] 0 0 0 0 1 0 0 0 0 0 ...
##  [list output truncated]
## – attr(*, "dim")= int [1:2] 2 1000
## – attr(*, "dimnames")=List of 2
##  ..$ : chr [1:2] "cv" "tocv"
##  ..$ : NULL
dim(samples_150)
## [1]  2 1000
```

##It is a list of 2000 elements, each element has 2 rows one for cardiovascular event and the other for time and 1000 columns or 1000 samples. Each row has 150 numbers

##To view the elements of cardiovascular events or 1st row only

```
str(samples_150[1,])
## List of 1000
## $ : num [1:150] 0 0 0 0 0 0 0 NA 0 0 ...
## $ : num [1:150] 1 0 0 0 0 0 0 0 0 0 ...
## $ : num [1:150] 0 0 0 0 0 0 0 0 0 0 ...
## $ : num [1:150] 1 0 0 0 0 0 0 0 0 0 ...
## $ : num [1:150] 0 0 0 0 0 0 0 0 0 0 ...
## $ : num [1:150] 0 0 NA 0 0 0 0 0 0 0 ...
## $ : num [1:150] 0 0 0 0 0 0 0 0 0 0 ...
## $ : num [1:150] 0 0 0 0 0 0 0 0 0 0 ...
## $ : num [1:150] NA 0 0 0 0 0 1 0 0 0 ...
## $ : num [1:150] 0 0 0 NA 0 0 0 0 0 0 ...
## $ : num [1:150] 0 0 0 0 0 0 NA 0 0 0 ...
## $ : num [1:150] 0 0 0 0 0 0 0 0 0 0 ...
## $ : num [1:150] 0 NA 0 0 0 0 0 0 0 0 ...
## $ : num [1:150] 0 0 0 0 0 0 0 0 0 0 ...
## $ : num [1:150] 0 0 0 0 0 0 0 0 0 0 ...
## $ : num [1:150] 0 0 NA 0 0 0 0 0 0 0 ...
## $ : num [1:150] 0 0 0 0 0 0 0 0 0 0 ...
## $ : num [1:150] 0 0 0 0 0 0 0 NA 0 0 ...
## $ : num [1:150] 0 0 0 0 0 0 0 0 0 NA ...
## $ : num [1:150] NA 1 0 0 0 0 NA 0 0 0 ...
## $ : num [1:150] 0 1 0 0 NA 0 NA 0 0 0 ...
## $ : num [1:150] 0 0 0 NA NA 0 0 0 0 0 ...
## $ : num [1:150] 0 0 0 1 0 0 0 0 0 0 ...
## $ : num [1:150] 0 0 NA 0 0 0 0 0 0 0 ...
## $ : num [1:150] 0 0 0 0 1 0 0 0 NA 0 ...
## $ : num [1:150] 0 0 0 0 0 0 0 0 0 0 ...
```

```
## $ : num [1:150] 0 0 0 0 1 0 0 0 0 0 ...
## $ : num [1:150] 0 NA 1 0 NA 0 0 0 0 0 ...
## $ : num [1:150] 0 0 0 0 0 0 0 0 1 0 ...
## $ : num [1:150] 0 0 0 0 0 0 0 0 0 0 ...
## $ : num [1:150] 1 0 0 0 1 0 0 0 0 0 ...
## $ : num [1:150] 0 0 0 0 NA 0 0 0 0 0 ...
## $ : num [1:150] 0 0 0 NA 0 0 0 0 0 0 ...
## $ : num [1:150] 1 0 0 0 NA 0 0 0 0 0 ...
## $ : num [1:150] 0 0 0 0 0 1 0 0 0 0 ...
## $ : num [1:150] 0 0 0 0 0 0 0 0 0 0 ...
## $ : num [1:150] 0 0 0 NA 0 0 0 0 0 0 ...
## $ : num [1:150] NA 0 0 0 0 0 0 0 0 0 ...
## $ : num [1:150] 0 0 0 0 0 NA 0 0 0 0 ...
## $ : num [1:150] 0 0 0 0 0 0 NA 0 0 0 ...
## $ : num [1:150] NA 0 0 1 0 0 0 0 0 0 ...
## $ : num [1:150] 0 0 0 0 0 0 0 0 1 0 ...
## $ : num [1:150] 0 0 0 0 0 0 0 0 0 0 ...
## $ : num [1:150] 0 0 0 0 0 1 0 0 0 0 ...
## $ : num [1:150] 0 0 NA 0 0 0 0 0 1 0 ...
## $ : num [1:150] 0 1 0 1 0 0 0 0 0 0 ...
## $ : num [1:150] 0 0 0 0 0 0 0 1 NA 0 ...
## $ : num [1:150] 0 0 0 0 0 0 0 0 0 0 ...
## $ : num [1:150] 0 0 0 0 0 0 0 0 0 0 ...
## $ : num [1:150] 0 0 0 0 1 0 0 0 0 0 ...
## $ : num [1:150] 0 0 0 NA 0 0 0 0 0 0 ...
## $ : num [1:150] 0 0 0 0 0 0 0 0 0 0 ...
## $ : num [1:150] 0 0 0 0 0 0 0 1 NA 0 ...
## $ : num [1:150] 0 0 0 0 0 0 0 0 0 0 ...
## $ : num [1:150] 0 0 0 0 NA 0 0 0 1 0 ...
## $ : num [1:150] 0 0 0 0 0 0 0 NA 0 0 ...
## $ : num [1:150] NA 0 0 0 0 0 0 0 0 0 ...
```

```
## $ : num [1:150] 0 0 0 0 0 0 0 0 0 0 ...
## $ : num [1:150] 0 0 0 0 0 0 0 0 NA 0 ...
## $ : num [1:150] 0 0 0 0 0 0 0 0 0 0 ...
## $ : num [1:150] 0 0 0 0 0 1 0 0 0 0 ...
## $ : num [1:150] 0 0 NA 0 0 0 0 0 0 0 ...
## $ : num [1:150] 0 0 0 0 0 0 1 0 0 0 ...
## $ : num [1:150] 0 0 0 0 0 0 0 0 0 0 ...
## $ : num [1:150] 0 0 0 0 0 0 0 0 0 0 ...
## $ : num [1:150] 0 0 0 1 0 0 0 0 0 0 ...
## $ : num [1:150] NA 0 0 NA 0 0 0 0 0 0 ...
## $ : num [1:150] 0 0 0 0 0 0 0 0 0 0 ...
## $ : num [1:150] 0 0 0 0 1 0 0 0 0 NA ...
## $ : num [1:150] 0 0 0 0 0 0 0 0 0 0 ...
## $ : num [1:150] 0 0 0 0 0 0 0 0 0 0 ...
## $ : num [1:150] 0 NA 1 1 0 0 0 0 0 0 ...
## $ : num [1:150] 0 NA 0 0 0 0 0 0 0 0 ...
## $ : num [1:150] 0 0 0 0 0 0 0 0 0 0 ...
## $ : num [1:150] 0 0 0 0 0 NA 0 1 0 0 ...
## $ : num [1:150] 0 0 0 0 0 0 0 0 0 0 ...
## $ : num [1:150] 0 0 0 0 0 0 0 0 0 0 ...
## $ : num [1:150] 0 0 0 0 0 NA 0 1 0 0 ...
## $ : num [1:150] 0 0 0 0 0 NA 0 0 0 0 ...
## $ : num [1:150] 0 0 0 0 0 0 0 0 0 NA ...
## $ : num [1:150] 0 0 0 0 0 0 0 0 0 0 ...
## $ : num [1:150] 0 NA 1 0 1 0 0 0 0 0 ...
## $ : num [1:150] 0 0 0 0 0 0 0 0 0 NA ...
## $ : num [1:150] 0 0 0 0 0 0 0 0 0 0 ...
## $ : num [1:150] 0 0 0 0 0 0 0 0 NA 0 ...
## $ : num [1:150] 0 0 0 0 0 0 0 0 0 0 ...
## $ : num [1:150] 0 0 0 0 0 0 0 NA 0 0 ...
## $ : num [1:150] 0 0 0 0 0 0 NA 0 1 0 ...
```

```
## $ : num [1:150] 0 0 0 0 0 0 0 0 0 NA ...
## $ : num [1:150] 0 0 NA 0 0 0 0 0 0 0 ...
## $ : num [1:150] 0 1 0 0 0 0 0 0 0 0 ...
## $ : num [1:150] 0 0 0 0 0 0 0 0 1 0 ...
## $ : num [1:150] 1 0 0 0 0 0 0 0 0 0 ...
## $ : num [1:150] NA 0 0 0 0 0 0 NA 0 0 ...
## $ : num [1:150] 0 0 0 NA 0 0 0 0 0 0 ...
## $ : num [1:150] 0 0 0 0 0 0 1 0 0 0 ...
## $ : num [1:150] 0 0 0 0 0 0 0 0 NA 0 ...
## $ : num [1:150] 0 0 0 0 0 0 0 0 0 0 ...
## $ : num [1:150] 0 0 NA 0 0 0 0 NA 0 0 ...
##  [list output truncated]
##To view the cardiovascular events of the 1st element
str(samples_150[1,][1])
## List of 1
## $ : num [1:150] 0 0 0 0 0 0 0 NA 0 0 ...
##To view the elements of time only
str(samples_150[2,])
## List of 1000
## $ : num [1:150] 3626 1703 247 1313 2619 ...
##  ..- attr(*, "label")= chr "Days to cardiovascular event or end of follow-up"
## $ : num [1:150] 1895 2603 1181 1646 3123 ...
##  ..- attr(*, "label")= chr "Days to cardiovascular event or end of follow-up"
## $ : num [1:150] 409 235 2678 1836 1959 ...
##  ..- attr(*, "label")= chr "Days to cardiovascular event or end of follow-up"
## $ : num [1:150] 278 295 3570 1446 681 ...
##  ..- attr(*, "label")= chr "Days to cardiovascular event or end of follow-up"
## $ : num [1:150] 1061 636 2451 933 3435 ...
```

```
## ..- attr(*, "label")= chr "Days to cardiovascular event or end of follow-up"
## $ : num [1:150] 500 709 NA 3397 3418 ...
## ..- attr(*, "label")= chr "Days to cardiovascular event or end of follow-up"
## $ : num [1:150] 2692 269 1958 2712 2305 ...
## ..- attr(*, "label")= chr "Days to cardiovascular event or end of follow-up"
## $ : num [1:150] 624.4 3624.5 2120.7 66.9 111.6 ...
## ..- attr(*, "label")= chr "Days to cardiovascular event or end of follow-up"
## $ : num [1:150] NA 729 3338 2086 1684 ...
## ..- attr(*, "label")= chr "Days to cardiovascular event or end of follow-up"
## $ : num [1:150] 1531 181 3153 NA 3162 ...
## ..- attr(*, "label")= chr "Days to cardiovascular event or end of follow-up"
## $ : num [1:150] 2983 3507 703 978 2712 ...
## ..- attr(*, "label")= chr "Days to cardiovascular event or end of follow-up"
## $ : num [1:150] 3152 2766 790 3552 2044 ...
## ..- attr(*, "label")= chr "Days to cardiovascular event or end of follow-up"
## $ : num [1:150] 510 NA 3394 1827 2421 ...
## ..- attr(*, "label")= chr "Days to cardiovascular event or end of follow-up"
## $ : num [1:150] 550 2064 1364 397 2414 ...
## ..- attr(*, "label")= chr "Days to cardiovascular event or end of follow-up"
## $ : num [1:150] 2132.8 1625.3 1296.5 50.9 3069.8 ...
## ..- attr(*, "label")= chr "Days to cardiovascular event or end of follow-up"
## $ : num [1:150] 2853 3201 NA 1695 1087 ...
```

..- attr(*, "label")= chr "Days to cardiovascular event or end of follow-up"

$: num [1:150] 2330 3296 197 3443 2781 ...

..- attr(*, "label")= chr "Days to cardiovascular event or end of follow-up"

$: num [1:150] 668 3344 1694 353 245 ...

..- attr(*, "label")= chr "Days to cardiovascular event or end of follow-up"

$: num [1:150] 2097 3096 849 507 1842 ...

..- attr(*, "label")= chr "Days to cardiovascular event or end of follow-up"

$: num [1:150] NA 110.2 1379 29.8 449.3 ...

..- attr(*, "label")= chr "Days to cardiovascular event or end of follow-up"

$: num [1:150] 1739 510 2291 774 NA ...

..- attr(*, "label")= chr "Days to cardiovascular event or end of follow-up"

$: num [1:150] 284 1865 179 NA NA ...

..- attr(*, "label")= chr "Days to cardiovascular event or end of follow-up"

$: num [1:150] 1724 2107 2640 645 110 ...

..- attr(*, "label")= chr "Days to cardiovascular event or end of follow-up"

$: num [1:150] 1285 2687 NA 2658 1897 ...

..- attr(*, "label")= chr "Days to cardiovascular event or end of follow-up"

$: num [1:150] 736 2332 2537 2973 196 ...

..- attr(*, "label")= chr "Days to cardiovascular event or end of follow-up"

$: num [1:150] 1668 1105 1550 2899 1774 ...

..- attr(*, "label")= chr "Days to cardiovascular event or end of follow-up"

$: num [1:150] 359 2119 1751 3397 1136 ...

..- attr(*, "label")= chr "Days to cardiovascular event or end of follow-up"

$: num [1:150] 2110 NA 198 312 NA ...

..- attr(*, "label")= chr "Days to cardiovascular event or end of follow-up"

$: num [1:150] 3066 3608 1271 814 985 ...

..- attr(*, "label")= chr "Days to cardiovascular event or end of follow-up"

$: num [1:150] 2572 1718 1747 3307 817 ...

..- attr(*, "label")= chr "Days to cardiovascular event or end of follow-up"

$: num [1:150] 1356 3030 3484 469 132 ...

..- attr(*, "label")= chr "Days to cardiovascular event or end of follow-up"

$: num [1:150] 2111 3317 2995 484 NA ...

..- attr(*, "label")= chr "Days to cardiovascular event or end of follow-up"

$: num [1:150] 1752 410 1055 NA 737 ...

..- attr(*, "label")= chr "Days to cardiovascular event or end of follow-up"

$: num [1:150] 1214 704 1720 2340 NA ...

..- attr(*, "label")= chr "Days to cardiovascular event or end of follow-up"

$: num [1:150] 1597 2914 1974 2024 493 ...

..- attr(*, "label")= chr "Days to cardiovascular event or end of follow-up"

$: num [1:150] 3618 2218 2100 32 1965 ...

..- attr(*, "label")= chr "Days to cardiovascular event or end of follow-up"

$: num [1:150] 3153 363 2422 NA 1013 ...

..- attr(*, "label")= chr "Days to cardiovascular event or end of follow-up"

$: num [1:150] NA 2748 460 469 1022 ...

```
## ..- attr(*, "label")= chr "Days to cardiovascular event or end of follow-up"
## $ : num [1:150] 3110 2812 828 2184 337 ...
## ..- attr(*, "label")= chr "Days to cardiovascular event or end of follow-up"
## $ : num [1:150] 1518 932 2664 484 2498 ...
## ..- attr(*, "label")= chr "Days to cardiovascular event or end of follow-up"
## $ : num [1:150] NA 3339.7 3225.1 37.1 2837.1 ...
## ..- attr(*, "label")= chr "Days to cardiovascular event or end of follow-up"
## $ : num [1:150] 2041 3445 1838 358 868 ...
## ..- attr(*, "label")= chr "Days to cardiovascular event or end of follow-up"
## $ : num [1:150] 521 636 2571 2621 1087 ...
## ..- attr(*, "label")= chr "Days to cardiovascular event or end of follow-up"
## $ : num [1:150] 952 2974 886 3271 3291 ...
## ..- attr(*, "label")= chr "Days to cardiovascular event or end of follow-up"
## $ : num [1:150] 3527 2142 NA 2543 3625 ...
## ..- attr(*, "label")= chr "Days to cardiovascular event or end of follow-up"
## $ : num [1:150] 1409 1356 1752 2538 215 ...
## ..- attr(*, "label")= chr "Days to cardiovascular event or end of follow-up"
## $ : num [1:150] 484 877 1429 3061 1334 ...
## ..- attr(*, "label")= chr "Days to cardiovascular event or end of follow-up"
## $ : num [1:150] 3057 2212 2272 3651 1368 ...
## ..- attr(*, "label")= chr "Days to cardiovascular event or end of follow-up"
## $ : num [1:150] 66.9 2287.3 2342.9 1960.9 3021.9 ...
```

```
## ..- attr(*, "label")= chr "Days to cardiovascular event or end of follow-up"
## $ : num [1:150] 2416 1385 2613 1711 196 ...
## ..- attr(*, "label")= chr "Days to cardiovascular event or end of follow-up"
## $ : num [1:150] 2837 1967 2348 NA 2064 ...
## ..- attr(*, "label")= chr "Days to cardiovascular event or end of follow-up"
## $ : num [1:150] 562 1341 292 1237 1375 ...
## ..- attr(*, "label")= chr "Days to cardiovascular event or end of follow-up"
## $ : num [1:150] 2914 501 3356 1140 237 ...
## ..- attr(*, "label")= chr "Days to cardiovascular event or end of follow-up"
## $ : num [1:150] 950 2068 1998 917 515 ...
## ..- attr(*, "label")= chr "Days to cardiovascular event or end of follow-up"
## $ : num [1:150] 1058 3394 657 558 NA ...
## ..- attr(*, "label")= chr "Days to cardiovascular event or end of follow-up"
## $ : num [1:150] 240 3640 170 292 338 ...
## ..- attr(*, "label")= chr "Days to cardiovascular event or end of follow-up"
## $ : num [1:150] NA 2572 3401 1918 2885 ...
## ..- attr(*, "label")= chr "Days to cardiovascular event or end of follow-up"
## $ : num [1:150] 2075.1 557.2 3054.9 55.9 32 ...
## ..- attr(*, "label")= chr "Days to cardiovascular event or end of follow-up"
## $ : num [1:150] 357 2402 605 1730 1943 ...
## ..- attr(*, "label")= chr "Days to cardiovascular event or end of follow-up"
## $ : num [1:150] 2934 1695 1460 709 742 ...
```

```
## ..- attr(*, "label")= chr "Days to cardiovascular event or end of follow-up"
## $ : num [1:150] 275 1947 3249 562 3605 ...
## ..- attr(*, "label")= chr "Days to cardiovascular event or end of follow-up"
## $ : num [1:150] 648 2930 NA 1838 2839 ...
## ..- attr(*, "label")= chr "Days to cardiovascular event or end of follow-up"
## $ : num [1:150] 2606.9 2302.3 41.4 2356.1 1340.7 ...
## ..- attr(*, "label")= chr "Days to cardiovascular event or end of follow-up"
## $ : num [1:150] 900 1697 2551 2366 3477 ...
## ..- attr(*, "label")= chr "Days to cardiovascular event or end of follow-up"
## $ : num [1:150] 1384 849 1523 1842 1488 ...
## ..- attr(*, "label")= chr "Days to cardiovascular event or end of follow-up"
## $ : num [1:150] 2010 2787.9 3204.1 94.1 1241.5 ...
## ..- attr(*, "label")= chr "Days to cardiovascular event or end of follow-up"
## $ : num [1:150] NA 2200 467 NA 1480 ...
## ..- attr(*, "label")= chr "Days to cardiovascular event or end of follow-up"
## $ : num [1:150] 2903 2122 3611 628 400 ...
## ..- attr(*, "label")= chr "Days to cardiovascular event or end of follow-up"
## $ : num [1:150] 2503.3 1437.2 376.9 1098.5 83.5 ...
## ..- attr(*, "label")= chr "Days to cardiovascular event or end of follow-up"
## $ : num [1:150] 2621 1930 2091 2957 3474 ...

## ..- attr(*, "label")= chr "Days to cardiovascular event or end of follow-up"
## $ : num [1:150] 1590 722 1749 1076 3451 ...
```

..- attr(*, "label")= chr "Days to cardiovascular event or end of follow-up"

$: num [1:150] 2074.1 NA 10.1 1479.5 1163.4 ...

..- attr(*, "label")= chr "Days to cardiovascular event or end of follow-up"

$: num [1:150] 842 NA 1207 3338 532 ...

..- attr(*, "label")= chr "Days to cardiovascular event or end of follow-up"

$: num [1:150] 3067 1668 2402 3461 1933 ...

..- attr(*, "label")= chr "Days to cardiovascular event or end of follow-up"

$: num [1:150] 2388 1773 577 3389 1598 ...

..- attr(*, "label")= chr "Days to cardiovascular event or end of follow-up"

$: num [1:150] 1049 2119 2905 1942 327 ...

..- attr(*, "label")= chr "Days to cardiovascular event or end of follow-up"

$: num [1:150] 2060 113 3410 2572 396 ...

..- attr(*, "label")= chr "Days to cardiovascular event or end of follow-up"

$: num [1:150] 3153 997 959 978 2837 ...

..- attr(*, "label")= chr "Days to cardiovascular event or end of follow-up"

$: num [1:150] 561.5 2148.8 55.6 3212.3 1888.1 ...

..- attr(*, "label")= chr "Days to cardiovascular event or end of follow-up"

$: num [1:150] 3153 1989 3108 2187 399 ...

..- attr(*, "label")= chr "Days to cardiovascular event or end of follow-up"

$: num [1:150] 2503 411 1934 3152 2666 ...

..- attr(*, "label")= chr "Days to cardiovascular event or end of follow-up"

$: num [1:150] 3236 NA 353.4 3567.3 56.4 ...

```
## ..- attr(*, "label")= chr "Days to cardiovascular event or end of follow-up"
## $ : num [1:150] 277 2639 559.4 64.5 2594.2 ...
## ..- attr(*, "label")= chr "Days to cardiovascular event or end of follow-up"
## $ : num [1:150] 3164 2919 2449 518 1104 ...
## ..- attr(*, "label")= chr "Days to cardiovascular event or end of follow-up"
## $ : num [1:150] 1375 2709 1045 939 1116 ...
## ..- attr(*, "label")= chr "Days to cardiovascular event or end of follow-up"
## $ : num [1:150] 575.5 66.9 2431.2 3045.6 1683.8 ...
## ..- attr(*, "label")= chr "Days to cardiovascular event or end of follow-up"
## $ : num [1:150] 2348 3625 636 3171 158 ...
## ..- attr(*, "label")= chr "Days to cardiovascular event or end of follow-up"
## $ : num [1:150] 601 131 2636 1776 264 ...
## ..- attr(*, "label")= chr "Days to cardiovascular event or end of follow-up"
## $ : num [1:150] 448 712 2529 3046 183 ...
## ..- attr(*, "label")= chr "Days to cardiovascular event or end of follow-up"
## $ : num [1:150] 1980 2110 NA 721 1998 ...
## ..- attr(*, "label")= chr "Days to cardiovascular event or end of follow-up"
## $ : num [1:150] 2440 425 1639 1075 2487 ...
## ..- attr(*, "label")= chr "Days to cardiovascular event or end of follow-up"
## $ : num [1:150] 1425 1105 110 1059 1699 ...

## ..- attr(*, "label")= chr "Days to cardiovascular event or end of follow-up"
```

```
## $ : num [1:150] 278 2988 729 749 620 ...
##  ..- attr(*, "label")= chr "Days to cardiovascular event or end of follow-up"
## $ : num [1:150] NA 669 1381 1122 2475 ...
##  ..- attr(*, "label")= chr "Days to cardiovascular event or end of follow-up"
## $ : num [1:150] 2828 1022 701 NA 406 ...
##  ..- attr(*, "label")= chr "Days to cardiovascular event or end of follow-up"
## $ : num [1:150] 2074 1027 3148 3207 1114 ...
##  ..- attr(*, "label")= chr "Days to cardiovascular event or end of follow-up"
## $ : num [1:150] 557 1947 3493 440 2427 ...
##  ..- attr(*, "label")= chr "Days to cardiovascular event or end of follow-up"
## $ : num [1:150] 2845 2979 3397 3513 1055 ...
##  ..- attr(*, "label")= chr "Days to cardiovascular event or end of follow-up"
## $ : num [1:150] 213 2498 NA 1638 2901 ...
##  ..- attr(*, "label")= chr "Days to cardiovascular event or end of follow-up"
##  [list output truncated]
##To view the follow-up times of the 1st element
str(samples_150[2,][1])
## List of 1
## $ : num [1:150] 3626 1703 247 1313 2619 ...
##  ..- attr(*, "label")= chr "Days to cardiovascular event or end of follow-up"
##To view the follow-up times of the last or 1000th element
str(samples_150[2,][1000])
## List of 1
```

```
## $ : num [1:150] 523 2305 950 NA 3474 ...
## ..- attr(*, "label")= chr "Days to cardiovascular event or end of follow-up"
##To obtain the sums of cardiovascular events only, use sapply where it sums every element of the list
sums_cv<-sapply(samples_150[1,],sum, na.rm = TRUE)
summary(sums_cv)
## Min. 1st Qu. Median Mean 3rd Qu. Max.
## 0.000 4.000 6.000 6.015 7.000 13.000
##To obtain the sums of time only, use sapply where it sums every element of the list
sums_time<-sapply(samples_150[2,],sum, na.rm = TRUE)
summary(sums_time)
## Min. 1st Qu. Median Mean 3rd Qu. Max.
## 205755 240129 248925 248752 257450 290964
##To obtain the incidence rate for each sample
rates_150<-sums_cv/sums_time
head(rates_150)
## [1] 3.026425e-05 2.538829e-05 3.378867e-05 3.591355e-05 2.672117e-05
## [6] 1.688767e-05
summary(rates_150)
## Min. 1st Qu. Median Mean 3rd Qu. Max.
## 0.000e+00 1.758e-05 2.356e-05 2.434e-05 3.017e-05 5.749e-05
##To view a histogram of incidence rates based on samples of size= 150
hist_150<-ggplot()+geom_histogram(aes(x = rates_150))
hist_150
```

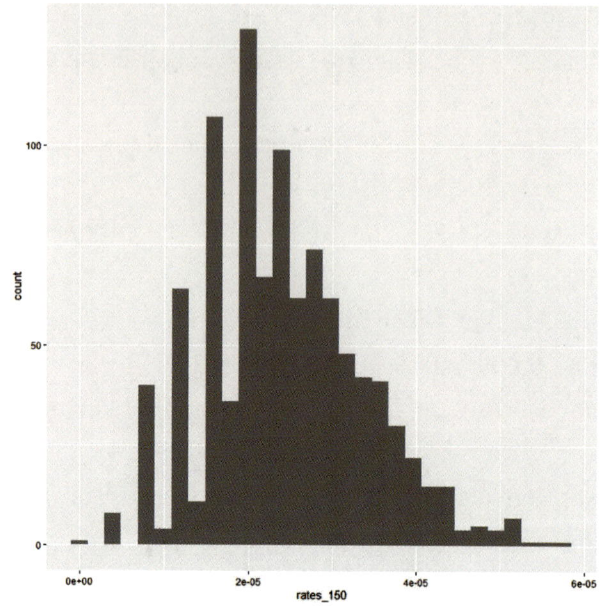

##To plot a boxplot of these incidence rates

```
box_150<-ggplot()+geom_boxplot(aes(y=rates_150))
box_150
```

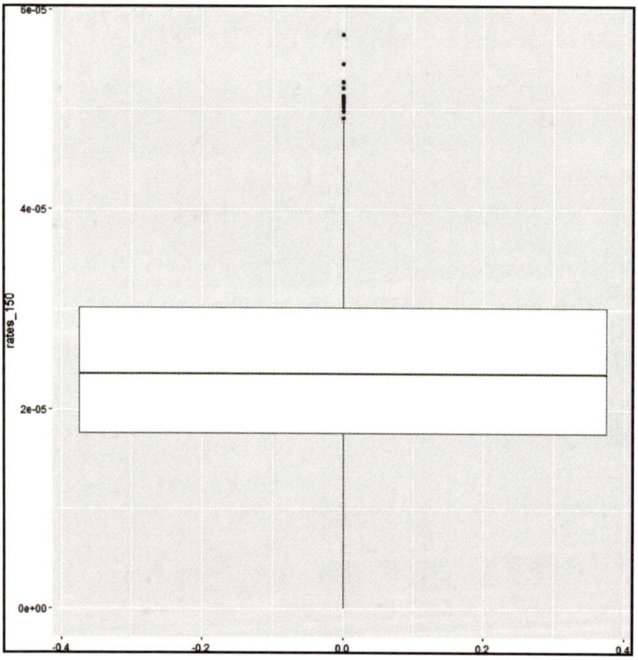

Plot all graphs side by side.

```
##Combine the 3 means in one dataframe
df<-data.frame(rates_20= rates_20, rates_50 = rates_50, rates_150 =
rates_150)
head(df)
##    rates_20   rates_50  rates_150
## 1 0.000000e+00 1.281609e–05 3.026425e–05
## 2 3.341490e–05 3.147441e–05 2.538829e–05
## 3 0.000000e+00 3.347042e–05 3.378867e–05
## 4 6.451571e–05 3.444698e–05 3.591355e–05
## 5 2.427611e–05 3.020894e–05 2.672117e–05
## 6 9.313591e–05 2.232570e–05 1.688767e–05
# gather the columns
df<- df %>% gather(key = “rate,” value = “value”)
table(df$rate)
##
## rates_150 rates_20 rates_50
##    1000    1000    1000
head(df)
##    rate    value
## 1 rates_20 0.000000e+00
## 2 rates_20 3.341490e–05
## 3 rates_20 0.000000e+00
## 4 rates_20 6.451571e–05
## 5 rates_20 2.427611e–05
## 6 rates_20 9.313591e–05
# arrange factor levels
df$rate <- factor(df$rate, levels = c(“rates_20,” “rates_50,””rates_150”))
ggplot(data = df, aes(x = value, fill = rate)) + geom_histogram()+facet_
wrap(~rate)
```

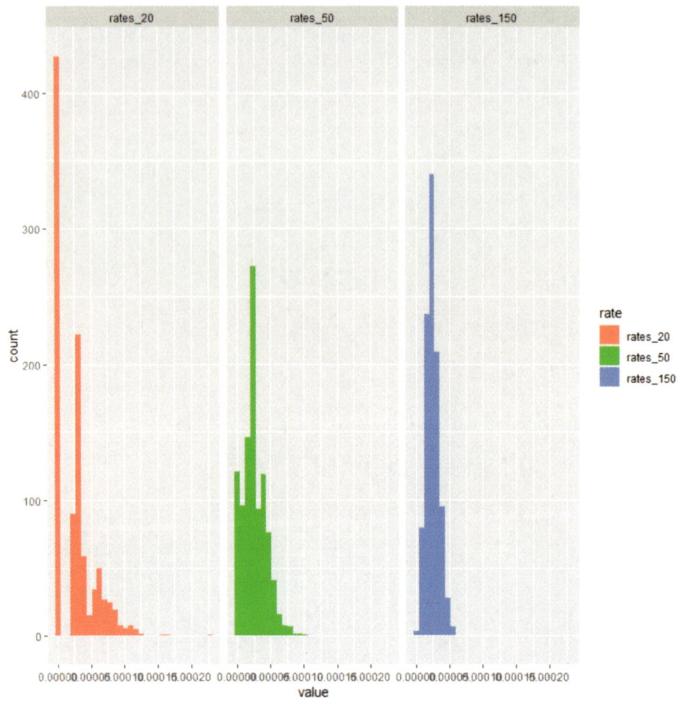


```
ggplot(data = df, aes(x = rate, y = value, col = rate))+geom_boxplot()
```

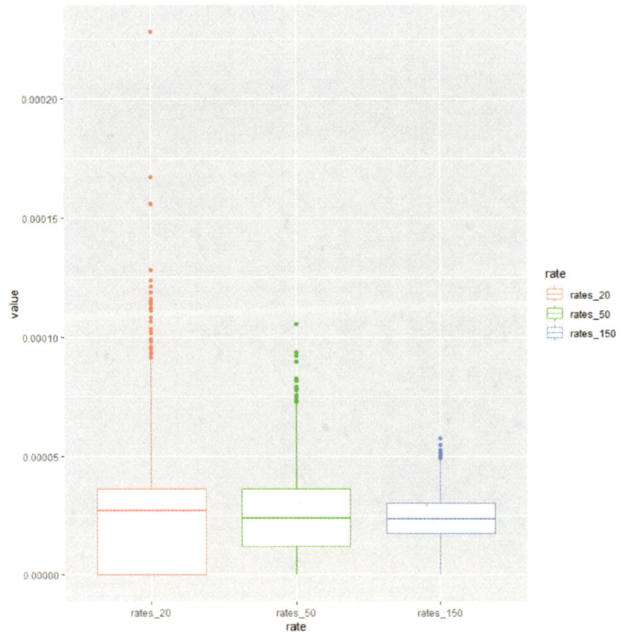

```
# to calculate the standard deviation for each rates type
df %>% group_by(rate) %>% summarise(sd = sd(value))
## # A tibble: 3 x 2
##  rate      sd
##  <fct>    <dbl>
## 1 rates_20 0.0000295
## 2 rates_50 0.0000173
## 3 rates_150 0.00000944
```

The same rules apply:

1. The sampling distribution for sample incidence rates (IRs) across random samples of the same size is approximately normal.
2. The variation in sample IRs decreases with increasing the sample size and the distribution shape becomes more normal. It decreases from 0.0000295 for 20 sample size to 0.00000944 for 150 sample size.
3. The mean of sample IRs is very close to the true population incidence rate = 2.4×10^{-5}.

6.5. THE CENTRAL LIMIT THEOREM (CLT)

It is a mathematical theory that unifies what we saw in the above examples. CLT states that the theoretical sampling distribution of a sample statistic (mean, proportion, IR) has the following properties:

1. Nearly normal or bell-shaped.
2. Will have on average the true population parameter or value that is being estimated.
3. The variability in the distribution depends on (a) the variability of values in our population () and (b) the sample size.

This variability in sample statistics (means, proportions, IRs) is called the standard error (SE) and is different from the variability of individual values in any single sample, which is called the standard deviation (s).

The general formula for the SE:

SE =

Since is unknown, it is replaced by the sample standard deviation (s) and called . The hat indicates that it is a sample estimate.

Standard error for means = /

Standard error for proportions= where p is the sample proportion and n is sample size

Standard error for incidence rate = / total follow up time.

These SEs will be used to calculate the confidence intervals (CI) for means, proportions, and IRs which is our next chapter.

CHAPTER 7

CONFIDENCE INTERVALS

CONTENTS

7.1. INTRODUCTION

Confidence interval (CI) is an interval that incorporates the uncertainty in our sample based estimates. CI is constructed based on some properties of the Normal distribution (see chapter 3).

In the normal distribution:

1. 68% of the values fall within 1 standard deviation (sd) from the mean;
2. 90% of the values fall within 1.65 sd from the mean;
3. 95% of the values fall within 2 sd (1.96 sd is exact value) from the mean;
4. 99% of the values fall within 2.58 sd from the mean;
5. 99.7% of the values fall within 3 sd from the mean.

7.2. CONFIDENCE INTERVAL (CI) FOR A SINGLE POPULATION PARAMETER (MEAN, PROPORTION, INCIDENCE RATE (IR))

7.2.1. Central Limit Theorem (CLT)

The central limit theorem (CLT) states that the theoretical sampling distribution of a sample statistic (mean, proportion, incidence rate (IR)) has the following properties:

1. Nearly normal or bell-shaped.
2. Will have on average the true population parameter or value that is being estimated.

These properties are used to construct CIs around our sample estimates. For example, 120 randomly selected men were measured for their blood pressure and the mean = 121 mmHg and sd = 11.5 mmHg. Standard error () for means=/

90% confidence interval (CI)= meanS

95% confidence interval (CI)= meanS OR more accurately meanS

99% confidence interval (CI)= meanS

All values that are multiplied by the standard error (SE) are from the normal distribution rule.

7.2.2. The Effect of Increasing Confidence and Increasing Sample Size on the Confidence Interval (CI) Width

Confidence intervals with different confidence levels are calculated for the mean blood pressure = 121 mmHg and sd = 11.5 mmHg obtained from a random sample of 120 men as follows:

```
##Calculating the standard error and assign its value to variable se
se<–11.5/sqrt(120)
se
## [1] 1.049802
##Calculating 90% CI and assign its value to variable CI_90
CI_90<-c(121+1.65*se,121-1.65*se)
CI_90
## [1] 122.7322 119.2678
##Calculating 95% CI and assign its value to variable CI_95
CI_95<-c(121+1.96*se,121-1.96*se)
CI_95
## [1] 123.0576 118.9424
##Calculating 99% CI and assign its value to variable CI_99
CI_99<-c(121+2.58*se,121-2.58*se)
CI_99
## [1] 123.7085 118.2915
```

Increasing the confidence level will make the CI wider for the same sample size.

Different sample sizes

```
## sample size 1000 with the same mean and sd as the 120 sample
se2<–11.5/sqrt(1000)
se2
## [1] 0.3636619
##Calculating 90% CI and assign its value to variable CI2_90
CI2_90<-c(121+1.65*se2,121-1.65*se2)
```

```
CI2_90
## [1] 121.6 120.4
##Calculating 95% CI and assign its value to variable CI2_95
CI2_95<-c(121+1.96*se2,121-1.96*se2)
CI2_95
## [1] 121.7128 120.2872
##Calculating 99% CI and assign its value to variable CI2_99
CI2_99<-c(121+2.58*se2,121-2.58*se2)
CI2_99
## [1] 121.9382 120.0618
```

Increasing the sample size will tighten the CI (narrower) for the same confidence level.

7.2.3. Interpretation of Confidence Intervals (CI)

CI gives us a range of possible values for the unknown population parameter (mean, proportion, or IR) from which the sample was taken.

For example, the 95% confidence level means that the true population parameter is not necessarily within this interval but for 95% of samples taken from this population, the CI generated from them will contain the true population parameter. We will never know whether we got one of these 95% samples or not, as in research, we only have one sample. We only know that this interval works 95% of the time.

In research, when taking a sample and constructing 95% CI, I do not know if I take one of the good 95% samples that yields interval containing the population value or one of the bad 5% samples that yields interval not containing the population value. As long as the 95% CI works 95% of the time, we can assume that the constructed interval contains the population value.

For the 99% CI, this interval will hold the true population value for 99% of samples taken from this population. For 99.9% CI, this interval will hold the true population value for 99.9% of samples taken from this population.

7.3. CALCULATION OF CONFIDENCE INTERVALS (CI)

Instead of doing complex computation for calculating the CI for a single population (mean, proportion, IR), R has excellent package called epiR that can do these calculations for us. This package was published in 2019-08-23.

7.3.1. Confidence Interval (CI) for Mean

Calculating and plotting the CIs for different random samples.

We will assume that regicor data is our population. We will take 100 random samples from the age column, each of size 50, and calculate the CI for the mean from each sample, then plotting these CIs to see how they cover the true population mean (which is the mean of whole age column in the dataset).

Argument, ctype = "mean.single," for the calculation of CI for means, requires a vector (single column of data) as input.

Generating 100 random samples from the age column:

```
##loading the necessary packages
library(tidyverse)
library(epiR)
library(compareGroups)
data("regicor")
set.seed(123)
samples_50<-replicate(100,sample(regicor$age,50))
dat<-data.frame(samples_50)
head(dat)
## X1 X2 X3 X4 X5 X6 X7 X8 X9 X10 X11 X12 X13 X14 X15 X16 X17
X18 X19 X20 X21
## 1 47 53 65 67 48 48 47 43 48 64 47 65 48 54 69 70 47 73 60 65 42
## 2 62 46 56 74 47 70 45 41 44 59 53 50 46 74 50 70 63 70 48 54 48
## 3 65 45 58 45 48 69 47 52 40 64 47 39 74 38 43 58 43 39 39 63 60
## 4 46 63 71 57 42 49 68 70 62 53 46 60 38 51 69 74 73 69 58 47 39
## 5 71 53 62 69 54 69 39 56 60 65 55 35 44 42 41 62 73 35 51 36 47
## 6 74 54 60 71 40 61 54 57 61 45 70 47 60 73 58 61 66 59 45 53 70
```

```
## X22 X23 X24 X25 X26 X27 X28 X29 X30 X31 X32 X33 X34 X35 X36
X37 X38 X39 X40
## 1 52 53 46 59 58 58 39 44 67 39 73 64 45 62 58 50 50 55 73
## 2 54 68 35 71 48 51 74 72 54 58 57 39 38 46 37 48 73 60 59
## 3 48 73 74 58 41 68 49 68 50 42 58 64 54 62 67 50 46 60 42
## 4 55 51 65 55 73 37 65 62 44 53 65 47 50 61 41 46 46 73 70
## 5 73 61 50 37 50 71 66 48 46 45 64 67 48 50 43 48 46 56 37
## 6 64 69 45 74 36 42 43 56 62 71 45 44 69 54 51 37 35 57 53
## X41 X42 X43 X44 X45 X46 X47 X48 X49 X50 X51 X52 X53 X54 X55
X56 X57 X58 X59
## 1 57 65 42 43 48 63 74 40 63 56 61 65 60 49 50 41 53 64 45
## 2 52 46 52 51 38 60 38 45 53 72 66 59 58 39 53 47 68 38 58
## 3 40 51 43 44 61 47 48 61 55 55 49 58 61 66 44 60 64 39 62
## 4 38 53 60 48 63 63 54 46 39 67 56 73 38 65 38 43 55 70 70
## 5 67 68 73 41 46 40 58 51 41 67 70 60 42 56 62 55 47 43 65
## 6 42 41 64 41 62 44 65 54 48 60 52 41 45 64 45 42 45 57 39
## X60 X61 X62 X63 X64 X65 X66 X67 X68 X69 X70 X71 X72 X73 X74
X75 X76 X77 X78
## 1 54 38 57 70 74 69 67 50 51 59 46 62 37 41 43 43 56 47 47
## 2 69 53 54 46 61 44 65 67 57 59 57 60 64 68 71 52 74 68 49
## 3 58 73 57 59 71 46 59 47 53 57 46 67 74 70 63 51 39 61 37
## 4 68 67 68 50 71 61 50 38 71 47 68 43 66 47 60 53 42 71 48
## 5 46 64 46 56 73 58 70 42 39 57 61 68 40 47 70 55 51 49 52
## 6 46 67 57 39 51 39 64 59 74 61 45 63 40 53 56 46 64 60 59
## X79 X80 X81 X82 X83 X84 X85 X86 X87 X88 X89 X90 X91 X92 X93
X94 X95 X96 X97
## 1 62 49 48 61 47 49 48 39 72 58 72 57 72 63 37 74 47 43 43
## 2 51 41 62 55 66 50 46 48 64 40 65 60 50 45 62 65 70 37 48
## 3 39 46 70 50 74 47 49 63 46 50 68 39 55 43 47 56 66 67 38
## 4 62 45 58 38 68 50 42 43 57 46 55 44 60 43 54 63 73 56 35
## 5 40 57 53 73 39 37 50 72 53 43 47 46 48 39 50 68 69 62 53
## 6 50 46 52 51 66 41 49 65 67 46 37 73 55 61 50 41 60 38 37
```

```
## X98 X99 X100
## 1 58 35 65
## 2 69 70 48
## 3 39 41 37
## 4 56 43 50
## 5 40 60 50
## 6 67 63 69
```

This dataframe consists of 100 columns (each corresponding to a single sample) and 50 rows which is the sample size.

Use the sapply function to apply the epi.conf to each column (sample).

```
x<-sapply(dat,epi.conf, ctype = "mean.single," conf.level = 0.95)
head(x)
## X1 X2 X3 X4 X5 X6 X7 X8
## est 54.54 53.52 56.3 53.56 54 57.02 52.2 54.1
## se 1.654236 1.509491 1.533903 1.505679 1.495572 1.520872 1.746366 1.645588
## lower 51.21569 50.48656 53.21751 50.53422 50.99454 53.96369 48.69055 50.79307
## upper 57.86431 56.55344 59.38249 56.58578 57.00546 60.07631 55.70945 57.40693
## X9 X10 X11 X12 X13 X14 X15 X16
## est 52.78 53.96 54.76 55.06 53.94 53.36 53.82 55.26
## se 1.546633 1.555493 1.59042 1.763672 1.733561 1.686624 1.59071 1.503142
## lower 49.67192 50.83412 51.56393 51.51577 50.45628 49.9706 50.62335 52.23932
## upper 55.88808 57.08588 57.95607 58.60423 57.42372 56.7494 57.01665 58.28068
## X17 X18 X19 X20 X21 X22 X23 X24
## est 56.94 52.4 52.92 53.98 54.58 55.12 54.78 54.9
## se 1.597807 1.520473 1.43081 1.550376 1.565727 1.352138 1.579277 1.606174
```

lower 53.72909 49.3445 50.04468 50.8644 51.43355 52.40278 51.60633 51.67227

upper 60.15091 55.4555 55.79532 57.0956 57.72645 57.83722 57.95367 58.12773

X25 X26 X27 X28 X29 X30 X31 X32

est 55.66 51.66 54.24 52.26 54.34 55.52 53.96 54.08

se 1.666381 1.510672 1.629714 1.625262 1.565863 1.546884 1.732041 1.349782

lower 52.31128 48.62419 50.96497 48.99391 51.19328 52.41142 50.47933 51.36751

upper 59.00872 54.69581 57.51503 55.52609 57.48672 58.62858 57.44067 56.79249

X33 X34 X35 X36 X37 X38 X39 X40

est 55.32 49.82 54.64 54.58 52.5 53.28 54.18 56.08

se 1.491866 1.491114 1.52175 1.642817 1.596488 1.543132 1.662231 1.795877

lower 52.32198 46.82349 51.58193 51.27863 49.29174 50.17896 50.83962 52.47105

upper 58.31802 52.81651 57.69807 57.88137 55.70826 56.38104 57.52038 59.68895

X41 X42 X43 X44 X45 X46 X47 X48

est 54.96 53.38 56.92 52.74 52.2 55.3 53.88 52.62

se 1.706879 1.431109 1.545986 1.480157 1.251285 1.479451 1.688289 1.593582

lower 51.5299 50.50408 53.81322 49.76551 49.68545 52.32693 50.48726 49.41758

upper 58.3901 56.25592 60.02678 55.71449 54.71455 58.27307 57.27274 55.82242

X49 X50 X51 X52 X53 X54 X55 X56

est 57.62 55.62 56.24 55.5 53.4 56.26 53.16 53.58

se 1.565675 1.463819 1.568194 1.554519 1.541534 1.44838 1.678522 1.541293

lower 54.47366 52.67834 53.0886 52.37608 50.30217 53.34937 49.78688 50.48266

```
## upper 60.76634 58.56166 59.3914 58.62392 56.49783 59.17063
56.53312 56.67734
##   X57   X58   X59   X60   X61   X62   X63   X64
## est 53.62 52.38 56.58 56.58 53.42 54.46 52.76 57.32
## se 1.446711 1.566718 1.497233 1.469719 1.676023 1.361635 1.637709
1.528088
## lower 50.71273 49.23156 53.5712 53.62649 50.05191 51.72369 49.4689
54.24919
## upper 56.52727 55.52844 59.5888 59.53351 56.78809 57.19631 56.0511
60.39081
##   X65   X66   X67   X68   X69   X70   X71   X72
## est 57.1   54.92 57.38 55.4   52.74 56.5   54.92 56.4
## se 1.553994 1.53406 1.434528 1.519667 1.519885 1.526968 1.491429
1.623803
## lower 53.97713 51.83719 54.49721 52.34611 49.68568 53.43144
51.92286 53.13685
## upper 60.22287 58.00281 60.26279 58.45389 55.79432 59.56856
57.91714 59.66315
##   X73   X74   X75   X76   X77   X78   X79   X80
## est 53.84 57.56 52.86 55.58 55.74 52.66 55.2   55.3
## se 1.629664 1.494796 1.323573 1.538112 1.563567 1.535062 1.571948
1.531239
## lower 50.56507 54.55609 50.20018 52.48905 52.59789 49.57518
52.04105 52.22286
## upper 57.11493 60.56391 55.51982 58.67095 58.88211 55.74482
58.35895 58.37714
##   X81   X82   X83   X84   X85   X86   X87   X88
## est 53.18 57.96 56.06 54.12 55.18 55.9   57.84 53.76
## se 1.416719 1.587955 1.785751 1.721803 1.656081 1.60846 1.484439
1.40511
## lower 50.333  54.76889 52.4714 50.65991 51.85198 52.66768 54.85691
50.93633
## upper 56.027  61.15111 59.6486 57.58009 58.50802 59.13232 60.82309
56.58367
```

```
## X89 X90 X91 X92 X93 X94 X95 X96
## est 55.24 56 58.48 54.92 55.14 54.02 55.8 54.34
## se 1.543534 1.55051 1.577712 1.705176 1.481315 1.65997 1.560481
1.668828
## lower 52.13815 52.88413 55.30947 51.49332 52.16319 50.68416
52.6641 50.98636
## upper 58.34185 59.11587 61.65053 58.34668 58.11681 57.35584
58.9359 57.69364
## X97 X98 X99 X100
## est 50.14 54.78 55.8 53.4
## se 1.530642 1.559249 1.517248 1.593354
## lower 47.06406 51.64657 52.75098 50.19804
## upper 53.21594 57.91343 58.84902 56.60196
```

The output of epi.conf for each column (sample):

1. *est* for the mean estimate;
2. *se* for the standard error;
3. *lower* for the lower confidence interval limit;
4. *upper* for the upper confidence interval limit.

Transpose the data for plotting so it now consists of 100 rows and 4 columns.

```
x2<-data.frame(t(x))
head(x2)
## est se lower upper
## X1 54.54 1.654236 51.21569 57.86431
## X2 53.52 1.509491 50.48656 56.55344
## X3 56.3 1.533903 53.21751 59.38249
## X4 53.56 1.505679 50.53422 56.58578
## X5 54 1.495572 50.99454 57.00546
## X6 57.02 1.520872 53.96369 60.07631
```

```
##Generating names of the samples from 1 to 100
nam2<-seq(1,100,1)
##Constructing a dataframe with the desired columns, sample name, sample
mean, lower confidence limit, upper confidence limit
x3<-data.frame(sample=nam2,mean = as.numeric(x2$est),
      lower=as.numeric(x2$lower), upper=as.numeric(x2$upper))
head(x3)
## sample mean  lower  upper
## 1  1 54.54 51.21569 57.86431
## 2  2 53.52 50.48656 56.55344
## 3  3 56.30 53.21751 59.38249
## 4  4 53.56 50.53422 56.58578
## 5  5 54.00 50.99454 57.00546
## 6  6 57.02 53.96369 60.07631
```

Plotting the CIs as error bars around their mean and plotting the population mean (regicor age mean) as a horizontal intercept.

```
ggplot(data = x3,aes(x = sample,y=mean))+geom_point()+
 geom_abline(slope = 0,intercept = mean(regicor$age))+
 geom_errorbar(aes(ymin=lower, ymax = upper))
```

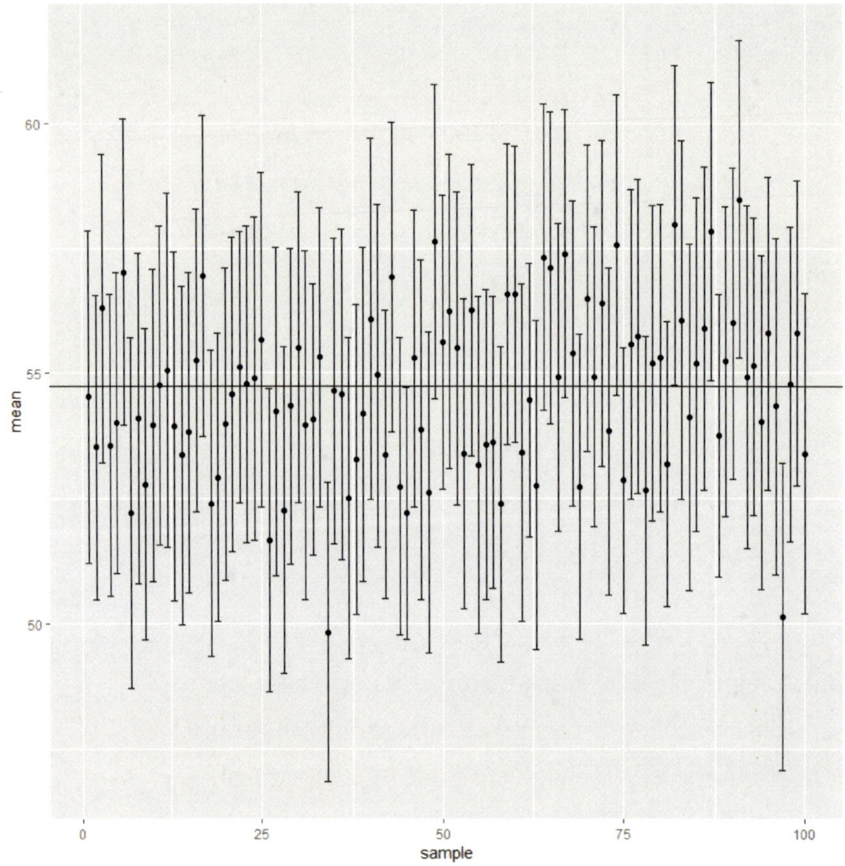

Some intervals do not cover the horizontal line (population mean).

Add another column to the dataset that determine if the interval covers the population mean (good) or not (bad). The interval will not cover the population mean if the lower limit is higher than the population mean or the upper limit is lower than the population mean.

```
x3<-x3 %>% mutate(color=ifelse(x3$lower>mean(regicor$age)|x3$upper
< mean(regicor$age),"bad,""good"))
head(x3)
## sample mean  lower  upper color
## 1   1 54.54 51.21569 57.86431 good
## 2   2 53.52 50.48656 56.55344 good
## 3   3 56.30 53.21751 59.38249 good
```

```
## 4   4 53.56 50.53422 56.58578 good
## 5   5 54.00 50.99454 57.00546 good
## 6   6 57.02 53.96369 60.07631 good
ggplot(data=x3, aes(x = sample,y=mean, color=color))+geom_point()+
 geom_abline(slope = 0,intercept = mean(regicor$age))+
 geom_errorbar(aes(ymin=lower, ymax = upper))
```

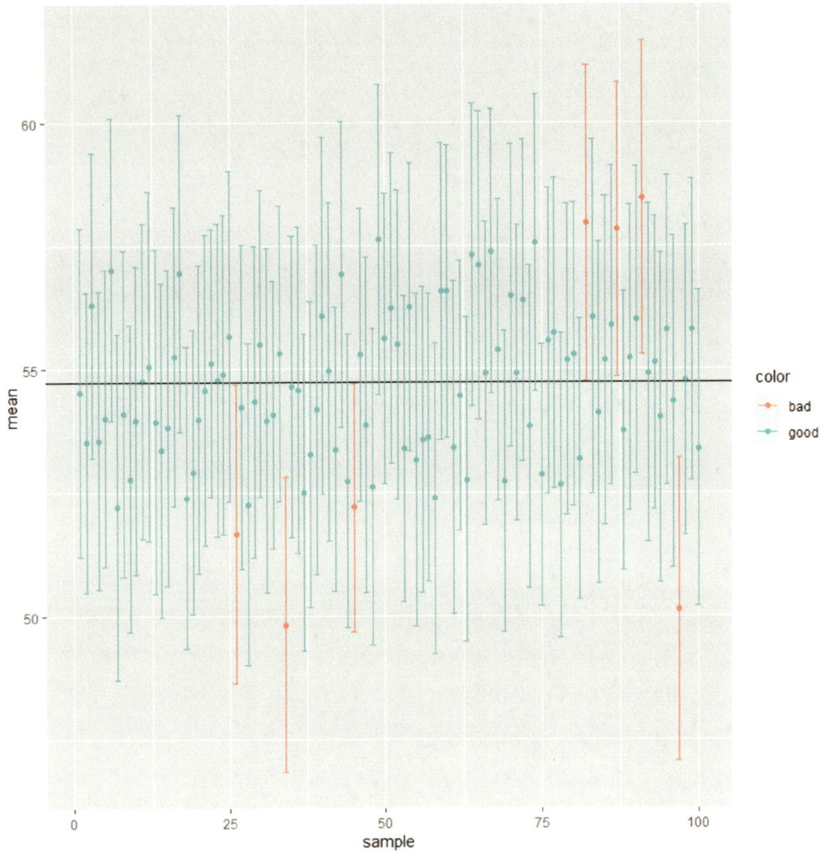

Some intervals of some samples do not cover the true mean and the remaining intervals cover the true mean. This is the meaning of a CI.

Calculating and plotting the CIs for a single sample mean with different confidence levels and different sample sizes.

Different sample sizes:

```
set.seed(123)
```

```
#sample size = 50
sample_50<-sample(regicor$age,50)
#sample size = 150
sample_150<-sample(regicor$age,150)
##As our dataset is a sample from a larger population, we can consider the
survey data is a sample of 2294 participants
sample_survey<- as.numeric(regicor$age)
##Calculating the 95% confidence interval for each sample
ci_50<-epi.conf(sample_50)
ci_50
##   est    se  lower  upper
## 1 54.54 1.654236 51.21569 57.86431
ci_150<-epi.conf(sample_150)
ci_150
##   est    se  lower  upper
## 1 54.42 0.8704582 52.69996 56.14004
ci_2294<- epi.conf(sample_survey)
ci_2294
##    est    se  lower  upper
## 1 54.73627 0.2306941 54.28388 55.18866
##Combining the results in a single dataframe
df<-rbind(ci_50,ci_150,ci_2294)
##Adding a column of the sample size
df<-data.frame(size=c(50,150,2294),df)
df
##  size   est    se  lower  upper
## 1   50 54.54000 1.6542364 51.21569 57.86431
## 2  150 54.42000 0.8704582 52.69996 56.14004
## 3 2294 54.73627 0.2306941 54.28388 55.18866
##Plotting the data to see the effect of increasing sample size
ggplot(data=df, aes(x = size,y=est,color=factor(size)))+
```

```
geom_point()+geom_errorbar(aes(ymin=lower, ymax = upper))
```

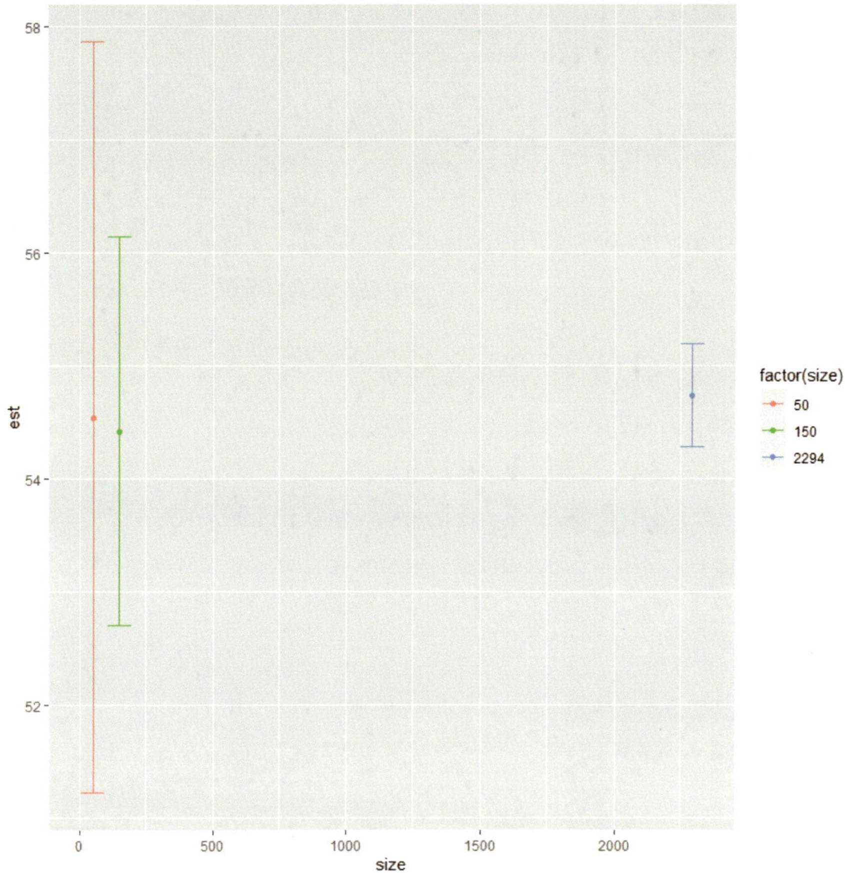

Note that with increasing the sample size, the CI becomes more narrower.

Different confidence levels for the same sample size.

```
ci_90<-epi.conf(sample_50,conf.level = 0.90)
ci_90
##   est    se lower  upper
## 1 54.54 1.654236 51.76659 57.31341
ci_95<-epi.conf(sample_50,conf.level = 0.95)
ci_95
##   est    se lower  upper
## 1 54.54 1.654236 51.21569 57.86431
ci_99<-epi.conf(sample_50,conf.level = 0.99)
```

```
ci_99
## est se lower upper
## 1 54.54 1.654236 50.10673 58.97327
##Combining the results in a single dataframe
df2<-rbind(ci_90,ci_95,ci_99)
##Adding a column of the confidence level
df2<-data.frame(level=c(90,95,99),df2)
df2
## level est se lower upper
## 1 90 54.54 1.654236 51.76659 57.31341
## 2 95 54.54 1.654236 51.21569 57.86431
## 3 99 54.54 1.654236 50.10673 58.97327
ggplot(data=df2, aes(x = level,y=est,color=factor(level)))+
 geom_point()+geom_errorbar(aes(ymin=lower, ymax = upper))
```

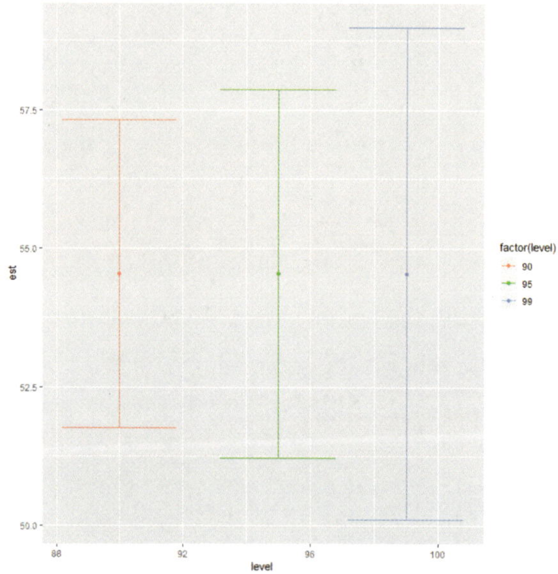

Increasing the confidence level, the interval widens.

For ci_50 (based on a sample size of 50), we have 95% confidence that the true mean age of the population surveyed lies in the interval from 51.2 years to 57.9 years. For ci_2294 (based on a sample size of 2294), we have 95% confidence that the true mean age of the population surveyed lies in the interval from 54.3 years to 55.2 years. For ci_90 (based on a sample size of 50), we have 90% confidence that the true mean age of the population surveyed lies in the interval from 51.8 years to 57.3 years. For ci_99 (based on a sample size of 50), we have 99% confidence that the true mean age of the population surveyed lies in the interval from 50.1 years to 59 years.

7.3.2. Confidence Interval (CI) for a Single Sample Proportion

In the regicor data, 2294 individuals were surveyed and 92 had cardiovascular events. Calculate the 95% CI for the population proportion of having cardiovascular events.

Argument ctype = "prop.single," for the calculation of CI for proportions, requires the data to be in two-column matrix with 1st column holding the positive cases and 2nd column holding the negative cases

```
table(regicor$cv)
##
## No Yes
## 2071 92
# rearranging the factor levels
regicor$cv<- factor(regicor$cv, levels = c("Yes,""No"))
table(regicor$cv)
##
## Yes No
## 92 2071
dat<- matrix(table(regicor$cv), ncol = 2)
dat
## [,1] [,2]
## [1,] 92 2071
epi.conf(dat, ctype = "prop.single")
## est se lower upper
## 1 0.04253352 0.004339095 0.03480899 0.05188008
sum(is.na(regicor$cv))
## [1] 131
```

The 95% CI is between 0.035 and 0.052 or 3.5% and 5.2%. The est is the estimated proportion = 92/2294= 0.04 but the estimated value was 0.043. The difference was due to the presence of 131 missing values for cv so the calculated proportion = 92/2163 = 0.043 or 4.3% which is the same result of epi.conf() function.

Based on a sample size of 2163, we have 95% confidence that the true proportion of cardiovascular events in the population surveyed lies in the interval between 0.035 and 0.052 or 3.5% and 5.2%.

7.3.3. Confidence Interval (CI) for a Single Sample Incidence Rate (IR)

The argument ctype= "inc.rate," for the calculation of CI for IRs, requires a two-column matrix; the first column specifies the number of positives and the second column specifies the total follow-up time.

There are two methods:

1. method="exact,"from Ulm (1990). A simple method to calculate the CI of a standardized mortality ratio. American Journal of Epidemiology 131: 373 – 375.
2. method="byar,"from Rothman (2002). Epidemiology An Introduction. Oxford University Press, London, pp. 130–143.

Use the regicor data to determine the 95% CI for the population IR of cardiovascular events. Incidence rate= IR = Total number of deaths (outcomes) in a sample/Total follow up time.

```
table(regicor$cv)

##

## Yes  No

##  92 2071


# replace No by 0 and Yes by 1

regicor$cv<- ifelse(regicor$cv=="No,"0,1)

table(regicor$cv)

##

##  0  1

## 2071  92
```

```
est_rate<- sum(regicor$cv, na.rm = TRUE)/sum(regicor$tocv, na.rm =
TRUE)
est_rate
## [1] 2.424021e–05
##This is our estimated incidence rate
##To obtain the confidence interval
cv <-sum(regicor$cv, na.rm = TRUE)
time<-sum(regicor$tocv, na.rm = TRUE)
dat <- matrix(c(cv, time), ncol = 2)
dat
##  [,1] [,2]
## [1,]  92 3795347
##Ulm method
epi.conf(dat, ctype = "inc.rate,"method= "exact")
##      est   lower   upper
## 1 2.424021e–05 1.954103e–05 2.972847e–05
##Rothman method
epi.conf(dat, ctype = "inc.rate,"method = "byar")
##      est   lower   upper
## 1 2.424021e–05 1.96588e–05 2.958362e–05
```

Note that the two methods yield nearly the same interval. The 95% CI for the mortality rate in this population is between 0.006 and 0.009 deaths/person day.

To calculate the IRs for both genders:

```
regicor %>% group_by(sex) %>% summarise(cv = sum(cv, na.rm
= TRUE),
          time = sum(tocv, na.rm = TRUE))
## # A tibble: 2 x 3
## sex   cv  time
## <fct> <dbl> <dbl>
## 1 Male   46 1817701.
```

```
## 2 Female  46 1977646.
dat_male <- matrix(c(46, 1817701), ncol = 2)
dat_male
##   [,1] [,2]
## [1,]  46 1817701
dat_female <- matrix(c(46, 1977646), ncol = 2)
dat_female
##   [,1] [,2]
## [1,]  46 1977646
epi.conf(dat_male, ctype = "inc.rate,"method= "exact")
##      est    lower    upper
## 1 2.530669e–05 1.852768e–05 3.375558e–05
epi.conf(dat_female, ctype = "inc.rate,"method= "exact")
##      est    lower    upper
## 1 2.325998e–05 1.702923e–05 3.102555e–05
##To determine the sample size for each gender
table(regicor$sex)
##
##  Male Female
##  1101  1193
```

Based on a sample of 1101 males and using the exact method, we have 95% confidence that the true IR of cardiovascular events in males from this population lies in the interval from 1.9×10^{-5} to 3.4×10^{-5} events/day.

Based on a sample of 1193 females and using the exact method, we have 95% confidence that the true IR of cardiovascular events in females from this population lies in the interval from 1.7×10^{-5} to 3.1×10^{-5} events/day.

When the CIs for different groups overlap as for males and females, this means that there is no real difference of cardiovascular events rate between males and females from this population. The observed difference in IRs between males and females (2.5×10^{-5} and 2.3×10^{-5}, respectively) was due to chance or sampling error. This is a non-statistical significant result.

CHAPTER 8

CONFIDENCE INTERVALS FOR COMPARING TWO OR MORE POPULATIONS

CONTENTS

8.1. INTRODUCTION

Researchers are usually interested in comparing the outcome between two or more populations using samples collected from these populations. It is not only important to estimate the outcome difference between two or more populations (mean difference, risk difference, risk ratio, odds ratio (OR), incidence rate ratio (IRR)) but also the uncertainty in these estimates as these estimates are subjected to sampling error.

8.1.1. Types of Comparisons Used in Research

1. Paired Study Design: Where for each person or observation in the 1st sample, there is a corresponding person or observation in the 2nd sample. The two samples are of equal size and it is commonly used for pretest, posttest data.
2. Unpaired Study Design: Where there is no linkage between the two samples and the two samples are not necessarily of the same size. The unpaired design is usually used for time to event data.

8.1.2. Examples

1. **Continuous Data, Paired:**
   - Systolic blood pressure measured before and after certain exercise in some patients. While unpaired study involves systolic blood pressure measured for two patient groups randomly assigned to two different treatments.
   - Weight measured before and after certain diet in some participants. While unpaired study involves weight measured for two groups of participants, one on a certain diet and one on their own diet regimen.
2. **Categorical Data, Paired:** patient images were classified as disease or not by a laboratory test and by independent investigators. While unpaired study involves patient images were classified as disease or not by a laboratory test where patients are from two groups, one receives treatment and the other receives placebo.
3. **Time to event Data, unpaired:** incidence rates (IRs) of death for two groups of patients, one with adenocarcinoma and the other with squamous cell carcinoma.

8.2. EXTENSION OF THE CENTRAL LIMIT THEOREM (CLT)

The difference between two quantities, whose distributions are normal, has itself a normal distribution:

1. The distribution of mean difference between two samples is normal and centered at the true mean difference between the two populations from which these samples were taken.
2. The distribution of proportion difference (risk difference) between two samples is normal and centered at the true proportion difference between the two populations from which these samples were taken.
3. For ratios as RR, OR, IRR, the distribution of log ratio between two samples is normal and centered at the true log ratio between the two populations from which these samples were taken.

Note that log(a/b)=log(a)-log(b), so for ratios, we compute log ratios and calculate the confidence intervals (CI) for log ratios then transform the results back to ratio scale.

The calculation of CIs for two or more populations difference is more complicated than for single population parameters and so we rely on epiR package for these calculations.

8.3. NULL VALUES

It is a value of measurement association between two populations that indicate that the population outcome quantities are equal and so there is no association between the outcome and the population. If the null value appears in the CI, this indicates that there is no association between the outcome and the population or a non-significant result.

The null value for:

1. Mean or proportion difference = zero
2. Relative risk, odds ratio, incidence rate ratio = 1
3. log(any ratio)= zero

8.4. CONFIDENCE INTERVAL (CI) FOR COMPARING MEANS BETWEEN TWO OR MORE POPULATIONS, MEAN DIFFERENCE

8.4.1. Paired Study

Argument, ctype = "mean.paired" of the epi.conf() function requires a two-column data frame; one column for each group:

- *Example 1:* Systolic blood pressure levels were measured in 16 middle-aged men before and after a standard exercise test. The values of the blood pressure levels were stored in the variables before (before the test) and after (after the test). The test requires a dataframe with the first column before test values and second column after test values.

```
library(tidyverse)
library(epiR)
before <- c(148,142,136,134,138,140,132,144,128,170,162,150,138,154,12
6,116)
after <- c(152,152,134,148,144,136,144,150,146,174,162,162,146,156,132
,126)
dat <- data.frame(before, after)
head(dat)
## before after
## 1 148 152
## 2 142 152
## 3 136 134
## 4 134 148
## 5 138 144
## 6 140 136
epi.conf(dat, ctype = "mean.paired," conf.level = 0.95)
## est se lower upper
## 1 6.625 1.491294 3.446382 9.803618
```

The mean of these 16 differences (after-before) = 6.6 mmHg and the 95% CI = 3.4 – 9.8 mmHg.

The null value (zero) does not appear in the CI and all its values are +ve meaning that there is a significant increase in the systolic blood pressure caused by this exercise test. This increase in blood pressure can be as low as 3.4 mmHg and as high as 9.8 mmHg.

- *Example 2:* Wiebe and Bortolotti,2002 http://www.biostathandbook.com/pairedttest.html examined color in the tail feathers of northern flickers. Some of the birds had one "odd" feather that was different in color or length from the rest of the tail feathers, presumably because it was regrown after being lost. They measured the yellowness of one odd feather on each of 16 birds and compared it with the yellowness of one typical feather from the same bird.

There are two yellowness measurements, one for the odd feather and the other for the typical feather.

```
typical<-c(–0.255, –0.213, –0.190, –0.185,
    –0.045, –0.025, –0.015, 0.003,
    0.015, 0.020, 0.023, 0.040,
    0.040, 0.050, 0.055, 0.058)
odd<-c(–0.324, –0.185, –0.299, –0.144,
   –0.027, –0.039, –0.264, –0.077,
   –0.017, –0.169, –0.096, –0.330,
   –0.346, –0.191, –0.128, –0.182)
dat <- data.frame(typical, odd)
epi.conf(dat, ctype = "mean.paired," conf.level = 0.95)
##    est    se  lower   upper
## 1 –0.137125 0.03373597 –0.2090315 –0.06521848
```

The mean of these 16 differences (odd – typical) = –0.14 and the 95% CI= –0.21 to –0.065

The null value (zero) does not appear in the CI and all its values are –ve meaning that there is a significant decrease in the yellowness index in the odd feather compared to typical feather. This decrease in the yellowness index can be as high as 0.21 and as low as 0.065.

8.4.2. Unpaired Study

Argument, ctype = "mean.unpaired" of the epi.conf() function requires a two-column data frame; the first column defining the groups must be a factor.

- *Example 1:* Use the regicor dataset from compareGroups package to determine the mean age difference between males and females.

```
library(compareGroups)
data("regicor")
##Viewing the mean age of each gender

regicor %>% group_by(sex) %>% summarise(mean_age= mean(age))
## # A tibble: 2 x 2
##  sex mean_age
##  <fct>  <dbl>
## 1 Male    54.8
## 2 Female  54.7

##Use the select function to create the desired dataframe

df <- select(regicor,sex,age)
head(df)
##     sex age
## 6101 Female 70
## 5762 Female 56
## 2992  Male 37
## 2611 Female 69
## 2762 Female 70
## 1516 Female 40
epi.conf(df, ctype = "mean.unpaired," conf.level = 0.95)
##     est    se  lower  upper
## 1 0.0932072 0.4618562 –0.8124927 0.9989071
```

The mean of these differences (males to females) = 0.09 years and the 95% CI = –0.81 to 1 years.

The null value (zero) does appear in the CI meaning that the age of females is not significantly lower than the age of males.

- *Example 2*: Use the same regicor data to determine the mean physical activity difference between different smoker groups. As we stated before, when comparing more than two groups, designate one group as a reference group (never smoker) and compare other groups to it.

So we will create two CIs:

1. For mean physical activity difference between never smokers and current smokers; and
2. For mean physical activity difference between never smokers and former smokers.

```
##Viewing the mean physical activity of each group

regicor %>% group_by(smoker) %>% summarise(mean_activity =
mean(phyact, na.rm = TRUE))
## # A tibble: 4 x 2
##  smoker          mean_activity
##  <fct>             <dbl>
## 1 Never smoker          396.
## 2 Current or former < 1y    390.
## 3 Former >= 1y          421.
## 4 <NA>              376.
## to remove NA values

regicor %>% filter(!is.na(smoker)) %>% group_by(smoker) %>%
 summarise(mean_activity = mean(phyact, na.rm = TRUE))
## # A tibble: 3 x 2
##  smoker          mean_activity
##  <fct>             <dbl>
## 1 Never smoker          396.
## 2 Current or former < 1y    390.
## 3 Former >= 1y          421.
```

```
##For CI of mean physical activity difference between never smokers and current smokers

df <- regicor %>% filter(smoker %in% c("Never smoker," "Current or former < 1y")) %>%
 filter(!is.na(smoker) & !is.na(phyact)) %>% select(smoker,phyact)
head(df)
##          smoker  phyact
## 1      Never smoker 304.2000
## 2      Never smoker 160.3000
## 3 Current or former < 1y 552.7912
## 4      Never smoker 522.0000
## 5 Current or former < 1y 386.9505
## 6      Never smoker 478.2761
table(df$smoker)
##
##     Never smoker Current or former < 1y      Former >= 1y
##         1169          568          0
df$smoker<- factor(df$smoker, levels = c("Never smoker," "Current or former < 1y"))
table(df$smoker)
##
##     Never smoker Current or former < 1y
##         1169          568
epi.conf(df, ctype = "mean.unpaired," conf.level = 0.95)
##    est    se  lower  upper
## 1 6.752007 19.88475 –32.24859 45.75261
```

The mean of physical activity difference (never smokers to current smokers) = 6.75 Kcal/week. The 95% CI = –32.2 to 45.8 Kcal/week.

The null value (zero) does appear in the CI meaning that the physical activity of never smokers is not significantly different from the physical activity of current smokers. In other words, the physical activity of the two groups are equivalent on the average.

```
##For CI of mean physical activity difference between never smokers and former smokers

df <- regicor %>% filter(smoker %in% c("Never smoker," "Former >= 1y")) %>%
 filter(!is.na(smoker) & !is.na(phyact)) %>% select(smoker,phyact)
head(df)
##      smoker phyact
## 1 Never smoker 304.2000
## 2 Never smoker 160.3000
## 3 Never smoker 522.0000
## 4 Never smoker 478.2761
## 5 Never smoker 324.0000
## 6 Former >= 1y 289.0000
table(df$smoker)
##

##      Never smoker Current or former < 1y      Former >= 1y
##          1169          0          422
df$smoker<- factor(df$smoker, levels = c("Never smoker," "Former >= 1y"))
table(df$smoker)
##
## Never smoker Former >= 1y
##      1169     422
epi.conf(df, ctype = "mean.unpaired," conf.level = 0.95)
##      est     se   lower  upper
## 1 –24.21501 21.35756 –66.10696 17.67695
```

The mean of physical activity difference (never smokers to former smokers) = –24.2 Kcal/week. The 95% CI = –66.1 to 17.7 Kcal/week.

The null value (zero) does appear in the CI meaning that the physical activity of never smokers is not significantly different from the physical activity of former smokers. In other words, the physical activity of the two groups are equivalent on the average.

8.5. CONFIDENCE INTERVAL (CI) FOR COMPARING PROPORTIONS BETWEEN TWO OR MORE POPULATIONS, PROPORTION DIFFERENCE

8.5.1. Paired Study

The argument, ctype= "prop.paired" requires a 4-column matrix.

- Column 1 is the number of positives for two groups.
- Column 2 is the number of positives for 1st group but negative for the 2nd group.
- Column 3 is the number of positives for the 2nd group which are negative for the 1st group.
- Column 4 is the number of negatives for two groups.

In a reliability exercise, 41 patients were randomly selected from those who had undergone a thalium-201 stress test. The 41 sets of images were classified as normal or not by the core thallium laboratory test and, independently, by clinical investigators from different centers. Of the 19 samples identified as ischemic by clinical investigators, 14 were identified as ischemic by the laboratory. Of the 22 samples identified as normal by clinical investigators, 0 were identified as ischemic by the laboratory. Is there a difference in the diagnosis between the laboratory test and the clinicians?

```
#column 1 is the number of positives for two groups, (14).

#column 2 is the number of positives for 1st group (clinicians) but negative for the 2nd group (lab), which is 5.

#column 3 is the number of positives for the 2nd group (lab), which are negative for the 1st group (clinicians), which is 0.

#columns 4 is the number of negatives for two groups (22).

#Proportion diagnosed as ischemic by 1st group (clinicians)=19/41=0.46

#Proportion diagnosed as ischemic by 2nd group (lab)=14/41=0.34

#Proportion difference estimated=0.46–0.34=0.12

dat <- matrix(c(14, 5, 0, 22), ncol = 4)
dat
```

```
## [,1] [,2] [,3] [,4]
## [1,] 14 5 0 22
epi.conf(dat, ctype = "prop.paired," conf.level = 0.95)
## est se lower upper
## 1 0.1219512 0.05110466 0.01148324 0.2264927
##If we set 1st group as lab and 2nd group as clinicians

dat2 <- as.matrix(cbind(14, 0, 5, 22))
dat2
## [,1] [,2] [,3] [,4]
## [1,] 14 0 5 22

epi.conf(dat2, ctype = "prop.paired," conf.level = 0.95)
## est se lower upper
## 1 –0.1219512 0.05110466 –0.2264927 –0.01148324
##The same proportion difference but with reversed sign
```

The estimated proportion difference = 0.12 or 12% and the 95% CI = 0.011 to 0.226 or 1.1% to 22.6%. The null value (zero) does not appear in the CI and all its values are +ve meaning that clinicians tend to overestimate the proportion of ischemic images over that of the lab test. This overestimation can be as low as 1% and as high as 22.6%. This means that clinicians diagnosis of ischemic images may not be reliable.

8.5.2. Unpaired Study

The argument, ctype = "prop.unpaired," requires a 4-column matrix.

- Column 1 is the number of positives for 1st group.
- Column 2 is the number of negatives for 1st group.
- Column 3 is the number of positives for the 2nd group.
- Column 4 is the number of negatives for the 2nd group.
- Using the regicor dataset, does the smoking status affects the death proportion?. As we stated before, when comparing more than two groups, designate one group as a reference group (never smoker) and compare other groups to it.

We have two proportion differences:

1. The proportion difference in deaths between never smokers and current smokers; and
2. The proportion difference in deaths between never smokers and former smokers.

The proportion difference in deaths between never smokers and current smokers.

```
##Viewing the numbers

table(regicor$smoker, regicor$death)
##

##              No Yes
## Never smoker      1052  86
## Current or former < 1y 488  69
## Former >= 1y      397  18
##First proportion difference

#column 1 is the number of deaths for never smokers which is 86.

#column 2 is the number of lives for never smokers which is 1052.

#column 3 is the number of deaths for current smokers which is 69.

#columns 4 is the number of lives for current smokers which is 488.

#Proportion of deaths in never smokers = 86/(86+1052) = 0.076

#Proportion with deaths in current smokers = 69/(69+488) = 0.124

#Proportion difference estimated = 0.076–0.124 = –0.048

dat <- matrix(c(86,1052,69,488), ncol = 4)
dat
##  [,1] [,2] [,3] [,4]
## [1,]  86 1052  69 488
epi.conf(dat, ctype = "prop.unpaired," conf.level = 0.95)
##     est    se   lower    upper
```

```
## 1 –0.04830674 0.01600749 –0.08136584 –0.01832961
```

The proportion difference in deaths between never smokers and current smokers.

The estimated proportion difference = –0.048 or –4.8% and the 95% CI = –0.08 to –0.018 or –8% to –1.8%.

The null value (zero) does not appear in the CI and all its values are –ve meaning that there is a real difference between never smokers and current smokers in death proportion with never smokers having lower proportion of deaths than current smokers by 4.8%. The % of deaths in never smokers is lower than that of current smokers by as low as 1.8% and as high as 8%.

In a group of 1000 never smokers and another group of 1000 current smokers from Girona (the place of regicor survey), we expect to see 48 (4.8% × 1000) fewer cases of deaths in the never smoker group compared to the current smoker group. These fewer cases can be anywhere from 18 to 81 fewer cases than the current smoker group.

The proportion difference in deaths between never smokers and former smokers.

```
##Viewing the numbers

table(regicor$smoker, regicor$death)
##
##              No Yes
## Never smoker      1052  86
## Current or former < 1y 488  69
## Former >= 1y      397  18
##First proportion difference

#column 1 is the number of deaths for never smokers which is 86.

#column 2 is the number of lives for never smokers which is 1052.

#column 3 is the number of deaths for former smokers which is 18.

#columns 4 is the number of lives for former smokers which is 397.

#Proportion of deaths in never smokers = 86/(86+1052) = 0.076

#Proportion with deaths in former smokers = 18/(18 + 397) = 0.043
```

```
#Proportion difference estimated = 0.076–0.043 = 0.033

dat <- matrix(c(86,1052,18,397), ncol = 4)
dat
##  [,1] [,2] [,3] [,4]
## [1,] 86 1052  18 397
epi.conf(dat, ctype = "prop.unpaired," conf.level = 0.95)
##    est    se   lower  upper
# 1 0.03219768 0.01270314 0.004307077 0.05525384
```

The proportion difference in deaths between never smokers and current smokers.

The estimated proportion difference = 0.032 or 3.2% and the 95% CI = 0.004 to 0.055 or 0.4% to 5.5%.

The null value (zero) does not appear in the CI and all its values are positive meaning that there is a real difference between never smokers and former smokers in death proportion with former smokers having lower proportion of deaths than never smokers by 3.2%. The % of deaths in the former smokers is lower than that of never smokers by as low as 0.4% and as high as 5.5%. However, we can argue that as the lower bound of CI approaches zero, this significant difference is not of practical usefulness.

In a group of 1000 never smokers and another group of 1000 former smokers from Girona (the place of regicor survey), we expect to see 32 (3.2% × 1000) fewer cases of deaths in the former smoker group compared to the never smoker group. These fewer cases can be anywhere from 4 to 55 fewer cases than the never smoker group.

8.6. CONFIDENCE INTERVAL (CI) FOR COMPARING PROPORTIONS BETWEEN TWO OR MORE POPULATIONS, RELATIVE RISK (RR) AND ODDS RATIO (OR)

The CIs for risk difference, RR, and OR should agree in terms of including or not including the Null value.

We can use the epi.2by2() function from the epiR package to calculate these CIs.

The epi.2by2() function requires 2 by 2 table format as follows:

1. The first row for exposure +ve and second row for exposure –ve
2. The first column for outcome +ve and second column for outcome –ve

Using the regicor dataset, does the smoking status affects the death proportion?. As we stated before, when comparing more than two groups, designate one group as a reference group (never smoker) and compare other groups to it.

We have two proportion differences:

1. The proportion difference in deaths between never smokers and current smokers.
2. The proportion difference in deaths between never smokers and former smokers.

The proportion difference in deaths between never smokers and current smokers.

```
##Viewing the numbers

table(regicor$smoker, regicor$death)
##
##                    No Yes
## Never smoker       1052  86
## Current or former < 1y 488  69
## Former >= 1y       397  18
#the number of current smokers (exposure+ve) having deaths (outcome+ve) is 69.

#the number of never smokers (exposure–ve) having deaths (outcome+ve) is 86.

#the number of current smokers (exposure+ve) having lives (outcome–ve) is 488.

#the number of never smokers (exposure–ve) having lives (outcome–ve) is 1052.

##The confidence interval for proportion difference
```

```
dat <- matrix(c(69,488,86,1052), ncol = 4)
dat
##   [,1] [,2] [,3] [,4]
## [1,]  69 488  86 1052
epi.conf(dat, ctype = "prop.unpaired," conf.level = 0.95)
##     est    se   lower   upper
## 1 0.04830674 0.01600749 0.01832961 0.08136584
```

##The confidence interval for relative risk and odds ratio

```
dat <- matrix(c(69,86,488,1052), ncol = 2)
dat
##   [,1] [,2]
## [1,]  69 488
## [2,]  86 1052
epi.2by2(dat = as.table(dat))
##        Outcome +  Outcome –  Total   Inc risk *   Odds
## Exposed +     69     488    557      12.39   0.1414
## Exposed –     86     1052   1138      7.56   0.0817
## Total       155     1540    1695      9.14   0.1006
##
## Point estimates and 95% CIs:
## -------------------------------------------------------------------
## Inc risk ratio               1.64 (1.21, 2.21)
## Odds ratio                  1.73 (1.24, 2.42)
## Attrib risk *               4.83 (1.69, 7.97)
## Attrib risk in population *         1.59 (–0.47, 3.65)
## Attrib fraction in exposed (%)        39.00 (17.64, 54.81)
## Attrib fraction in population (%)      17.36 (5.50, 27.73)
## -------------------------------------------------------------------
## Test that OR = 1: chi2(1) = 10.503 Pr>chi2 = 0.00
## Wald confidence limits
```

CI: confidence interval

* Outcomes per 100 population units

Comparing death proportions between current smokers and never smokers.

The estimated proportion difference (Attrib risk) = 4.83% and the 95% CI = 1.69% to 7.97%. The absolute decrease in deaths for never smokers compared to current smokers can be as low as 1.7% and as high as 8%. Note that it is the same result as that of epi.conf() function.

The estimated RR (Incidence risk ratio)= 1.64 and the 95% CI= 1.21 to 2.21. For a current smoker individual in Girona, he will have a higher risk of death by (1.64- 1=0.64) or 64% compared to a never smoker individual. This increase in risk can be as low as (1.21–1=0.21) or 21% and as high as (2.21–1 = 1.21) or 121%. In other words, the death risk for current smokers is 1.64 times the death risk for never smokers.

The estimated odds ratio = 1.73 and the 95% CI = 1.24 to 2.42. For a current smoker individual in Girona, he will have a higher odds of death by (1.73 – 1= 0.73) or 73% compared to a never smoker individual. This increase in odds can be as low as (1.24–1=0.24) or 24% and as high as (2.42–1 = 1.42) or 142%. In other words, the odds of death for current smokers is 1.73 times the death odds for never smokers.

The proportion difference in deaths between never smokers and former smokers.

```
##Viewing the numbers

table(regicor$smoker, regicor$death)
##
##                  No Yes
## Never smoker      1052  86
## Current or former < 1y 488  69
## Former >= 1y      397  18
#the number of former smokers (exposure+ve) having deaths (outcome+ve) is 18.

#the number of never smokers (exposure–ve) having deaths (outcome+ve) is 86.
```

#the number of former smokers (exposure+ve) having lives (outcome–ve) is 397.

#the number of never smokers (exposure–ve) having lives (outcome–ve) is 1052.

##The confidence interval for proportion difference

```
dat <- matrix(c(18,397,86,1052), ncol = 4)
dat
##  [,1] [,2] [,3] [,4]
## [1,]  18 397  86 1052
epi.conf(dat, ctype = "prop.unpaired," conf.level = 0.95)
##     est     se    lower    upper
## 1 –0.03219768 0.01270314 –0.05525384 –0.004307077
```

##The confidence interval for relative risk and odds ratio

```
dat <- matrix(c(18,86,397,1052), ncol = 2)
dat
##  [,1] [,2]
## [1,]  18 397
## [2,]  86 1052
epi.2by2(dat = as.table(dat))
##       Outcome +  Outcome –  Total   Inc risk *   Odds
## Exposed +     18     397    415      4.34   0.0453
## Exposed –     86     1052    1138      7.56   0.0817
## Total      104     1449    1553      6.70   0.0718
##
## Point estimates and 95% CIs:
## -------------------------------------------------------------------
## Inc risk ratio            0.57 (0.35, 0.94)
## Odds ratio              0.55 (0.33, 0.93)
## Attrib risk *            –3.22 (–5.71, –0.73)
## Attrib risk in population *       –0.86 (–2.84, 1.12)
```

```
## Attrib fraction in exposed (%)      -74.23 (-185.95, -6.16)
## Attrib fraction in population (%)    -12.85 (-23.03, -3.51)
## -------------------------------------------------------------
## Test that OR = 1: chi2(1) = 5.046 Pr>chi2 = 0.02
## Wald confidence limits
## CI: confidence interval
## * Outcomes per 100 population units
```

Comparing death proportions between former smokers and never smokers.

The estimated proportion difference (Attrib risk) = –3.22% and the 95% CI = –5.71% to –0.73%. The absolute decrease in deaths for former smokers compared to never smokers can be as low as 0.73% and as high as 5.71%. Note that it is nearly the same result as that of epi.conf() function. The difference because the epi.conf() function uses Altman et al. method (2000). Statistics with Confidence, second edition. British Medical Journal, London, pp. 28, 29 and pp. 45–56.). While the epi.2by2() function uses the Wald (1943). Tests of statistical hypotheses concerning several parameters when the number of observations is large. Transactions of the American Mathematical Society (*54*, 426–482) and score methods (Miettinen and Nurminen, 1985). Comparative analysis of two rates (Statistics in Medicine, *4*, 213–226).The estimated RR (Incidence risk ratio)= 0.57 and the 95% CI= 0.35 to 0.94. For a former smoker individual in Girona, he will have a lower risk of death by (1- 0.57 = 0.43) or 43% compared to a never smoker individual. This decrease in death risk can be as low as (1–0.94= 0.06) or 6% and as high as (1–0.35 = 0.65) or 65%. In other words, the death risk for former smokers is 0.57 (about half) times the death risk for never smokers.

The estimated odds ratio = 0.55 and the 95% CI = 0.33 to 0.93. For a former smoker individual in Girona, he will have a lower odds of death by (1 – 0.55= 0.45) or 45% compared to a never smoker individual. This decrease in odds can be as low as (1–0.93=0.07) or 7% and as high as (1–0.33 = 0.67) or 67%. In other words, the odds of death for former smokers is 0.55 times the death odds for never smokers.

Note that risk difference, RR, and OR agree in terms of association direction. As former smokers are associated with a decreased death proportion, so risk difference will be negative and RR and OR will be less than 1.

8.7. CONFIDENCE INTERVAL (CI) FOR COMPARING INCIDENCE RATE (IR) BETWEEN TWO OR MORE POPULATIONS, INCIDENCE RATE RATIOS (IRRS)

We will use the epi.2by2() function from the epiR package with method="cohort.time" to calculate these CIs

The epi.2by2 function requires 2 by 2 table format as follows:

1. The first row for exposure +ve and second row for exposure –ve;
2. The first column for outcome +ve and second column for the total follow-up time.

8.7.1. Regicor Data

Using regicor data, does gender affects the mortality rate?

```
# replace "No" with 0 and "Yes" with 1 for death column

table(regicor$death)
##
##  No Yes
## 1975 173
regicor$death<- ifelse(regicor$death == "No,"0,1)
table(regicor$death)
##
##  0  1
## 1975 173
##Viewing the deaths and survival time for each gender

regicor %>% group_by(sex) %>% summarise(death= sum(death, na.rm = TRUE),
              time=sum(todeath, na.rm = TRUE))
## # A tibble: 2 x 3
##  sex  death   time
##  <fct> <dbl>  <dbl>
## 1 Male   87 1727482.
## 2 Female  86 1969882.
```

```
##The number of deaths in males = 87

##The survival time in males = 1727482

##The number of deaths in females = 86

##The survival time in females = 1969882

dat <- (matrix(c(87,86,1727482,1969882), ncol = 2))
dat
##   [,1] [,2]
## [1,]  87 1727482
## [2,]  86 1969882
epi.2by2(dat = as.table(dat), method = "cohort.time")
##        Outcome +  Time at risk    Inc rate *
## Exposed +      87     1727482      0.00504
## Exposed –      86     1969882      0.00437
## Total       173     3697364      0.00468
##
## Point estimates and 95% CIs:
## -------------------------------------------------------------------
## Inc rate ratio                 1.15 (0.85, 1.57)
## Attrib rate *                0.00 (–0.00, 0.00)
## Attrib rate in population *          0.00 (–0.00, 0.00)
## Attrib fraction in exposed (%)        13.31 (–18.16, 36.41)
## Attrib fraction in population (%)      6.70 (0.72, 12.87)
## -------------------------------------------------------------------
## Wald confidence limits
## CI: confidence interval
## * Outcomes per 100 units of population time at risk
```

The estimated Incidence rate ratio = 1.15 and the 95% CI = 0.85 to 1.57.

The null value (1 for ratios) does appear in the CI, meaning that there is no real difference between the two gender groups in the mortality rate, even though males have higher IR of death (0.00504) than that of females (0.00437). The observed ratio, that is larger than 1 (1.15), was due to sampling variability or chance and not due to a real difference.

Using regicor data, does history of hypertension (histhtn) affects the mortality rate.

```
##Viewing the deaths and survival time for each gender

regicor %>% group_by(histhtn) %>% summarise(death= sum(death, na.rm = TRUE),

                time=sum(todeath, na.rm = TRUE))
## # A tibble: 3 x 3
##  histhtn death   time
##  <fct>  <dbl>  <dbl>
## 1 Yes    76 1144045.
## 2 No     97 2538407.
## 3 <NA>    0  14911.
##The number of deaths in individuals with hypertension history = 76

##The survival time in individuals with hypertension history = 1144045

##The number of deaths in individuals with no hypertension history = 97

##The survival time in individuals with no hypertension history = 2538407

dat <- (matrix(c(76, 97, 1144045,2538407), ncol = 2))
dat
##  [,1] [,2]
## [1,]  76 1144045
## [2,]  97 2538407
epi.2by2(dat = as.table(dat), method = "cohort.time")
##        Outcome +  Time at risk   Inc rate *
## Exposed +     76     1144045     0.00664
## Exposed -     97     2538407     0.00382
```

```
## Total      173    3682452    0.00470
##
## Point estimates and 95% CIs:
## -------------------------------------------------------------------
## Inc rate ratio                    1.74 (1.27, 2.37)
## Attrib rate *                     0.00 (0.00, 0.00)
## Attrib rate in population *       0.00 (–0.00, 0.00)
## Attrib fraction in exposed (%)    42.48 (21.28, 57.84)
## Attrib fraction in population (%) 18.66 (14.51, 22.99)
## -------------------------------------------------------------------
## Wald confidence limits
## CI: confidence interval
## * Outcomes per 100 units of population time at risk
```

The estimated Incidence rate ratio = 1.74 and the 95% CI= 1.27 to 2.37.

The null value (1 for ratios) does not appear in the CI, meaning that there is a real difference between the two groups in the mortality rate.

The IR of death in individuals with hypertension history was 1.74 times the IR of death in individuals with no hypertension history. Individuals, from Girona, with hypertension history have (1.74–1=0.74) 74% greater risk of death compared to individuals with no such history. After accounting for sampling variability, this increase in death risk can be as low as 27% (1.27–1=0.27) and as high as 137% (2.37–1= 1.37).

8.7.2. Veteran Data

Using veteran dataset from the survival package, does cancer cell type affect the mortality rate?

The celltype in this dataset has four categories: squamous; smallcell; adeno; and large. As we stated before, we must design a reference group and compare all other groups to it. If we assume that our reference group was squamous so we have three IRRs:

- Incidence rate ratio of smallcell to squamous;
- Incidence rate ratio of adeno to squamous; and
- Incidence rate ratio of large to squamous.

1. Incidence rate ratio of smallcell to squamous:

```
library(survival)
data("veteran")
##Viewing the deaths and survival time for each group

veteran %>% group_by(celltype) %>% summarise(death=
sum(status),time=sum(time))
## # A tibble: 4 x 3
##  celltype death time
##  <fct>  <dbl> <dbl>
## 1 squamous  31 7007
## 2 smallcell  45 3440
## 3 adeno   26 1731
## 4 large   26 4485
##The number of deaths in smallcell = 45

##The survival time in smallcell = 3440

##The number of deaths in squamous = 31

##The survival time in squamous = 7007

dat <- matrix(c(45,31,3440,7007), ncol = 2)
dat
##  [,1] [,2]
## [1,]  45 3440
## [2,]  31 7007
epi.2by2(dat = as.table(dat), method = "cohort.time")
##       Outcome +  Time at risk   Inc rate *
## Exposed +     45     3440      1.308
## Exposed -     31     7007      0.442

## Total      76     10447     0.727
##
## Point estimates and 95% CIs:
```

```
## ---------------------------------------------------------------
## Inc rate ratio              2.96 (1.83, 4.83)
## Attrib rate *              0.87 (0.45, 1.28)
## Attrib rate in population *        0.29 (0.06, 0.51)
## Attrib fraction in exposed (%)       66.18 (45.37, 79.31)
## Attrib fraction in population (%)     39.19 (31.03, 47.56)
## ---------------------------------------------------------------
## Wald confidence limits
## CI: confidence interval
## * Outcomes per 100 units of population time at risk
## Smallcell is significantly larger than squamous in mortality rate
```

The estimated Incidence rate ratio = 2.96 and the 95% CI = 1.83 to 4.83.

The null value (1 for ratios) does not appear in the CI, meaning that there is a real difference between the two groups in the mortality rate. The IR of death in the smallcell group was 2.96 times the IR of death in the squamous group. Smallcell type cancer was associated with (2.96–1=1.96) 196% greater risk of death compared to squamous cancer type. After accounting for sampling variability, this increase in death risk can be as low as 83% (1.83–1=0.83) and as high as 383% (4.83–1=3.83).

2. **Incidence Rate Ratio of Adeno to Squamous:**

```
##Viewing the deaths and survival time for each group

veteran %>% group_by(celltype) %>% summarise(death= sum(status),time=sum(time))
## # A tibble: 4 x 3
##  celltype death time
##  <fct>  <dbl> <dbl>
## 1 squamous   31 7007
## 2 smallcell  45 3440
## 3 adeno    26 1731
## 4 large    26 4485
```

```
##The number of deaths in adeno = 26

##The survival time in adeno = 1731

##The number of deaths in squamous = 31

##The survival time in squamous = 7007

dat <- matrix(c(26,31, 1731, 7007), ncol = 2)
dat
##   [,1] [,2]
## [1,]  26 1731
## [2,]  31 7007
epi.2by2(dat = as.table(dat), method = "cohort.time")

##       Outcome +  Time at risk    Inc rate *

## Exposed +     26      1731       1.502
## Exposed –     31      7007       0.442
## Total         57      8738       0.652

##
## Point estimates and 95% CIs:
## ----------------------------------------------------------------
## Inc rate ratio                      3.40 (1.94, 5.91)
## Attrib rate *                       1.06 (0.46, 1.66)
## Attrib rate in population *         0.21 (–0.02, 0.44)
## Attrib fraction in exposed (%)      70.55 (48.35, 83.08)
## Attrib fraction in population (%)   32.18 (25.70, 39.16)
## ----------------------------------------------------------------
## Wald confidence limits
## CI: confidence interval
## * Outcomes per 100 units of population time at risk
## Adeno is significantly larger than squamous in mortality rate
```

The estimated Incidence rate ratio = 3.40 and the 95% CI= 1.94 to 5.91.

The null value (1 for ratios) does not appear in the CI, meaning that there is a real difference between the two groups in the mortality rate. The IR of death in the adeno cancer group was 3.40 times the IR of death in the squamous group. Adeno type cancer was associated with (3.40–1=2.40) 240% greater risk of death compared to squamous cancer type. After accounting for sampling variability, this increase in death risk can be as low as 94% (1.94–1=0.94) and as high as 491% (5.91–1=4.91).

3. Incidence Rate Ratio of Large to Squamous:

```
##Viewing the deaths and survival time for each group

veteran %>% group_by(celltype) %>% summarise(death= sum(status),time=sum(time))
## # A tibble: 4 x 3
##  celltype death time
##  <fct>  <dbl> <dbl>
## 1 squamous   31 7007
## 2 smallcell  45 3440
## 3 adeno    26 1731
## 4 large    26 4485
##The number of deaths in large = 26

##The survival time in large = 4485

##The number of deaths in squamous = 31

##The survival time in squamous = 7007

dat <- matrix(c(26,31,4485, 7007), ncol = 2)
dat
##  [,1] [,2]
## [1,]  26 4485
## [2,]  31 7007
epi.2by2(dat = as.table(dat), method = "cohort.time")
##      Outcome +  Time at risk   Inc rate *
## Exposed +     26     4485      0.580
```

```
## Exposed –     31     7007     0.442
## Total      57     11492     0.496
##
## Point estimates and 95% CIs:
## -------------------------------------------------------------------
## Inc rate ratio                    1.31 (0.75, 2.28)
## Attrib rate *                     0.14 (–0.13, 0.41)
## Attrib rate in population *            0.05 (–0.15, 0.26)
## Attrib fraction in exposed (%)         23.68 (–33.82, 56.15)
## Attrib fraction in population (%)      10.80 (2.28, 19.98)
## -------------------------------------------------------------------
## Wald confidence limits
## CI: confidence interval
## * Outcomes per 100 units of population time at risk
##large is not significantly different from squamous
```

The estimated Incidence rate ratio = 1.31 and the 95% CI= 0.75 to 2.28. The null value (1 for ratios) does appear in the CI, meaning that there is no real difference between the two groups in the mortality rate, even though the large cancer group has higher IR (0.58) than that of the squamous cancer group (0.442).

The observed ratio, that is more than 1 (1.31), was due to sampling variability or chance and not due to a real difference.

Final conclusion, the most lethal cancer type is adeno (IRR=3.40) followed by smallcell (IRR=2.96) then large and squamous which are equivalent in their mortality rates.

CHAPTER 9

HYPOTHESIS TESTING FOR COMPARING MEANS

CONTENTS

9.1. INTRODUCTION TO HYPOTHESIS TESTING

It is not only important to estimate the outcome difference between two or more populations (mean difference, risk difference, risk ratio, OR, incidence rate ratio (IRR)) but also the uncertainty in these estimates as these estimates are subjected to sampling error.

There are two approaches to estimate this uncertainty:

1. Confidence intervals (CI), discussed in chapters 7 and 8, where you let your data take you to the truth (population parameters) by adding uncertainty bounds around estimates that are based on the data.

2. Hypothesis testing, where you start with two exclusive possibilities for the unknown truth(population parameters). Then, use the data of the sample to choose between these two possibilities for the truth. The two possibilities are: Null hypothesis, Ho, and Alternative hypothesis, Ha.

Null hypothesis, Ho: There is no difference between the two populations and the difference = zero or Null value.

Alternative hypothesis, Ha: There is difference between the two populations so the difference zero.

9.1.1. Different Forms of Hypothesis Testing

1. Mean difference:

   Ho: or

   Ha: or

2. Risk difference:

   Ho: p1 = p2 or p1-p2 = 0

   Ha: p1 or

3. Relative risk:

   Ho: or

   Ha: or log

4. Odds ratio:

   Ho: or

   Ha: or

5. Incidence rate ratio:

   Ho: or

   Ha: or

9.1.2. Steps of Hypothesis Testing

1. Specify the two competing hypotheses, null, and alternative.
2. Assume the Null hypothesis to be the truth or a zero difference. Also, we know how estimates vary around the Null value; For example, 95% of all estimates fall within 2 standard errors (SEs) from this Null value.
3. Compute how far our sample estimate is from the zero difference under the Null hypothesis.
4. Translate this distance into a p-value. p-value is the probability of our sample results under the Null hypothesis. p-value is used for differentiating real differences from sampling errors in the data collected. Generally in research, the cut off used is 0.05 and it corresponds to 95% CI. This 0.05 is called the rejection level, level, or significance level.
5. Make a decision, accept Ha or fail to reject Ho.

9.1.3. Different p-Values

If p-value < 0.05, so it is a statistically significant result at 0.05 level, and we reject the Null hypothesis. Our sample data are unlikely under the Ho, they have a probability of less than 0.05.

If p-value >= 0.05, so it is a statistically nonsignificant result at 0.05 level, and we fail to reject the Null hypothesis. We say fail to reject the Null hypothesis because if we have a p-value of 0.25. This means that our sample data have a probability of 25% under the Null hypothesis which is considered a large percentage. In your opinion, you may consider it small and accept Ha.

This has led some statisticians to consider abandoning the use of p-values for significance. See this article from [Nature Journal] https://www.nature.com/articles/d41586-019-00874-8 for an interesting discussion.

There are numerous tests, explored in the following sections, that carry out these steps for us and provide the p-value required for a decision.

9.1.4. Relation Between 95% Confidence Intervals (CI), Null Value, and p-Value

When p-value < 0.05, so it is a statistically significant result at 0.05 level. This means that our sample results have a probability of less than 5% under the Ho, and the 95% CI for the estimate will not include the Null value.

When p-value >= 0.05, so it is a statistically nonsignificant result at 0.05 level. This means that our sample results have a probability of more than 5% under the Ho, and the 95% CI for the estimate will include the Null value.

9.2. HYPOTHESIS TESTING FOR COMPARING MEANS BETWEEN TWO POPULATIONS

9.2.1. Paired t-Test

Paired t-test is used for paired data and uses the t-distribution to calculate the p-value. t-distribution is like the standard normal distribution (with mean = 0) but with more flatter tails for small sample sizes.

Assumptions for paired t-test

1. Paired data;
2. Large data >30 or check the normality of differences with Shapiro-Wilk test.

   If the data failed the normality test, we can use the paired Wilcoxon test
3. Equal variances of the two groups. This can be checked with F-test. The default value in t-test is unequal variances

Generally, all these tests require a dataframe with the first column for grouping factor and second column for numerical values

We will use the same examples of chapter 8 so you can compare between CI results and hypothesis testing results

- *Example 1:* Systolic blood pressure levels were measured in 16 middle-aged men before and after a standard exercise test. The values of the blood pressure levels were stored in the variables before (before the test) and after (after the test).

Check normality of the difference

```
library(tidyverse)
```

```
before <- c(148,142,136,134,138,140,132,144,128,170,162,150,138,154,12
6,116)
after <- c(152,152,134,148,144,136,144,150,146,174,162,162,146,156,132
,126)
d<-after-before
shapiro.test(d)
##
## Shapiro-Wilk normality test
##
## data: d
## W = 0.98613, p-value = 0.9941
```

Shapiro-Wilk test gives a p-value >0.05, which means that the differences are normally distributed.

Checking equal variance in the two groups.

```
##Creating the dataframe

dat <- data.frame(before, after)
dat<-dat %>% gather(key= type,value = Bl.p.)
dat
##   type Bl.p.
## 1 before  148
## 2 before  142
## 3 before  136
## 4 before  134
## 5 before  138
## 6 before  140
## 7 before  132
## 8 before  144
## 9 before  128
## 10 before  170
## 11 before  162
```

```
## 12 before  150
## 13 before  138
## 14 before  154
## 15 before  126
## 16 before  116
## 17 after  152
## 18 after  152
## 19 after  134
## 20 after  148
## 21 after  144
## 22 after  136
## 23 after  144
## 24 after  150
## 25 after  146
## 26 after  174
## 27 after  162
## 28 after  162
## 29 after  146
## 30 after  156
## 31 after  132
## 32 after  126
##Checking equal variance

var.test(Bl.p. ~ type, data = dat)
##
## F test to compare two variances
##
## data: Bl.p. by type
## F = 0.82155, num df = 15, denom df = 15, p-value = 0.7084
## alternative hypothesis: true ratio of variances is not equal to 1
```

95 percent confidence interval:

0.2870465 2.3513626

sample estimates:

ratio of variances

0.8215537

F-test gives a p-value >0.05, which means that the two variances are equal: Numerical summaries and visual displays.

##Visualizing the data

```
ggplot(data = dat, aes(x = type,y=Bl.p., col = type))+geom_boxplot()
```

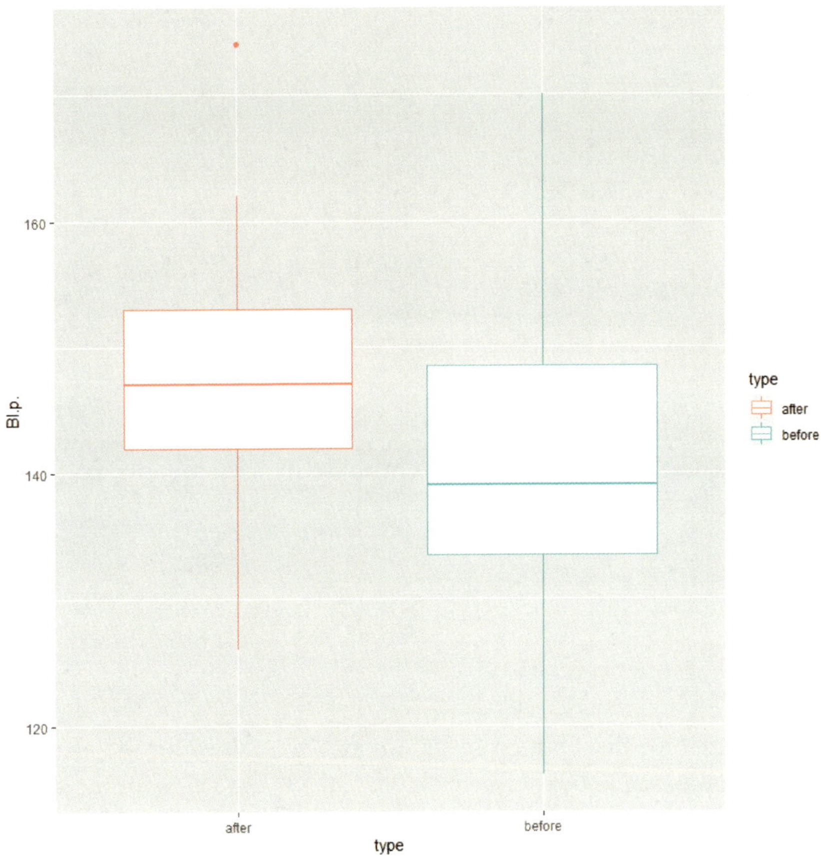


```
ggplot(data = dat, aes(x = Bl.p., col = type))+geom_density()
```

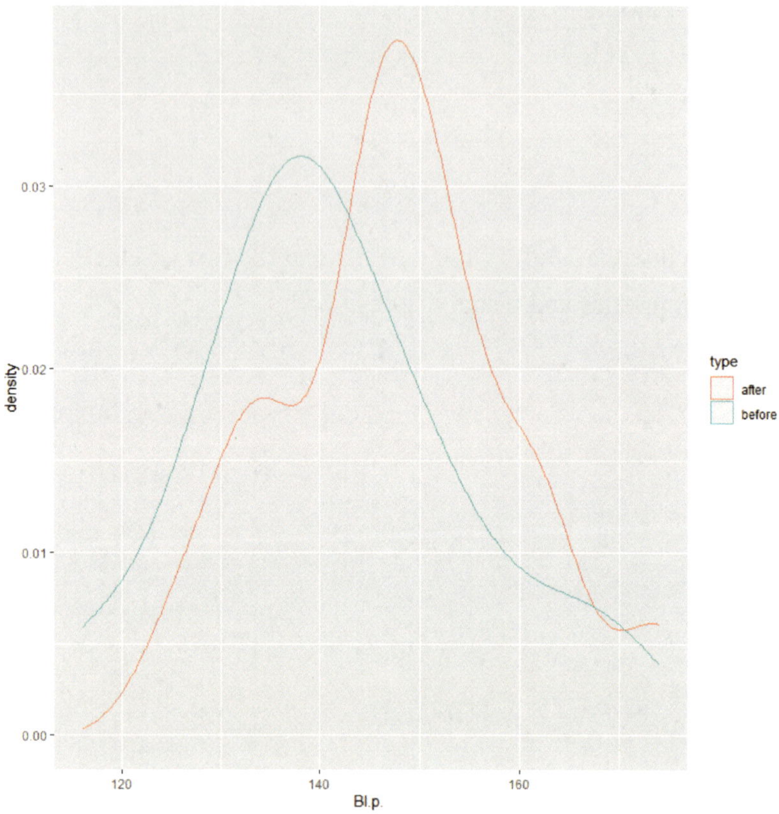

##Calculating the mean for each group

```
dat %>% group_by(type) %>% summarise(mean = mean(Bl.p.))
## # A tibble: 2 x 2
##  type  mean
##  <chr> <dbl>
## 1 after  148.
## 2 before 141.
```

We see that the distribution of blood pressure values for the after test is shifted up compared to the before test. Also, the mean value for the after test is greater than the mean value for the before test.

Conducting the t.test

alternative hypothesis "two.sided" which is default

```
t.test(Bl.p. ~ type, data = dat, paired=TRUE, var.equal=TRUE)
```

```
##
## Paired t-test
##
## data: Bl.p. by type
## t = 4.4425, df = 15, p-value = 0.0004748
## alternative hypothesis: true difference in means is not equal to 0
## 95 percent confidence interval:
## 3.446382 9.803618
## sample estimates:
## mean of the differences
##        6.625
## alternative hypothesis "greater" i.e. after greater than before

t.test(Bl.p. ~ type, data = dat, paired=TRUE, alternative="greater," var.equal=TRUE)
##
## Paired t-test
##
## data: Bl.p. by type
## t = 4.4425, df = 15, p-value = 0.0002374
## alternative hypothesis: true difference in means is greater than 0
## 95 percent confidence interval:
## 4.010686   Inf
## sample estimates:
## mean of the differences
##        6.625
levels(factor(dat$type))
## [1] "after" "before"
```

In R, the "after" level is the first level for the column "type" so we specify the alternative as "greater" meaning that after test values are greater than before test values.

Assuming that is the mean for after test values and is the mean for before test values

For two-sided test,

Null hypothesis, Ho: or

Alternative hypothesis, Ha: or

For one-sided test, alternative="greater"

Null hypothesis, Ho: or

Alternative hypothesis, Ha: or

t-test gives a p-value <0.05 in both cases (two-sided and one-sided test), which means the after test blood pressure measurements are greater than before test measurements by 6.6 on the average.

Alternative hypothesis ="two.sided" is generally used in research.

- *Example 2:* Wiebe and Bortolotti,2002 http://www.biostathandbook.com/pairedttest.html examined color in the tail feathers of northern flickers. Some of the birds had one "odd" feather that was different in color or length from the rest of the tail feathers, presumably because it was regrown after being lost. They measured the yellowness of one odd feather on each of 16 birds and compared it with the yellowness of one typical feather from the same bird. Therefore, there are two yellowness measurements, one for the odd feather and the other for the typical feather.

Check normality of the difference.

```
typical<-c(–0.255, –0.213, –0.190, –0.185,
    –0.045, –0.025, –0.015, 0.003,
    0.015, 0.020, 0.023, 0.040,
    0.040, 0.050, 0.055, 0.058)

odd<-c(–0.324, –0.185, –0.299, –0.144,
  –0.027, –0.039, –0.264, –0.077,
  –0.017, –0.169, –0.096, –0.330,
  –0.346, –0.191, –0.128, –0.182)
d2<-typical-odd
shapiro.test(d2)
```

```
##
## Shapiro-Wilk normality test
##
## data: d2
## W = 0.93987, p-value = 0.3474
```

Shapiro-Wilk test gives a p-value >0.05, which means that the differences are normally distributed.

Checking equal variance in the two groups.

```
##Creating the dataframe

dat2 <- data.frame(typical,odd)
dat2<-dat2 %>% gather(key=type,value = yellowness)
dat2
##    type yellowness
## 1 typical   –0.255
## 2 typical   –0.213
## 3 typical   –0.190
## 4 typical   –0.185
## 5 typical   –0.045
## 6 typical   –0.025
## 7 typical   –0.015
## 8 typical   0.003
## 9 typical   0.015
## 10 typical   0.020
## 11 typical   0.023
## 12 typical   0.040
## 13 typical   0.040
## 14 typical   0.050
## 15 typical   0.055
## 16 typical   0.058
## 17   odd   –0.324
## 18   odd   –0.185
```

```
## 19  odd  -0.299
## 20  odd  -0.144
## 21  odd  -0.027
## 22  odd  -0.039
## 23  odd  -0.264
## 24  odd  -0.077
## 25  odd  -0.017
## 26  odd  -0.169
## 27  odd  -0.096
## 28  odd  -0.330
## 29  odd  -0.346
## 30  odd  -0.191
## 31  odd  -0.128
## 32  odd  -0.182
var.test(yellowness ~ type, data = dat2)
##
## F test to compare two variances
##
## data: yellowness by type
## F = 1.0641, num df = 15, denom df = 15, p-value = 0.9059
## alternative hypothesis: true ratio of variances is not equal to 1
## 95 percent confidence interval:
## 0.3717825 3.0454840
## sample estimates:
## ratio of variances
##      1.064076
```

F-test gives a p-value >0.05, which means that the two variances are equal.

Numerical summaries and visual displays.

```
##Visualizing the data

ggplot(data=dat2, aes(x=type,y=yellowness, col=type))+geom_boxplot()
```

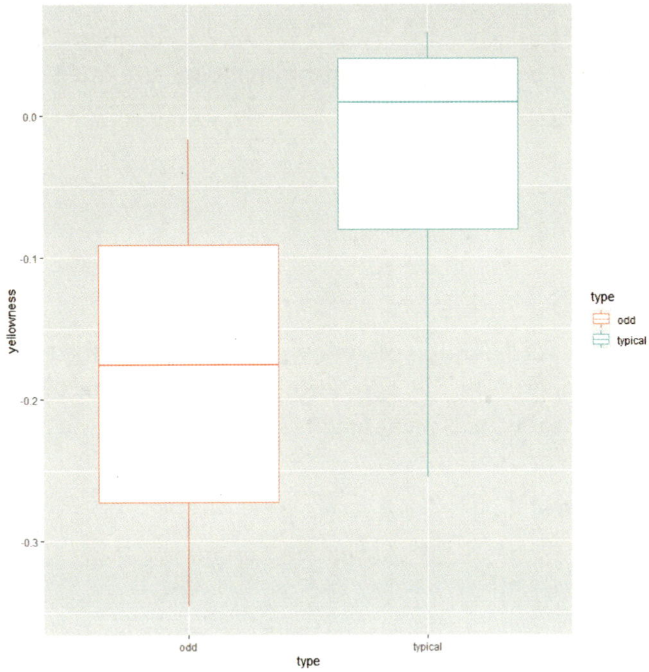


```
ggplot(data = dat2, aes(x = yellowness, col = type))+geom_density()
```

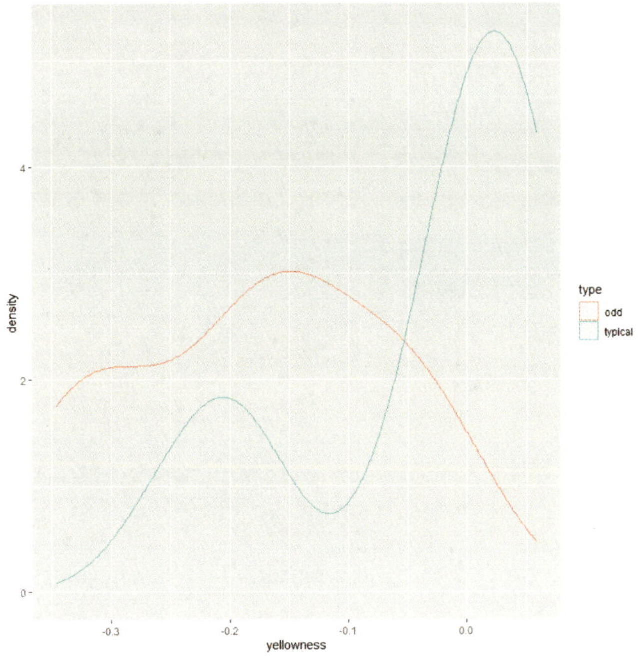

```
##Calculating the mean for each group

dat2 %>% group_by(type) %>% summarise(mean = mean(yellowness))
## # A tibble: 2 x 2
##  type   mean
##  <chr>  <dbl>
## 1 odd   –0.176
## 2 typical –0.039
```

We see that the distribution of yellowness indices for the typical feathers is shifted up compared to the odd feathers. Also, the mean value for the yellowness indices for the typical feathers is greater than the mean value for the odd feathers.

Conducting the t.test

```
##Conducting the test for an alternative hypothesis "two.sided" which is default

t.test(yellowness ~ type, data = dat2, paired=TRUE, var.equal=TRUE)
##
## Paired t-test

##
## data: yellowness by type
## t = –4.0647, df = 15, p-value = 0.001017
## alternative hypothesis: true difference in means is not equal to 0
## 95 percent confidence interval:
## –0.20903152 –0.06521848
## sample estimates:
## mean of the differences
##        –0.137125
##Conducting the test for an alternative hypothesis "less" i.e. odd less than typical

t.test(yellowness ~ type, data = dat2, paired=TRUE, alternative="less," var.equal=TRUE)
##
```

```
## Paired t-test
##
## data: yellowness by type
## t = –4.0647, df = 15, p-value = 0.0005083
## alternative hypothesis: true difference in means is less than 0
## 95 percent confidence interval:
##    -Inf –0.07798415
## sample estimates:
## mean of the differences
##      –0.137125
levels(factor(dat2$type))
## [1] "odd"  "typical"
```

In R, the "odd" level is the first level for the column "type" so we specify the alternative as "less" meaning that odd feathers indices are less than typical feathers indices.

Assuming is the mean of yellow indices for odd feathers and is the mean of yellow indices for typical feathers

For two-sided test,

Null hypothesis, Ho: or

Alternative hypothesis, Ha: or

For one-sided test, alternative="less"

Null hypothesis, Ho: or

Alternative hypothesis, Ha: or

t-test gives a p-value <0.05 in both cases (two-sided and one-sided), which means that the yellow indices of the odd feathers are less than the yellow indices of the typical feathers by 0.14 on the average.

9.2.2. Unpaired t-Test

Unpaired t-test is used for unpaired data and uses the t-distribution to calculate the p-value.

Assumptions:

1. Unpaired data.
2. Large data >30 or check the normality of data with Shapiro-Wilk test.

   If the data failed the normality test, we can use the two-samples Wilcoxon rank test.
3. Equal variances of the two groups. This can be checked with F-test. The default value in t-test is unequal variances.

The test requires a dataframe with the first column for grouping factor and second column for numerical values

- *Example 1:* Use the regicor dataset from compareGroups package to determine the mean age difference between males and females.

Check normality

```
library(compareGroups)
data("regicor")
table(regicor$sex)
##
## Male Female
## 1101 1193
```

As this is a large sample (1101 males and 1193 females), we can ignore the normality test according to CLT (Central Limit Theorem).

Checking equal variance in the two groups:

```
var.test(age ~ sex, data = regicor)
##
## F test to compare two variances
##
## data: age by sex
## F = 1.0125, num df = 1100, denom df = 1192, p-value = 0.8336
## alternative hypothesis: true ratio of variances is not equal to 1
## 95 percent confidence interval:
## 0.9017386 1.1370755
## sample estimates:
```

ratio of variances

1.012455

F-test gives a p-value >0.05, which means that the two variances are equal.

Numerical summaries and visual displays.

##Visualizing the data

```
ggplot(data = regicor, aes(x = sex, y= age, col = sex))+geom_boxplot()
```

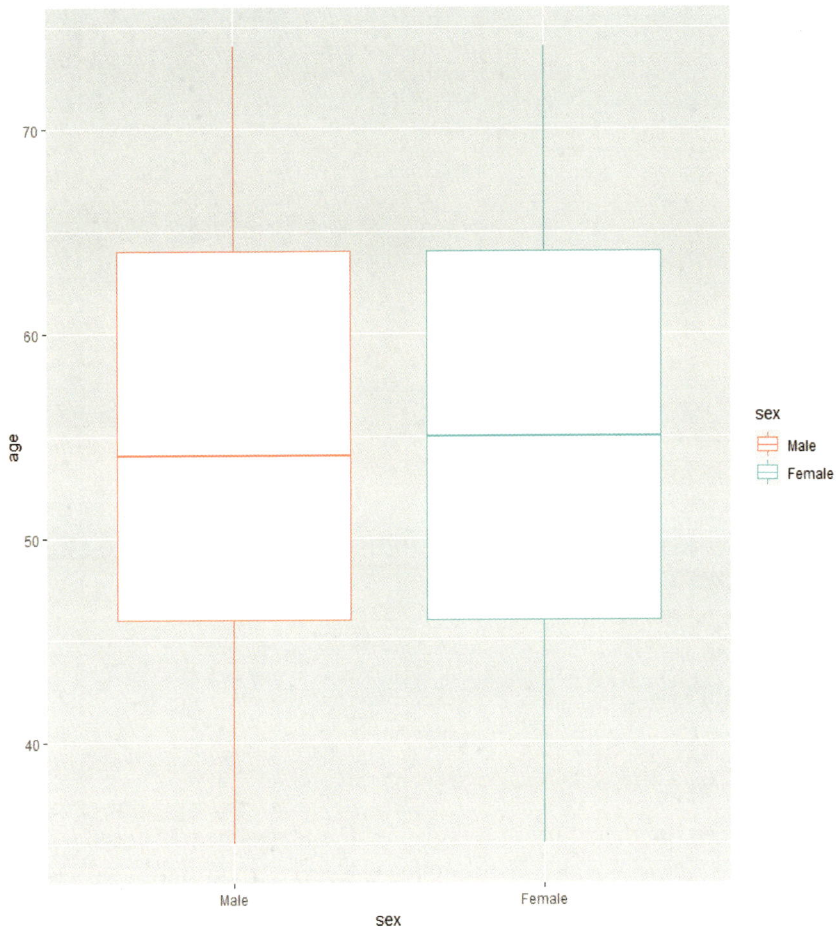


```
ggplot(data = regicor, aes(x = age, col = sex))+geom_density()
```

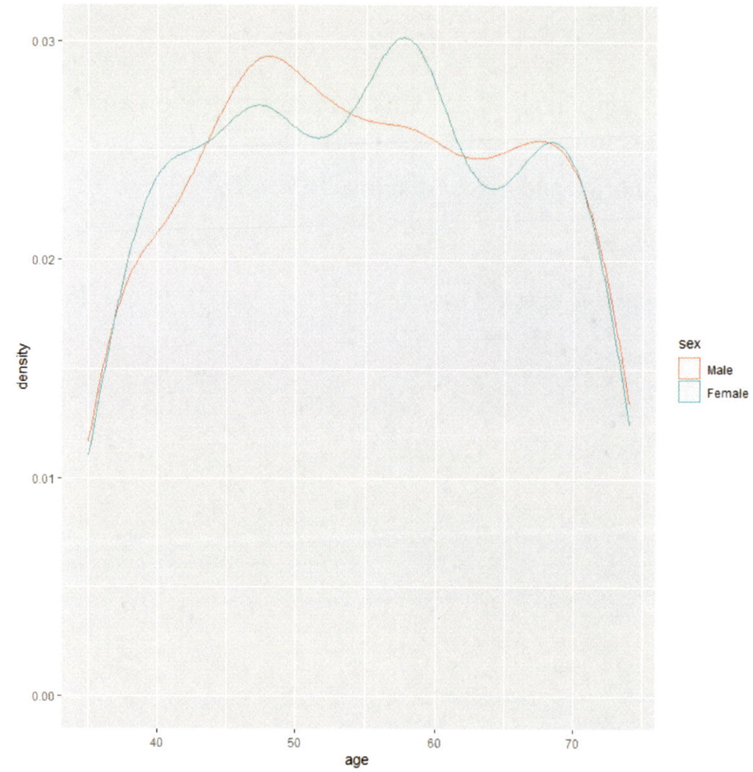


```
##Calculating the mean for each group

regicor %>% group_by(sex) %>% summarise(mean = mean(age))
## # A tibble: 2 x 2

##  sex   mean
##  <fct> <dbl>
## 1 Male  54.8
## 2 Female 54.7
```

We see that the distribution of ages for females is nearly equal to that of males. Also, the mean value for females is nearly equal the mean value for males.

Conducting the test for an alternative hypothesis "two.sided" which is default.

```
t.test(age ~ sex, data = regicor, var.equal = TRUE)
##
## Two Sample t-test
##
## data: age by sex
## t = 0.20181, df = 2292, p-value = 0.8401
## alternative hypothesis: true difference in means is not equal to 0
## 95 percent confidence interval:
## -0.8124927 0.9989071
## sample estimates:
##  mean in group Male mean in group Female
##       54.78474       54.69153
##Conducting the test for an alternative hypothesis "greater" or males older than females
t.test(age ~ sex, data = regicor,alternative="greater," var.equal = TRUE)
##
## Two Sample t-test
##
## data: age by sex
## t = 0.20181, df = 2292, p-value = 0.42
## alternative hypothesis: true difference in means is greater than 0
## 95 percent confidence interval:
## -0.6667859    Inf
## sample estimates:
##  mean in group Male mean in group Female
##       54.78474       54.69153
```

t-test gives a p-value larger than 0.05 in both cases, which means that the age of females is not significantly lower than the age of males.

- *Example 2:* Use the same regicor data to determine the mean physical activity difference between males and females.

Check normality.

```
table(regicor$sex)
##
## Male Female
## 1101  1193
```

As this is a large sample (1101 males and 1193 females), we can ignore the normality test according to CLT.

Checking equal variance in the two groups:

```
var.test(phyact ~ sex, data = regicor)
##
## F test to compare two variances
##
## data: phyact by sex
## F = 1.4668, num df = 1059, denom df = 1145, p-value = 2.167e–10
## alternative hypothesis: true ratio of variances is not equal to 1
## 95 percent confidence interval:
## 1.303378 1.651077
## sample estimates:
## ratio of variances
##     1.46676
```

F-test gives a p-value < 0.05, which means that the two variances are not equal.

Numerical summaries and visual displays.

```
##Visualizing the data

ggplot(data = regicor, aes(x = sex, y= phyact, col = sex))+geom_boxplot()
```

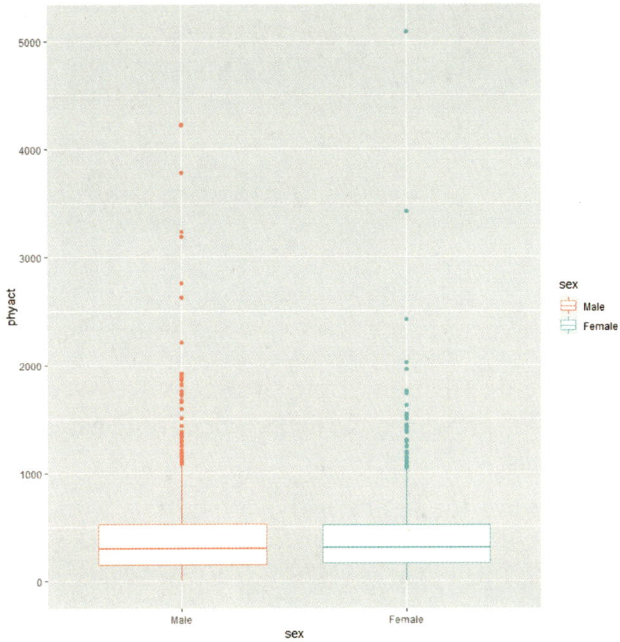

ggplot(data = regicor, **aes**(x = phyact, col = sex))+**geom_density**()

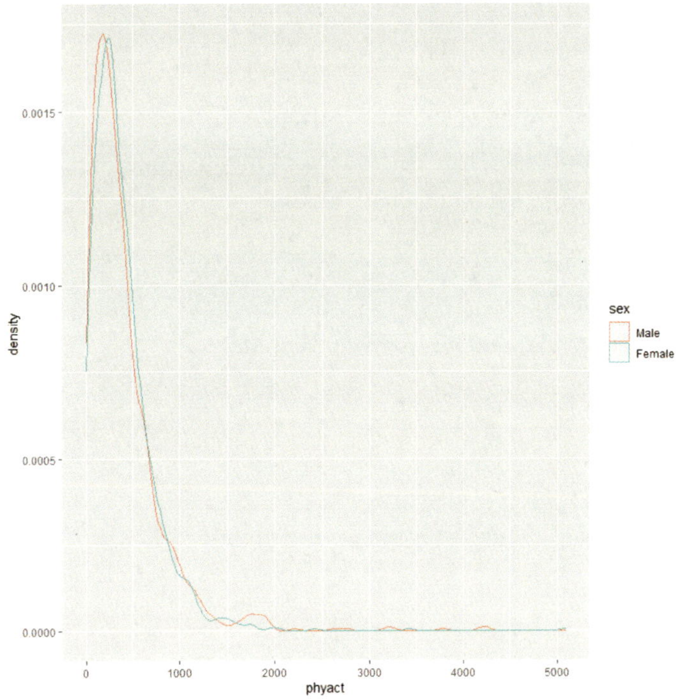

```
##Calculating the mean for each group

regicor %>% group_by(sex) %>% summarise(mean = mean(phyact,
na.rm = TRUE))
## # A tibble: 2 x 2
##  sex   mean
##  <fct> <dbl>
## 1 Male  407.
## 2 Female 392.
```

We see that the distribution of physical activity for females is nearly equal to that of males. However, the mean value for females is less than the mean value for males.

Conducting the test for an alternative hypothesis "two.sided" which is default.

```
t.test(phyact ~ sex, data = regicor, var.equal = FALSE)
##
## Welch Two Sample t-test

##
## data: phyact by sex
## t = 0.89949, df = 2058.6, p-value = 0.3685
## alternative hypothesis: true difference in means is not equal to 0
## 95 percent confidence interval:
## –17.69165 47.67095
## sample estimates:
##  mean in group Male mean in group Female
##       406.6184       391.6288
##Conducting the test for an alternative hypothesis "greater" or males do
greater physical activity than females

t.test(phyact ~ sex, data = regicor, var.equal = FALSE, alternative =
"greater")
##
## Welch Two Sample t-test
```

```
##
## data: phyact by sex
## t = 0.89949, df = 2058.6, p-value = 0.1842
## alternative hypothesis: true difference in means is greater than 0
## 95 percent confidence interval:
## –12.43358   Inf
## sample estimates:
##  mean in group Male mean in group Female
##       406.6184       391.6288
```

t-test gives a p-value larger than 0.05 in both cases, which means that the physical activity of females is not significantly lower than the physical activity of males.

The interpretation of two.sided or one.sided test will depend on your interested question. If you want to test that the two groups are not similar to each other, use the "two.sided" approach. If you want to test that one group is larger than (or smaller than) the other, use the "one-sided" approach. Other examples of one-sided approach, you want to see if the dug effect increase weight or decrease blood pressure than the placebo. You are not interested if the drug is different from the placebo.

9.3. HYPOTHESIS TESTING FOR COMPARING MEANS BETWEEN TWO POPULATIONS, NON-PARAMETRIC TESTS

Non-parametric tests are used when your data deviate from normality or you have a small sample size (<30).

9.3.1. Paired Study, Wilcoxon Signed-Rank Test

There is no assumptions for this test. The Wilcoxon signed-rank test compares medians instead of means

- *Example 1:* We want to compare the weights of 10 mice before and after certain treatment.

Creating the data.

```
# Weight of the mice before treatment

before <-c(200.1, 190.9, 192.7, 213, 241.4, 196.9, 172.2, 185.5, 205.2,
193.7)
# Weight of the mice after treatment

after <-c(392.9, 393.2, 345.1, 393, 434, 427.9, 422, 383.9, 392.3, 352.2)
## Create the dataframe

dat3 <- data.frame(before,after)
dat3<-dat3 %>% gather(key=type,value = weight)
dat3
##   type weight
## 1 before 200.1
## 2 before 190.9
## 3 before 192.7
## 4 before 213.0
## 5 before 241.4
## 6 before 196.9
## 7 before 172.2
## 8 before 185.5
## 9 before 205.2
## 10 before 193.7
## 11 after 392.9
## 12 after 393.2
## 13 after 345.1
## 14 after 393.0
## 15 after 434.0
## 16 after 427.9
## 17 after 422.0
## 18 after 383.9
## 19 after 392.3
## 20 after 352.2
```

Numerical summaries and visual displays.

```
##Visualizing the data
ggplot(data = dat3, aes(x = type,y=weight, col = type))+geom_boxplot()
```

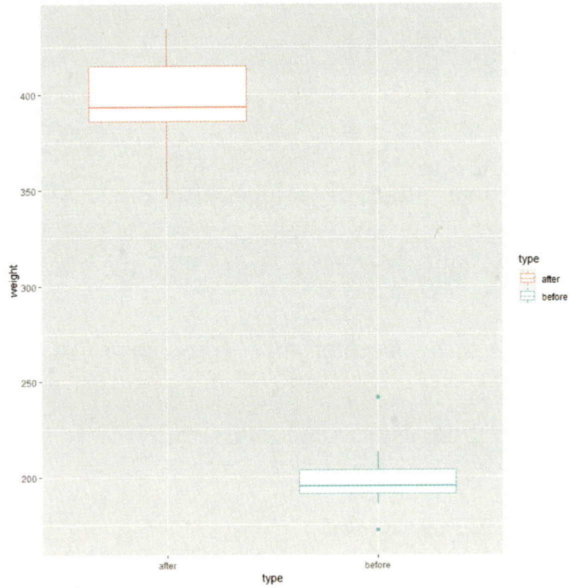


```
ggplot(data = dat3, aes(x = weight, col = type)) + geom_density()
```

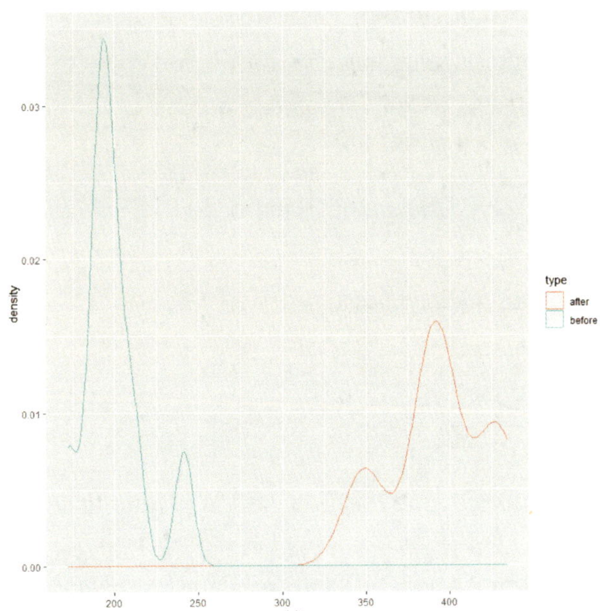

```
##Calculating the median for each group

dat3 %>% group_by(type) %>% summarise(median = median(weight))
## # A tibble: 2 x 2
##  type  median
##  <chr>  <dbl>
## 1 after  393.
## 2 before  195.
```

We see that the distribution of weights after the test is shifted up when compared to the distribution of weights before the test. Also, the median value for the after weights is greatly larger than the median value for the before weights.

Conducting the test for an alternative hypothesis "two.sided" which is default.

```
wilcox.test(weight ~ type, data = dat3,paired=TRUE)
##
## Wilcoxon signed rank exact test
##
## data: weight by type
## V = 55, p-value = 0.001953
## alternative hypothesis: true location shift is not equal to 0
##Conducting the test for an alternative hypothesis "greater" or after weights heavier than before weights

wilcox.test(weight ~ type, data = dat3, paired=TRUE,alternative="greater")
##
## Wilcoxon signed rank exact test
##
## data: weight by type
## V = 55, p-value = 0.0009766

## alternative hypothesis: true location shift is greater than 0
```

Wilcoxon signed-rank test gives a p-value <0.05 in both cases, which means that the mice have gained a significant weight after this treatment.

9.3.2. Unpaired Study, Two-Samples Wilcoxon Rank Sum Test or "Mann-Whitney" Test

There is no assumptions for this test. The Wilcoxon rank sum test compares medians instead of means

- *Example 1:* Permeability constants of the human chorioamnion (a placental membrane) at term and between 12 to 26 weeks gestational age (preterm). The alternative of interest is greater permeability of the human chorioamnion for the term pregnancy.

```
term <- c(0.80, 0.83, 1.89, 1.04, 1.45, 1.38, 1.91, 1.64, 0.73, 1.46)
preterm <- c(1.15, 0.88, 0.90, 0.74, 1.21)
## Conducting the test for an alternative hypothesis "greater"

wilcox.test(term, preterm, alternative="greater")
##
## Wilcoxon rank sum exact test
##
## data: term and preterm
## W = 35, p-value = 0.1272
## alternative hypothesis: true location shift is greater than 0
```

Wilcoxon rank sum test gives a p-value larger than 0.05, which means that the permeability constants of the human term chorioamnion are the same as the permeability constants of the human preterm chorioamnion.

- *Example 2:* Suppose we want to compare the length of hospital stay for patients with the same diagnosis at two different hospitals.

```
hospital1<- c(21, 10, 32, 60, 8, 44, 29, 5, 13, 26, 33)
hospital2<- c(86, 27, 10, 68, 87, 76, 125, 60, 35, 73, 96, 44, 238)
median(hospital1)
## [1] 26
median(hospital2)
## [1] 73
##Conducting the test for an alternative hypothesis "two.sided" which is default

wilcox.test(hospital1, hospital2)
```

```
##
## Wilcoxon rank sum test with continuity correction
##
## data: hospital1 and hospital2
## W = 17.5, p-value = 0.001925
## alternative hypothesis: true location shift is not equal to 0
```

Wilcoxon rank sum test gives a p-value < 0.05 which means that the length of stay in the hospital2 is longer than the length of stay in hospital1 for patients with the same diagnosis.

CHAPTER 10

HYPOTHESIS TESTING FOR PROPORTIONS AND TIME TO EVENT DATA

CONTENTS

In this chapter, we will limit our discussion to comparing two populations only.

10.1. COMPARING PROPORTIONS BETWEEN TWO POPULATIONS USING CHI-SQUARE TEST

The general formula for hypothesis testing that compare proportions:

Ho: or

Ha: or

The Chi-square test use the approach of contingency table of the data to test that. The contingency table is a table of data rows (one variable) cross classified by columns (the other variable).

10.1.1. Steps of Hypothesis Testing Performed by the Chi-Square Test

1. The Chi-square test uses the contingency table of data to calculate an expected table. The expected table contain the theoretical data values that would be expected when there is no relation between the rows (predictor) and the columns (outcome), i.e., the null hypothesis is true, Ho: .
2. The test calculates the discrepancies between each observed and expected value and aggregates them.
3. The aggregated value is compared to a Chi-square distribution under the null hypothesis and calculates a p-value.
4. p-value is the probability of our sample results under the Null hypothesis
5. Make a decision, accept Ha or fail to reject Ho

If p-value < 0.05, so it is a statistically significant result at 0.05 level. Reject the Null hypothesis and conclude that our sample data are unlikely under the Ho, null hypothesis, they have a probability of less than 0.05.

If p-value >= 0.05, so it is a statistically nonsignificant result at 0.05 level and we fail to reject the Null hypothesis.

10.1.2. Assumptions of the Test

1. Unpaired data.

2. The normal approximation to the binomial distribution is valid. The normal approximation can be shown to be approximately true if no expected value in the table is less than 5 (sometimes known as "the rule of five").

When we run the test in R and there is an expected value less than 5, R will throw an error (Warning message: In chisq.test(x, y) : Chi-squared approximation may be incorrect) indicating that this test is not suitable. In that case, Fisher Exact test is a suitable alternative.

It is not important to assign which variable (predictor or outcome) to columns or rows. The Chi-square test requires a matrix of rows and columns. The Chi-square test has no versions for "less" or "greater" alternative hypotheses

- *Example 1:* A researcher reported adverse effects in 85 patients receiving either terbinafine or placebo treatment for dermatophyte onychomycosis. Out of 56 patients receiving terbinafine, 5 patients experienced adverse effects. Out of 29 patients receiving a placebo, none experienced adverse effects. Is there more adverse effects due to the drug?

```
#the number of positives for 1st group (drug) which is 5.

#the number of negatives for 1st group (drug) which is 51.

#the number of positives for the 2nd group (placebo), which is 0.

#the number of negatives for the 2nd group (placebo), which is 29.

#Note that the values in a matrix are arranged column wise

dat<-matrix(c(5,51,0,29),nrow = 2)
dat
##  [,1] [,2]
## [1,] 5 0
## [2,] 51 29
#To add names to columns and rows

dimnames(dat)<-list(SE=c("SE+,""SE-"),drug_type=c("drug,""placebo"))
dat
##  drug_type
```

```
## SE  drug placebo
##  SE+  5    0
##  SE-  51   29
```

##We have assigned the side effects to rows and drug type to columns

##Run the test

```
test<-chisq.test(dat)
```

##Note the warning message because some of the expected values are less than 5 so Fisher Exact test should be used

##You can access the expected values by typing

```
test$expected
##    drug_type
## SE      drug  placebo
##  SE+ 3.294118 1.705882
##  SE- 52.705882 27.294118
```

```
##To see the test results
test
##
## Pearson's Chi-squared test with Yates' continuity correction
##
## data: dat
## X-squared = 1.3747, df = 1, p-value = 0.241
```

#To see that the results will not change if we assign drug type to rows and side effects to columns

```
dat2<-matrix(c(5,0,51,29), nrow = 2)
dat2
##    [,1] [,2]
## [1,]  5  51
## [2,]  0  29
chisq.test(dat2)
```

```
##
## Pearson's Chi-squared test with Yates' continuity correction
##
## data: dat2
## X-squared = 1.3747, df = 1, p-value = 0.241
chisq.test(dat)
##
## Pearson's Chi-squared test with Yates' continuity correction
##
## data: dat
## X-squared = 1.3747, df = 1, p-value = 0.241
##Same p-value
```

The Chi-square test is not suitable for our data as some expected values are less than 5 so Fisher Exact test should be used.

- *Example 2:* All women with at least one birth in an international study are classified as either having breast cancer or not according to age at first birth, either ≤ 29 or ≥ 30 years. 683 of 1498 (45.5%) women with breast cancer had an age at first birth ≥ 30 years, while 2537 of 8747 (29%) women with breast cancer had an age at first birth ≤ 29 years.

Source: Based on WHO Bulletin, 43, 209–221, 1970.

Research question: Is there a significant difference or simply due to chance?

```
#the number of breast cancer for 1st group (≥30) which is 683.

#the number of no breast cancer for 1st group (≥30) which is 815(=1498–683).

#the number of breast cancer for 2nd group (≤ 29) which is 2537.

#the number of no breast cancer for 2nd group (≤ 29) which is 6210(=8747–2537).

#Note that the values in a matrix are arranged column wise
```

```
dat<-matrix(c(683,815,2537,6210),nrow = 2)
dat
##   [,1] [,2]
## [1,] 683 2537
## [2,] 815 6210
#To add names to columns and rows

dimnames(dat)<-list(cancer=c("C+,""C-"), age_group=c(">=30,""<=29"))
dat
##    age_group
## cancer >=30 <=29
##   C+ 683 2537
##   C- 815 6210
##We have assigned the cancer status to rows and age group to columns

##Run the test

test<-chisq.test(dat)
test
##
## Pearson's Chi-squared test with Yates' continuity correction

##
## data: dat
## X-squared = 162.56, df = 1, p-value < 2.2e–16
#Note that no warning is produced

#To see expected values

test$expected
##    age_group
## cancer   >=30   <=29
##   C+ 470.8209 2749.179
```

```
## C- 1027.1791 5997.821
```

The results are statistically significant. Therefore, there is an association between having breast cancer and having a first child after the age of 30. In other words, age of first birth is an important risk factor for breast cancer.

10.2. COMPARING PROPORTIONS BETWEEN TWO POPULATIONS USING FISHER EXACT TEST

Fisher exact test is used when the expected value in any cell of the contingency table is less than 5. For tables in which use of the Chi-square test is appropriate, the two tests give very similar results although the p-value will be different as the Fisher exact test uses the hypergeometric distribution to calculate its p-value.

Fisher exact test has versions for "less" or "greater" alternative hypotheses for the 2 by 2 cross table, but Fisher test use the same matrix formula as the Chi-square test.

- *Example 1:* A researcher reported adverse effects in 85 patients receiving either terbinafine or placebo treatment for dermatophyte onychomycosis. Out of 56 patients receiving terbinafine, 5 patients experienced adverse effects. Out of 29 patients receiving a placebo, none experienced adverse effects. Is there more adverse effects due to the drug?

```
#the number of positives for 1st group (drug) which is 5.

#the number of negatives for 1st group (drug) which is 51.

#the number of positives for the 2nd group (placebo), which is 0.

#the number of negatives for the 2nd group (placebo), which is 29.

#Note that the values in a matrix are arranged column wise

dat<-matrix(c(5,51,0,29),nrow = 2)
dat
##  [,1] [,2]
## [1,] 5 0
## [2,] 51 29
```

```
#To add names to columns and rows

dimnames(dat)<-list(SE=c("SE+,""SE-"),drug_type=c("drug,""placebo"))
dat
##   drug_type
## SE  drug placebo
##  SE+  5    0
##  SE-  51   29
##We have assigned the side effects to rows and drug type to columns

##Run the test with alternative hypothesis="two.sided," the default

fisher.test(dat)
##
## Fisher's Exact Test for Count Data
##
## data: dat
## p-value = 0.1606
## alternative hypothesis: true odds ratio is not equal to 1
## 95 percent confidence interval:
## 0.4832707    Inf
## sample estimates:
## odds ratio
##    Inf
##Run the test with alternative hypothesis="greater," as we are interested
to see if drug produces more adverse effects than placebo\

fisher.test(dat, alternative = "greater")
##
## Fisher's Exact Test for Count Data
##
## data: dat
## p-value = 0.1165
```

```
## alternative hypothesis: true odds ratio is greater than 1
## 95 percent confidence interval:
## 0.6477159   Inf
## sample estimates:
## odds ratio
##   Inf
```

Using different alternative hypotheses, p-value is greater than 0.05. Therefore, there is no relation between the drug use and the incidence of adverse effects.

- *Example 2:* Suppose a retrospective study is done among men ages 50–54 in a specific county who died over a 1-month period. The investigators try to include approximately an equal number of men who died from cardiovascular disease (CVD) and men who died from other causes. Of 35 people who died from CVD, 5 were on a high-salt diet before they died, whereas of 25 people who died from other causes, 2 were on such a diet. There is a difference in the proportions, 5/35=0.14 while 2/25=0.08 so we may ask a question:

Is there a relation between high-salt intake and death from CVD?

```
dat<-matrix(c(5,30,2,23),nrow = 2)
dat
##   [,1] [,2]
## [1,]  5  2
## [2,]  30  23
#To add names to columns and rows

dimnames(dat)<-list(Salt=c("Salt+,""Salt-"), Death_
cause=c("CVD,""other"))
dat
##    Death_cause
## Salt  CVD other
##  Salt+  5   2
##  Salt- 30  23
```

```
##We have assigned the salt intake to rows and death cause to columns

##Run the test with alternative hypothesis="two.sided," the default

fisher.test(dat)
##
## Fisher's Exact Test for Count Data

##
## data: dat
## p-value = 0.6882
## alternative hypothesis: true odds ratio is not equal to 1

## 95 percent confidence interval:
##  0.278957 21.620483
## sample estimates:
## odds ratio

##  1.897126
##Run the test with alternative hypothesis="greater," as we are interested
to see if high salt-intake produces more CVD related deaths

fisher.test(dat,alternative = "greater")
##
## Fisher's Exact Test for Count Data
##
## data: dat
## p-value = 0.3747
## alternative hypothesis: true odds ratio is greater than 1
## 95 percent confidence interval:
## 0.3572533    Inf
## sample estimates:
## odds ratio
##  1.897126
```

The two proportions in this example are not significantly different with either a one-sided or two-sided test, and we cannot say, on the basis of this limited amount of data, that there is a significant association between high salt intake and death from cardiovascular events.

10.3. COMPARING PROPORTIONS BETWEEN TWO POPULATIONS USING MCNEMAR TEST (PAIRED DATA)

McNemar's test is used to compare two binomial proportions from matched samples.

10.3.1. Assumptions

1. Paired data or matched pairs.
2. The normal approximation to the binomial distribution is valid. This occurs when the number of discordant pairs is ≥20.

The McNemar test requires a data matrix as before.

- *Example 1:* The Approval of the President's performance in measured in two surveys, performed one month apart, for a random sample of 1600 voting-age Americans. The data are presented as below.
  - Reference: Alan Agresti (1990). Categorical data analysis. New York: Wiley. Pages 350–354.
  - Question: Is there a significant change in approvals during this 1 month period?

#the number of approvals for two groups, 794.

#the number of approvals for 1st group (1st survey) but disapprove for the 2nd group (2nd survey), which is 150.

#the number of approvals for the 2nd group but disapprove for the 1st group which is 86.

#the number of disapprovals for two groups (570).

#Proportion approvals for 1st group = (794+150)/1600=0.59

#Proportion approvals 2nd group = (794+86)/1600=0.55

```
#Note the number of discordant pairs=150+86= 236 which is larger than 20

Performance <- matrix(c(794, 86, 150, 570),nrow = 2)
dimnames(Performance) <- list("1st Survey" = c("Approve," "Disapprove"),"2nd Survey" = c("Approve," "Disapprove"))
Performance
##      2nd Survey
## 1st Survey  Approve Disapprove
## Approve    794    150
## Disapprove   86    570
mcnemar.test(Performance)
##
## McNemar's Chi-squared test with continuity correction
##
## data: Performance
## McNemar's chi-squared = 16.818, df = 1, p-value = 4.115e–05
```

There is a significant decrease in the approval ratings during this month period.

- *Example 2:* Suppose we want to compare two different chemotherapy regimens for breast cancer after mastectomy. The two treatment groups should be as comparable as possible on other prognostic factors. To accomplish this goal, a matched study is set up such that a random member of each matched pair gets treatment A whereas the other member gets treatment B. The patients are assigned to pairs matched on age (within 5 years) and clinical conditions. The patients are followed for 5 years, with survival as the outcome variable.

There are 90 pairs in which both patients died within 5 years, 510 pairs in which both patients survived for 5 years, 16 pairs in which the treatment A patient survived and the treatment B patient died, and 5 pairs in which the treatment B patient survived and the treatment A patient died.

For 600 pairs (90 + 510), the outcomes of the two treatments are the same, whereas for 21 pairs (16 + 5), the outcomes of the two treatments are different.

Which treatment is the drug of choice?

Treatment A has a survival rate= (510+16)/621=0.85

Treatment B has a survival rate=(510+5)/621=0.83

```
##510 pairs in which both patients survived for 5 years

##16 pairs in which the treatment A patient survived and the treatment B patient died

##5 pairs in which the treatment B patient survived and the treatment A patient died

##90 pairs in which both patients died within 5 years

#Note the number of discordant pairs=16+5= 21 which is larger than 20, So McNemar test can be used

outcome <- matrix(c(510,5,16,90),nrow = 2)
dimnames(outcome)<-list(A=c("Survived,""Died"), B=c("Survived,""Died"))
outcome
##     B
## A     Survived Died
## Survived  510  16
## Died        5  90
mcnemar.test(outcome)
##
## McNemar's Chi-squared test with continuity correction
##
## data: outcome
## McNemar's chi-squared = 4.7619, df = 1, p-value = 0.0291
```

We conclude that the treatment A member of the pair is significantly more likely to survive for 5 years than the treatment B member. Thus, when all other things being equal (such as toxicity, cost, etc.), treatment A would be the treatment of choice.

- *Example 3 (from Chapter 8):* In a reliability exercise, 41 patients were randomly selected from those who had undergone a thalium-201 stress test. The 41 sets of images were classified as normal or not by the core thallium laboratory test and, independently, by clinical investigators from different centers. Of the 19 samples identified as ischemic by clinical investigators, 14 were identified as ischemic by the laboratory. Of the 22 samples identified as normal by clinical investigators, 0 were identified as ischemic by the laboratory. Is there a difference in the diagnosis between the laboratory test and the clinicians?

Here the number of discordant observations=5 which is less than 20 so McNemar test cannot be used. Instead, Exact McNemar test from exact2x2 package can be used.

```
library(exact2x2)
#the number of positives or ischemic for two groups, (14).

#the number of positives for 1st group (clinicians) but negative for the 2nd group (lab), which is 5.

#the number of positives for the 2nd group (lab), which are negative for the 1st group (clinicians), which is 0.

#the number of negatives for two groups (22).

images <- matrix(c(14,0,5,22),nrow = 2)
dimnames(images)<-list(clinicians=c("+,""-"), lab=c("+,""-"))
images
##      lab

## clinicians + -
##    + 14 5
##    – 0 22
#two sided test which is default

exact2x2(images,paired = TRUE)
##
## Exact McNemar test (with central confidence intervals)
```

```
##
## data: images
## b = 5, c = 0, p-value = 0.0625
## alternative hypothesis: true odds ratio is not equal to 1
## 95 percent confidence interval:
## 0.9163559   Inf
## sample estimates:
## odds ratio
##    Inf
##If we are interested in that clinicians have high false positive rate than lab, alternative="greater"
exact2x2(images,paired = TRUE,alternative = "greater")
##
## Exact McNemar-type test
##
## data: images
## b = 5, c = 0, p-value = 0.03125
## alternative hypothesis: true odds ratio is greater than 1
## 95 percent confidence interval:
## 1.218674   Inf
## sample estimates:
## odds ratio
##    Inf
```

We conclude that clinicians have significantly higher tendency to overestimate the proportion of ischemic images over that of the lab test. The two-sided test is nonsignificant for the exact McNemar test but it is significant when using epi.conf() function of the epiR package. This is because they use different approaches to calculate their p-value; so, it is important to name the statistical tests used in your analysis in the materials and methods section.

10.4. COMPARING TIME TO EVENT DATA BETWEEN TWO POPULATIONS USING LOG-RANK TEST

Ho:

Ha:

In log-rank test compares the distance between the Kaplan Meier curves of the two or more samples and calculates a p-value that determines if there is a significant difference between the different survival of different samples.

10.4.1. Steps of Hypothesis Testing Done by Log-Rank Test

1. The log-rank test compares the number of events observed at each event time in two samples to the number of expected events in two samples under the null hypothesis, Ho.
2. Calculate the discrepancy between each observed and expected 2X2 table at each event time.
3. Aggregate these discrepancies across all event times and compare it to the distribution of these discrepancies under the null hypothesis using Chi-square distribution to get a p-value.
4. Using the p-value to accept Ha or fail to reject Ho.

The survdiff() function of the survival package performs the log-rank test and requires:

1. formula in the form, survival object ~ grouping factor;
2. dataframe containing the variables.

- *Example 1:* We will use survival package in R with its accompanying dataset veteran. See chapter 5.

Is there a significant difference between the two treatments, standard, and test, in the survival curves?

```
library(survival)
library(survminer)
data("veteran")
#Surv function takes the time and status column to build the survival object

km<-Surv(veteran$time,veteran$status)
##To plot the survival curves for the two treatments

trt_fit<-survfit(km ~trt, data = veteran)
```

```
ggsurvplot(trt_fit,data = veteran)
```

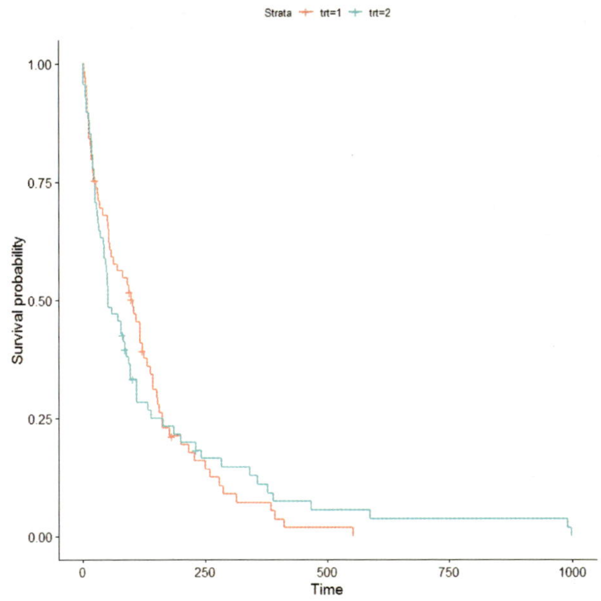


```
ggsurvplot(trt_fit,data = veteran, conf.int = TRUE,pval = TRUE,pval.
method = TRUE)
```

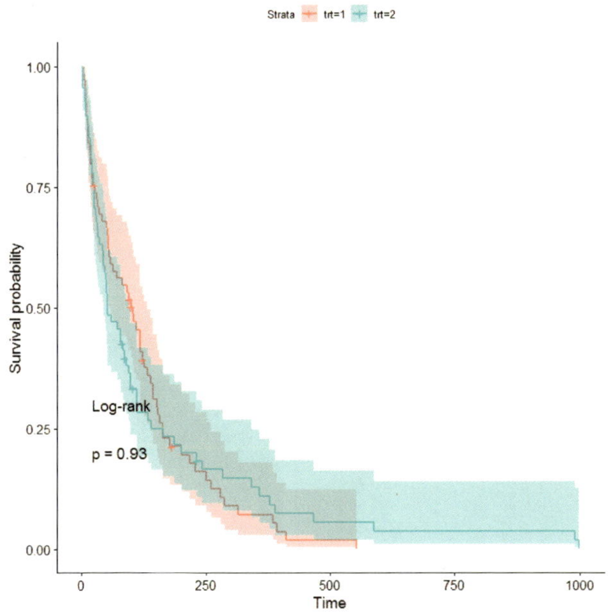

```
##To run the log-rank test

survdiff(formula = km ~ trt, data = veteran)
## Call:
## survdiff(formula = km ~ trt, data = veteran)
##
##    N Observed Expected (O-E)^2/E (O-E)^2/V
## trt=1 69    64    64.5  0.00388  0.00823
## trt=2 68    64    63.5  0.00394  0.00823
##
## Chisq= 0 on 1 degrees of freedom, p= 0.9
```

There is no difference between the two treatments in survival or mortality rates. The survdiff() function gives the same p-value as the plot. The non-significance is evident from the plot where the confidence intervals (CI) of two survival curves are overlapped.

- *Example 2:* Does prior therapy affect the mortality rate?

```
##To plot the survival curves for the two groups

prior_fit<-survfit(km ~prior, data = veteran)
ggsurvplot(prior_fit)
```

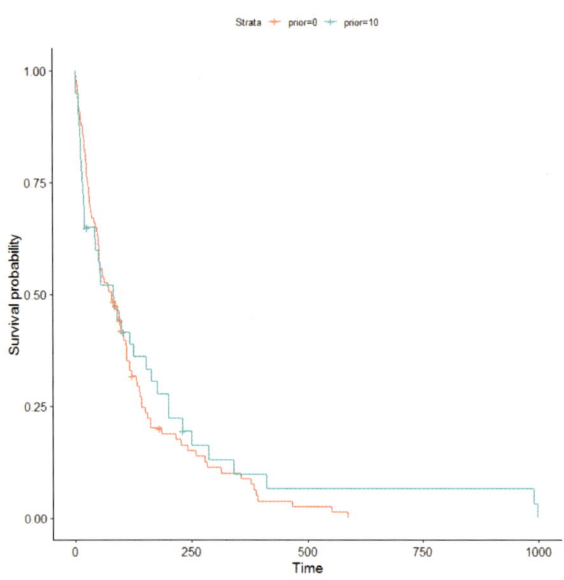

```
ggsurvplot(prior_fit, conf.int = TRUE,pval = TRUE,pval.method = TRUE)
```

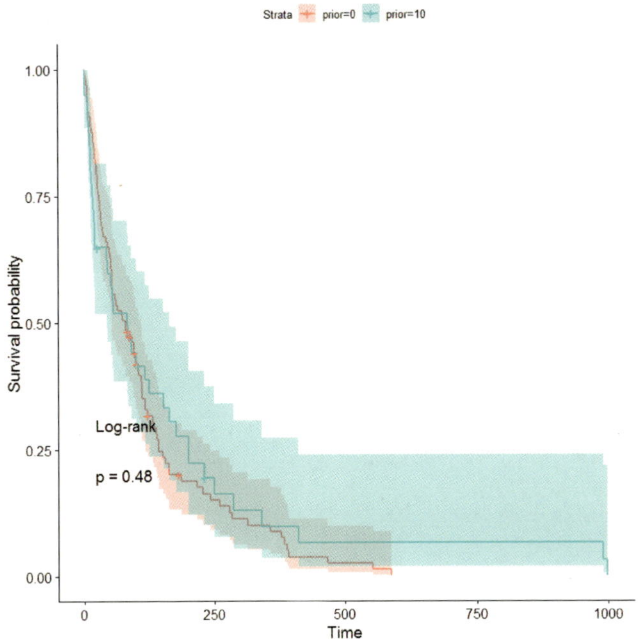


```
##To run the log-rank test

survdiff(formula = km ~ prior, data = veteran)
## Call:
## survdiff(formula = km ~ prior, data = veteran)
##
##         N Observed Expected (O-E)^2/E (O-E)^2/V
## prior=0 97    91   87.4    0.150    0.501
## prior=10 40    37   40.6    0.323    0.501
##
## Chisq= 0.5 on 1 degrees of freedom, p= 0.5
```

There is no significant difference between the two groups (using prior therapy or not) in the survival curves.

CHAPTER 11

HYPOTHESIS TESTING FOR MORE THAN TWO POPULATIONS

CONTENTS

11.1. THE PROBLEM OF MULTIPLE COMPARISONS IN STATISTICAL TESTS

When you perform a large number of statistical tests (between different populations), some will have P values less than 0.05 purely by chance, even if all your null hypotheses are really true. Therefore, these p-values need to be adjusted using several methods.

The methods available in R are:

Bonferroni correction ("bonferroni"), Holm ("holm"), Hochberg ("hochberg"), Hommel ("hommel"), Benjamini, and Hochberg ("BH" or its alias "fdr," fdr=false discovery rate), and Benjamini and Yekutieli ("BY").

When you run a statistical test (comparing means, proportions, or survival curves) for more than two populations and get a significant result, you will do a pairwise test to see which two populations are significantly different from each other. This pairwise testing will use the p-value adjustment methods stated above.

11.2. COMPARING MEANS BETWEEN MORE THAN TWO POPULATIONS USING ANALYSIS OF VARIANCE (ANOVA) TEST

Ho: means are equal for all categories

Ha: at least one mean is different from one other mean.

There are several steps of doing the ANOVA test and assumptions that must be met. If the test fails the assumptions, Kruskal-Wallis test will be done instead. The ANOVA assumptions involve residuals which are difference between the observed value and the mean value (expected value) for that group.

11.2.1. ANOVA Assumptions

1. The residuals should be homoscedastic, i.e., homogenous variance of numerical values across the factor levels, tested by Levene's test.
2. The residuals should be normally distributed, tested by Shapiro-Wilk test.
3. The residuals show no outliers, tested by outlier test.

Also, R provides a set of plots (Diagnostic plots) that help in diagnosis of these problems.

- *Example 1*: For the regicor data, what is the mean difference in age among different smoking statuses? Is there a significant difference?

Steps of running ANOVA test and Kruskal-Wallis test.

- Step 1: Check the format of the variables, Y (age) must be numeric and X factor (smoking status).

```
library(tidyverse)
library(car)
library(compareGroups)
data("regicor")
str(regicor$age)
## int [1:2294] 70 56 37 69 70 40 66 53 43 70 ...
## – attr(*, "label")= Named chr "Age"
##  ..- attr(*, "names")= chr "age"

str(regicor$smoker)
## Factor w/ 3 levels "Never smoker,"..: 1 1 2 1 NA 2 1 1 3 3 ...
## – attr(*, "label")= Named chr "Smoking status"
##  ..- attr(*, "names")= chr "smoker"
```

The two columns are in the desired format.

- Step 2: Summary statistics (mean) and visual displays.

```
regicor %>% filter(!is.na(smoker)) %>% group_by(smoker) %>%
 summarise(mean = mean(age)) %>% arrange(desc(mean))
## # A tibble: 3 x 2

##  smoker        mean
##  <fct>        <dbl>
## 1 Never smoker       56.5
## 2 Former >= 1y       55.3
## 3 Current or former < 1y 50.6
```

```
# Plot a box plot

regicor %>% filter(!is.na(smoker)) %>%
 ggplot(aes(x = smoker,y= age,fill= smoker))+geom_boxplot()
```

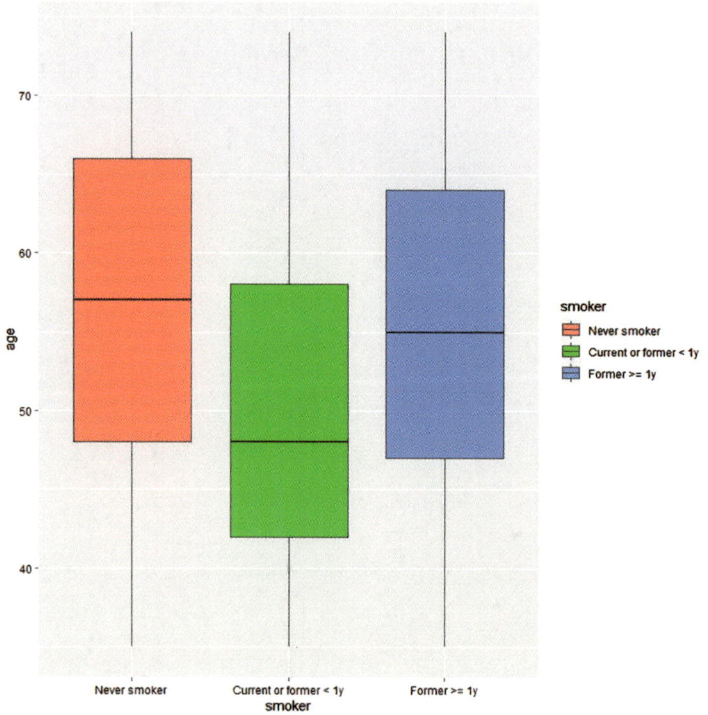

We see that never smokers have the highest mean age followed by the former smokers and lastly the current smokers.

- Step 3: Compute the one-way ANOVA test (one-way as there is one grouping factor, smoking status).

```
dat<- regicor %>% filter(!is.na(smoker))
anov_smok <-aov(age ~ smoker, data = dat)
summary(anov_smok)
##           Df Sum Sq Mean Sq F value Pr(>F)
## smoker     2  14032    7016   60.81 <2e–16 ***
## Residuals 2230 257302   115
## ---
## Signif. codes: 0 '***' 0.001 '**' 0.01 '*' 0.05 '.' 0.1 ' ' 1
```

The mean age was significantly different among the different smoking groups.

- Step 4: Run a pairwise analysis and plot the results.

```
TukeyHSD(anov_smok)
## Tukey multiple comparisons of means
## 95% family-wise confidence level
##
## Fit: aov(formula = age ~ smoker, data = dat)
##
## $smoker
##                 diff    lwr    upr    p adj
## Current or former < 1y-Never smoker –5.899225 –7.163596 –4.6348548 0.0000000
## Former >= 1y-Never smoker –1.166397 –2.571409 0.2386154 0.1259249
## Former >= 1y-Current or former < 1y 4.732828 3.146686 6.3189710 0.0000000
plot(TukeyHSD(anov_smok))
```

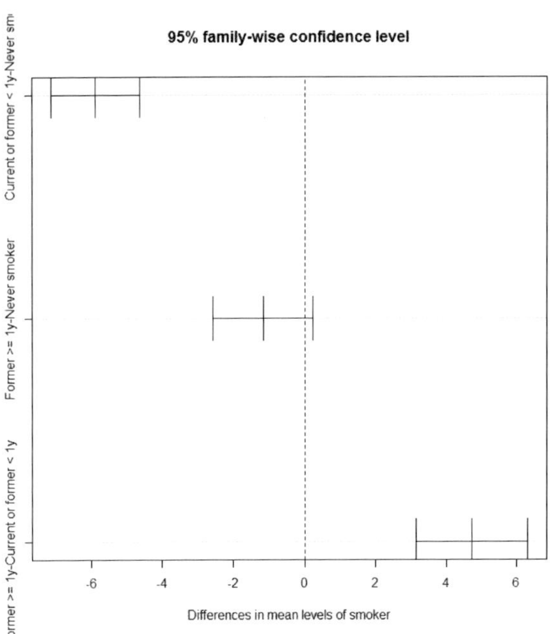

Current smokers are significantly younger than never smokers and former smokers. However, never smokers and former smokers are statistically equivalent in their mean age.

- Step 5: Run residual analysis.

The residuals should be homoscedastic, normally distributed and with no outliers.

```
# homoscedastic

plot(anov_smok, which = 1)
```

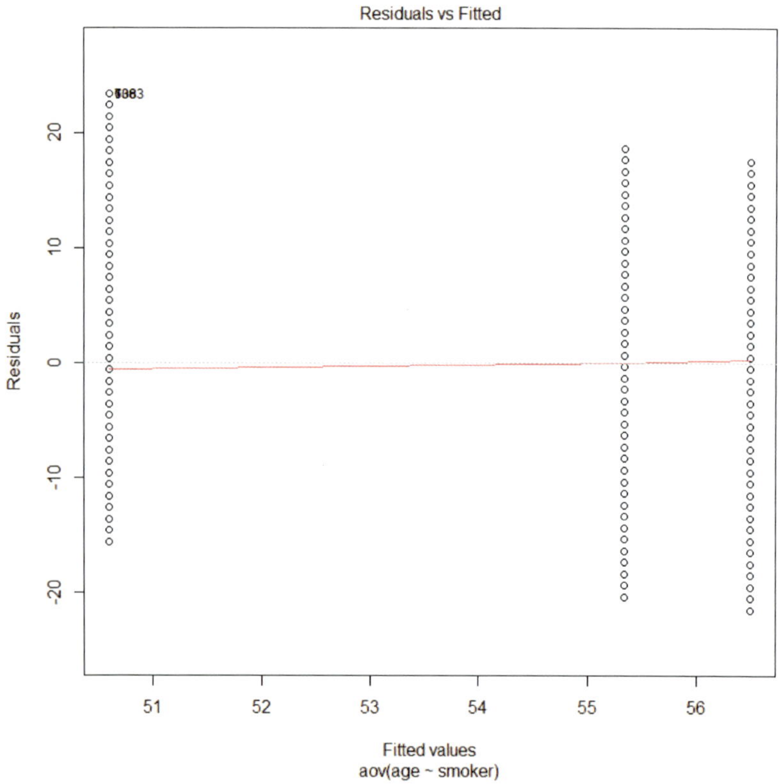


```
leveneTest(anov_smok)
## Levene's Test for Homogeneity of Variance (center = median)
##       Df F value Pr(>F)
## group  2 0.8197 0.4407
##     2230
```

```
##The residuals should be normally distributed

plot(anov_smok,which = 2)
```

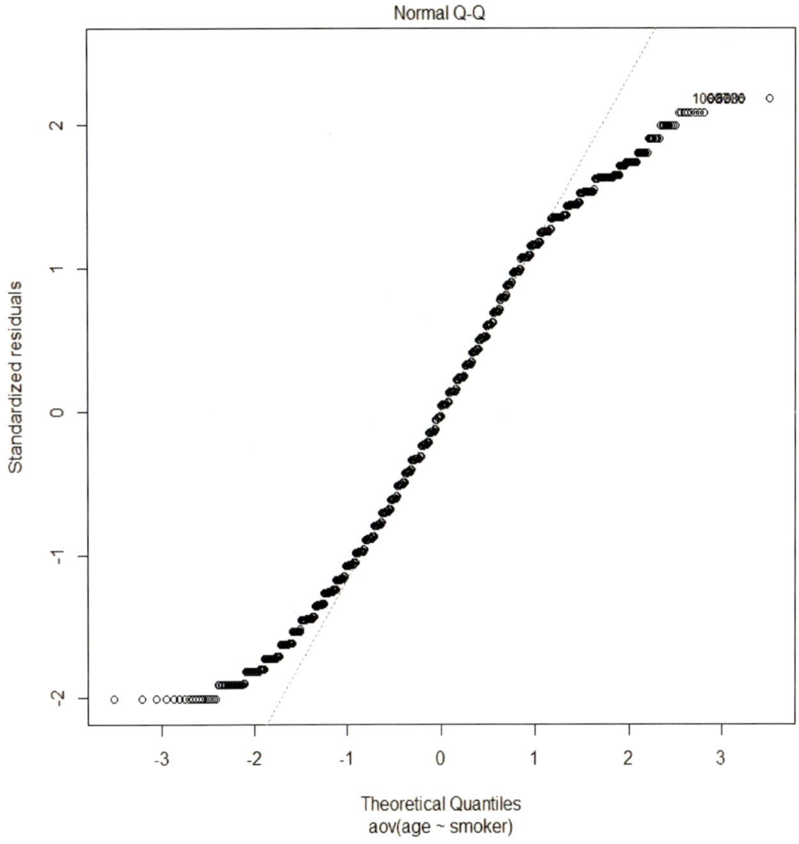


```
shapiro.test(anov_smok$residuals)
##
## Shapiro-Wilk normality test
##
## data: anov_smok$residuals
## W = 0.97798, p-value < 2.2e–16
```

```
##The residuals show no outliers

plot(anov_smok,which = 4)
```

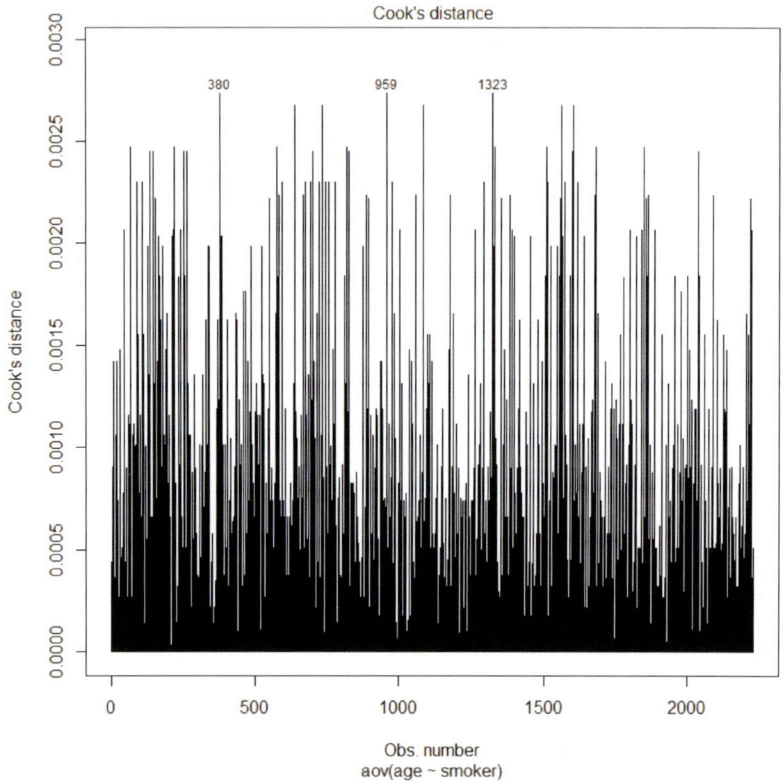


```
outlierTest(anov_smok)
## No Studentized residuals with Bonferroni p < 0.05
## Largest |rstudent|:

##    rstudent unadjusted p-value Bonferroni p
## 638 2.181933      0.029219       NA
```

The Levene's test gives a p-value greater than 0.05 which means that the variance is homogenous across the different smoking groups.

Residuals do not follow their normal line and the Shapiro-Wilk test gives a p-value less than 0.05 which indicates that the residuals are not normally distributed.

There are no significant outliers in these residuals.

So the assumptions are not met so run Kruskal Wallis test for medians instead.

- Step 6: Run Kruskal Wallis test for medians instead.

```
kw_smok <-kruskal.test(age ~ smoker,data = dat)
kw_smok
##
## Kruskal-Wallis rank sum test
##
## data: age by smoker
## Kruskal-Wallis chi-squared = 117.78, df = 2, p-value < 2.2e–16
# pairwise testing

pairwise.wilcox.test(dat$age, dat$smoker)
##
## Pairwise comparisons using Wilcoxon rank sum test with continuity correction
##
## data: dat$age and dat$smoker
##
##          Never smoker Current or former < 1y
## Current or former < 1y < 2e–16   –
## Former >= 1y     0.039    2.8e–12
##
## P value adjustment method: holm
```

The age means were significantly different by the ANOVA test. However the assumptions are not met so we run Kruskal-Wallis test.

The Kruskal-Wallis test and pairwise comparison has shown that current smokers are significantly younger than never smokers and former smokers. Also, former smokers are significantly younger than never smokers.

11.3. COMPARING MEANS BETWEEN MORE THAN TWO POPULATIONS USING KRUSKAL-WALLIS TEST

Kruskal-Wallis test is commonly used for data that do not meet the assumptions of ANOVA test or a small sample size for each level (less than 30)

- *Example:* Mucociliary efficiency from the rate of removal of dust in normal subjects, subjects with obstructive airway disease, and subjects with asbestosis.

Reference: Myles and Douglas (1973).

```
x <- c(2.9, 3.0, 2.5, 2.6, 3.2) # normal subjects
y <- c(3.8, 2.7, 4.0, 2.4, NA)  # with obstructive airway disease
z <- c(2.8, 3.4, 3.7, 2.2, 2.0) # with asbestosis
df<- data.frame(normal = x, obs = y, asb = z)
df
##  normal obs asb
## 1  2.9 3.8 2.8
## 2  3.0 2.7 3.4
## 3  2.5 4.0 3.7
## 4  2.6 2.4 2.2
## 5  3.2 NA 2.0
df<-gather(df, key = condition, value = efficiency)
df
##  condition efficiency
## 1   normal    2.9
## 2   normal    3.0
## 3   normal    2.5
## 4   normal    2.6
## 5   normal    3.2
## 6    obs    3.8
## 7    obs    2.7
## 8    obs    4.0
```

```
## 9   obs   2.4
## 10  obs   NA
## 11  asb   2.8
## 12  asb   3.4
## 13  asb   3.7
## 14  asb   2.2
## 15  asb   2.0
# run test

kw_cond <-kruskal.test(efficiency ~ condition, data = df)
kw_cond
##
## Kruskal-Wallis rank sum test
##
## data: efficiency by condition
## Kruskal-Wallis chi-squared = 0.77143, df = 2, p-value = 0.68
```

The test was not significant so we conclude that there is no difference among the different health conditions in mucociliary efficiency.

11.4. COMPARING PROPORTIONS BETWEEN MORE THAN 2 POPULATIONS USING CHI-SQUARE TEST

Ho: The proportions at one variable are the same for different values of other variables.

Ha: At least, two proportions are different from each other.

11.4.1. Assumptions

1. Unpaired data.
2. The normal approximation to the binomial distribution is valid. The normal approximation can be shown to be approximately true if no expected value in the table is less than 5 (sometimes known as "the rule of five"). Otherwise, Fisher Exact test will be used.

Note: It is important to assign predictor to rows to be able to do pairwise comparison.

- *Example 1:* In the regicor dataset, is there a relation between smoking status and the proportion of cardiovascular events. In this case, the smoking status is the predictor so assigned to rows and cardiovascular events is the outcome so assigned to columns.

Step 1: Viewing the numbers and proportions.

```
library(rcompanion)
library(tidyverse)
library(scales)
library(compareGroups)
data("regicor")
dat<- regicor %>% filter(!is.na(smoker) & !is.na(cv))
table(dat$smoker, dat$cv)
##
##                          No Yes
## Never smoker           1099  37
## Current or former < 1y  506  47
## Former >= 1y            419   8
prop.table(table(dat$smoker, dat$cv),1)
##
##                                No        Yes
## Never smoker           0.96742958 0.03257042
## Current or former < 1y 0.91500904 0.08499096
## Former >= 1y           0.98126464 0.01873536
```

3.3 % of never smokers had cardiovascular events compared to 8.5% of current smokers and 1.9% of former smokers.

Step 2: Plot the cross tab.

```
##The two variables must be factors and we can control their level orders
str(dat$smoker)
## Factor w/ 3 levels "Never smoker,"..: 1 1 2 1 2 1 1 3 3 3 ...
str(dat$cv)
```

```
## Factor w/ 2 levels "No,""Yes": 1 1 1 1 1 1 1 1 1 1 ...
# data for bar plots
dat2<- dat %>% group_by(smoker) %>% count(cv) %>%
 mutate(percentage = percent(n/sum(n)))
dat2
## # A tibble: 6 x 4
## # Groups:  smoker [3]
##   smoker          cv    n percentage
##   <fct>           <fct> <int> <chr>
## 1 Never smoker      No    1099 97%
## 2 Never smoker      Yes   37 3%
## 3 Current or former < 1y No   506 92%
## 4 Current or former < 1y Yes   47 8%
## 5 Former >= 1y      No   419 98%
## 6 Former >= 1y      Yes    8 2%
ggplot(data = dat,aes(x = smoker, fill= cv))+
 geom_bar(position = "fill," width = 0.5)+
 geom_text(data = dat2,aes(y=n,label=percentage),
  position=position_fill(vjust=0.5),size=4)+ylab("percentage")+scale_y_
continuous(labels = percent_format())
```

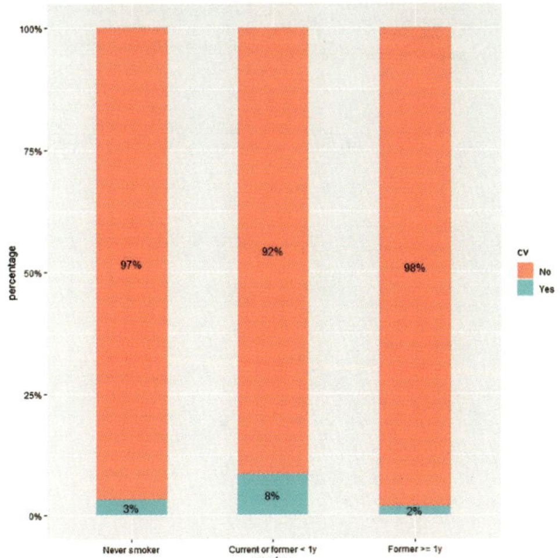

Step 3: Construct the matrix for test.

```
dat<-matrix(table(dat$smoker, dat$cv),nrow = 3)
dat
##   [,1] [,2]
## [1,] 1099  37
## [2,] 506  47
## [3,] 419  8
dimnames(dat)<-list(smoking =c("never,""current,""former"),

        cv_event =c("no,""yes"))
dat
##     cv_event
## smoking   no yes
##  never  1099 37
##  current 506 47
##  former  419  8
```

Step 4: Running the Chi-square test.

```
test<-chisq.test(dat)
test
##
## Pearson's Chi-squared test
##
## data: dat
## X-squared = 32.451, df = 2, p-value = 8.982e–08
test$expected
##     cv_event
## smoking     no   yes
##  never  1086.6087 49.39130
##  current 528.9565 24.04348
##  former  408.4348 18.56522
```

Note that there is no warning as all values of expected table are larger than 5.

Step 5: Pairwise comparison of the Chi-square test.

```
pairwiseNominalIndependence(dat,fisher = FALSE, gtest = FALSE,chisq
= TRUE)
##     Comparison p.Chisq p.adj.Chisq
## 1 never : current 5.87e–06  1.76e–05
## 2  never : former 1.98e–01  1.98e–01
## 3 current : former 1.50e–05  2.25e–05
```

The Chi-square and pair-wise tests has shown that current smokers have significantly higher proportion of cardiovascular events than both never and former smokers. Never smokers and former smokers are equivalent in their proportion of cardiovascular events.

11.5. COMPARING PROPORTIONS BETWEEN MORE THAN 2 POPULATIONS USING FISHER EXACT TEST

Fisher exact test is used when the expected value in any cell of the contingency table is less than 5. For tables in which the use of the Chi-square test is appropriate, the two tests give very similar results although the p-value will be different as the Fisher exact test uses the hypergeometric distribution to calculate its p-value.

Fisher exact test has versions for "less" or "greater" alternative hypotheses. Fisher test use the same matrix formula as the Chi-square test.

- *Example 1:* Fredericks (2012) wanted to know whether checking termite monitoring stations frequently would scare termites away and make it harder to detect termites. He checked the stations (small bits of wood in plastic tubes, placed in the ground near termite colonies) either every day, every week, every month, or just once at the end of the three-month study, and recorded how many had termite damage by the end of the study.

The number of damaged stations that are checked daily is 1, while the number of undamaged stations that are checked daily is 24.

The number of damaged stations that are checked weekly is 5, while the number of undamaged stations that are checked weekly is 20.

The number of damaged stations that are checked monthly is 14, while the number of undamaged stations that are checked monthly is 11.

The number of damaged stations that are checked quarterly is 11, while the number of undamaged stations that are checked quarterly is 14.

The research question was: Is there a relation between the frequency of checking stations and damage?

Step 1: Viewing the numbers and proportions.

```
dat<-matrix(c(1,5,14,11,24,20,11,14),nrow = 4)
dat
##   [,1] [,2]
## [1,]  1  24
## [2,]  5  20
## [3,]  14  11
## [4,]  11  14
dimnames(dat)<-list(frequency=c("daily,""weekly,""monthly,""quater
ly"), damage=c("damaged,""undamaged"))
dat
##      damage
## frequency damaged undamaged
##  daily     1    24
##  weekly     5    20
##  monthly    14    11
##  quarterly   11    14
dat2<- dat %>% data.frame() %>% mutate(frequency=row.names(dat))
%>% gather(key=damage, value = value, -frequency)
dat2
##  frequency  damage value
## 1   daily  damaged   1
## 2  weekly  damaged   5
## 3  monthly  damaged  14
## 4 quarterly  damaged  11
```

```
## 5  daily undamaged  24
## 6  weekly undamaged  20
## 7  monthly undamaged  11
## 8 quarterly undamaged  14
dat3<- dat2 %>% group_by(frequency) %>% mutate(percentage = percent(value/sum(value)))
dat3
## # A tibble: 8 x 4
## # Groups:  frequency [4]
##   frequency damage  value percentage
##   <chr>   <chr>   <dbl> <chr>
## 1 daily   damaged    1 4%
## 2 weekly  damaged    5 20%
## 3 monthly  damaged   14 56%
## 4 quarterly damaged   11 44%
## 5 daily   undamaged  24 96%
## 6 weekly  undamaged  20 80%
## 7 monthly  undamaged  11 44%
## 8 quarterly undamaged  14 56%
```

4% of daily examined stations are damaged compared to 20% of weekly examined, 56% of monthly examined, and 44% of quarterly examined stations.

Step 2: Plotting the data.

```
ggplot(data = dat2,aes(x = frequency, y= value, fill = damage))+
 geom_col(position = "fill," width = 0.5)+
 geom_text(data = dat3,aes(y= value,label=percentage),
    position = position_fill(vjust = 0.5),size=4) +
 ylab("percentage")+scale_y_continuous(labels = percent_format())
```

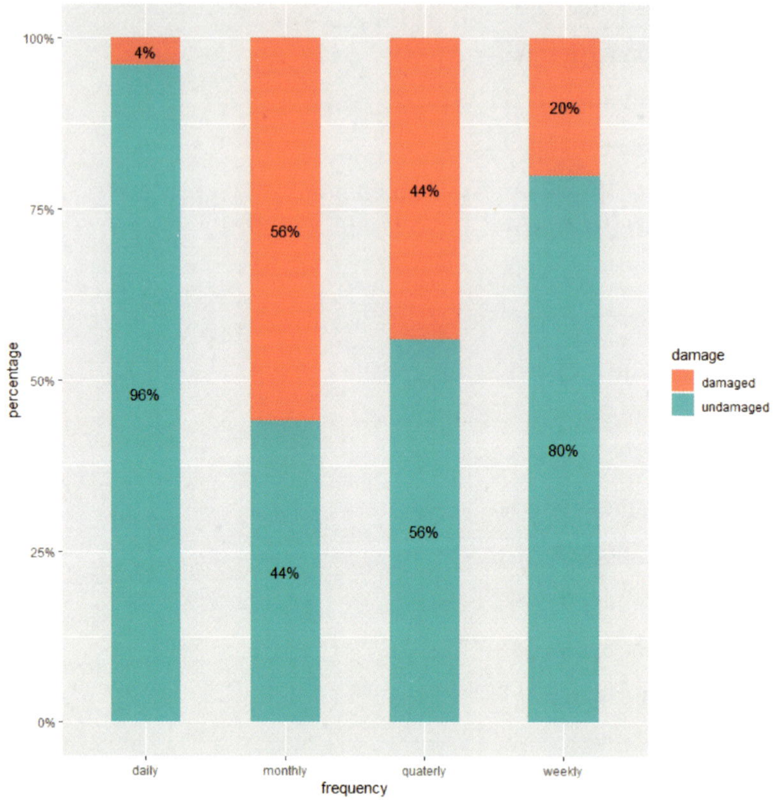

Step 3: Run the test and pairwise comparison.

```
test<-fisher.test(dat)
test
##
## Fisher's Exact Test for Count Data
##
## data: dat
## p-value = 0.0001228
## alternative hypothesis: two.sided
pairwiseNominalIndependence(dat,fisher = TRUE, gtest = FALSE,chisq
= FALSE)
##      Comparison p.Fisher p.adj.Fisher
## 1    daily : weekly 0.189000   0.227000
```

```
## 2  daily : monthly 0.000102   0.000612
## 3  daily : quarterly 0.001920   0.005760
## 4  weekly : monthly 0.018600   0.037200
## 5 weekly : quarterly 0.128000   0.192000
## 6 monthly : quarterly 0.572000   0.572000
```

Fisher Exact test and pair-wise tests were used and showed the following:

1. Daily and weekly inspection are equivalent in their percentage damage.
2. Daily inspection produces significantly less damage than monthly and quarterly inspections.
3. Weekly inspection produces less damage than monthly inspection but equivalent damage to quarterly inspection.
4. Monthly and quarterly inspection are equivalent in the percentage damage.

11.6. COMPARING SURVIVAL CURVES BETWEEN MORE THAN TWO POPULATIONS USING LOG-RANK TEST

Ho: All survival curves are equivalent for all groups or populations.

Ha: At least two populations are different in their survival curves.

The log-rank test compares the distance between the Kaplan Meier curves and calculates a p-value that determines if there is a significant difference between the different survival curves. The survdiff() function of the survival package performs the log-rank test and requires:

1. Formula in the form, survival object ~ grouping factor.
2. Dataframe containing the variables.

- *Example 1:* Is there a significant difference in survival curves according to cancer cell type in veteran dataset?

Step 1: Build the survival object.

```
library(survival)
library(survminer)
data("veteran")
km<-Surv(veteran$time,veteran$status)
```

Step 2: Plot the survival curves for different cancer cell types.

```
cancer_fit<-survfit(km ~celltype, data = veteran)
ggsurvplot(cancer_fit,data = veteran)
```

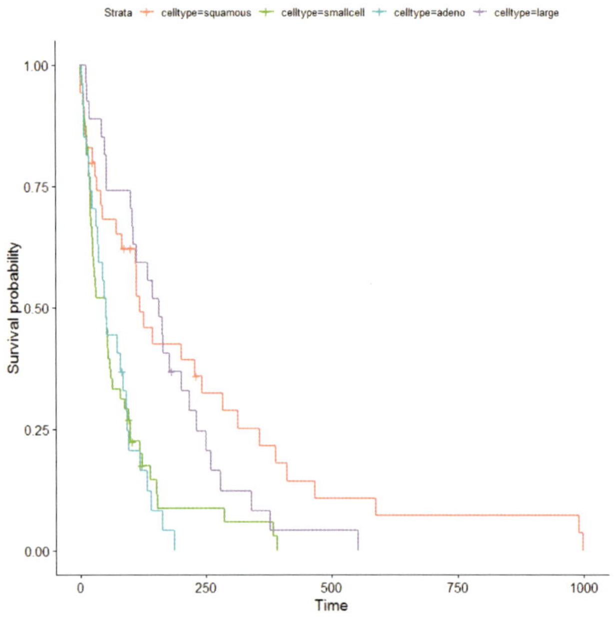


```
ggsurvplot(cancer_fit,pval = TRUE,pval.method = TRUE)
```

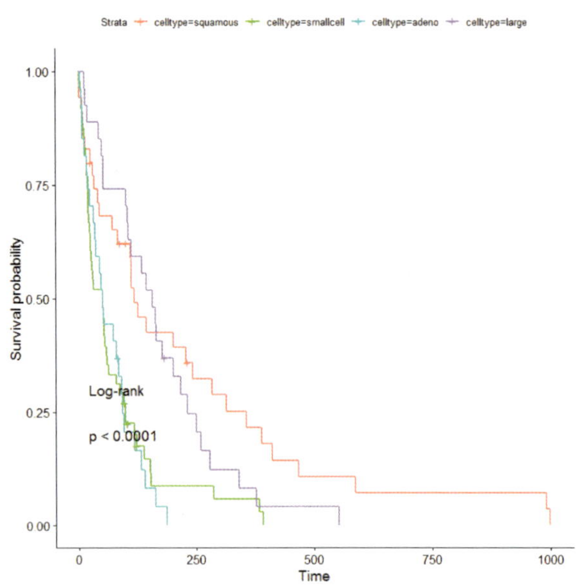

```
##Step 3: Run the log-rank test

survdiff(formula = km ~ celltype, data = veteran)
## Call:
## survdiff(formula = km ~ celltype, data = veteran)
##
##          N Observed Expected (O-E)^2/E (O-E)^2/V
## celltype=squamous 35    31   47.7   5.82   10.53
## celltype=smallcell 48    45   30.1   7.37   10.20
## celltype=adeno   27    26   15.7   6.77   8.19
## celltype=large   27    26   34.5   2.12   3.02
##
## Chisq= 25.4 on 3 degrees of freedom, p= 1e–05
##Step 4: Pairwise comparison

pairwise_survdiff(Surv(time, status) ~ celltype, data = veteran)
##
## Pairwise comparisons using Log-Rank test
##
## data: veteran and celltype
##
##      squamous smallcell adeno
## smallcell 0.00134 –     –
## adeno   0.00134 0.75565  –
## large   0.43731 0.00331  0.00016
##
## P value adjustment method: BH
```

Log_rank test and pairwise comparison has shown that:

1. Large cell and squamous cell cancers are equivalent in their survival rates.
2. Large cell cancer has higher survival rate than small cell and adeno cell cancers.

3. Adeno cell cancer has significantly lower survival rate (more deaths) than squamous cell cancer.
4. Small cell and Adeno cell cancers are equivalent in their survival rates.
5. Small cell cancer has significantly lower survival rate (more deaths) than squamous cell cancer.

- *Example 2:* Is there a difference in survival curves according to diagnosis time (from diagnosis to treatment)? Diagnosis time is a numerical value so we cut it into 4 quantiles, each containing nearly 25% of the observations. We are expecting that increasing time from diagnosis to randomization (treatment) will decrease the survival rate but does data provide evidence for that?

```
##Define the boundary limits for quantiles, we put 0 to include 1 values of diagtime

summary(veteran$diagtime)
## Min. 1st Qu. Median Mean 3rd Qu. Max.
## 1.000 3.000 5.000 8.774 11.000 87.000
veteran$quant<-cut(veteran$diagtime, breaks = c(0,3,5,11,87), labels=c("1,""2,""3,""4"))
table(veteran$quant)
##
## 1 2 3 4
## 42 33 30 32
##To plot the survival curves for each quantile

quantile_fit<-survfit(km ~quant, data = veteran)
ggsurvplot(quantile_fit,data = veteran)
```

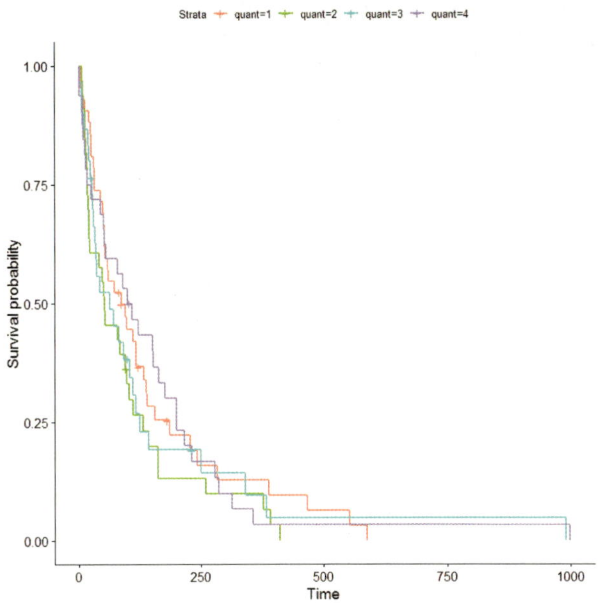


```
ggsurvplot(quantile_fit,pval = TRUE,pval.method = TRUE)
```

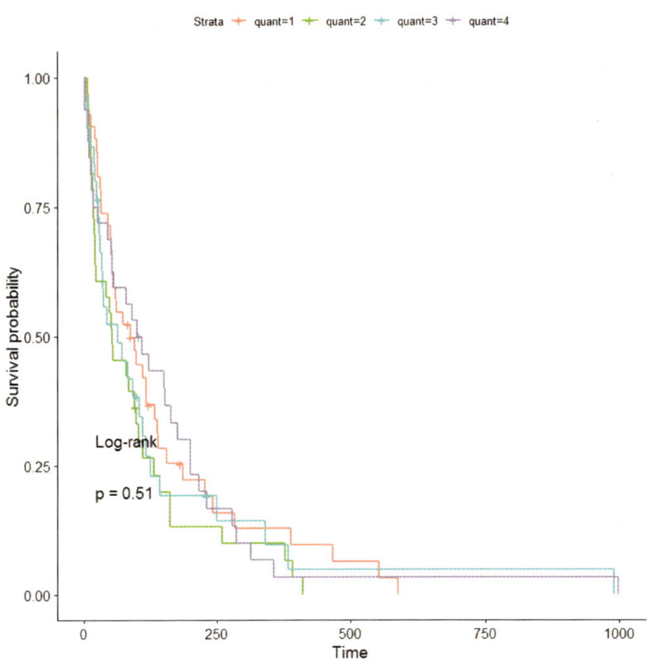

```
##To run the log-rank test

survdiff(formula = km ~ quant, data = veteran)
## Call:
## survdiff(formula = km ~ quant, data = veteran)
##
##          N Observed Expected (O-E)^2/E (O-E)^2/V
## quant=1 42       38     42.1    0.3968     0.605
## quant=2 33       32     25.9    1.4239     1.825
## quant=3 30       27     25.5    0.0828     0.105
## quant=4 32       31     34.4    0.3443     0.493
##
## Chisq= 2.3 on 3 degrees of freedom, p= 0.5
##Pairwise comparison although it is not needed

pairwise_survdiff(Surv(time, status) ~ quant, data = veteran)
##
## Pairwise comparisons using Log-Rank test
##
## data: veteran and quant
##
##   1    2    3
## 2 0.74 -    -
## 3 0.76 0.76 -
## 4 0.93 0.76 0.76
##
## P value adjustment method: BH
```

Running log-rank test and pairwise testing has showed that there is no difference between the diagnosis time quantiles and survival rates.

CHAPTER 12

SIMPLE AND MULTIPLE LINEAR REGRESSION

CONTENTS

12.1. AN OVERVIEW OF SIMPLE REGRESSION

Simple regression is a general framework for relating any outcome type (continuous, binary, or time to event) to a single predictor by a linear equation. We called it simple because we only consider one predictor

All tests covered in previous lectures can be done by simple regression models

1. Comparing means between two or more groups, using t-test or ANOVA, can be done using simple linear regression
2. Comparing proportions between two or more groups, using Chi-square or Fisher exact test, can be done using simple logistic regression
3. Comparing survival curves between two or more groups, using log rank test, can be done using simple Cox proportional hazards (PH) regression

In addition, regression models can be extended to allow using multiple predictors, and so called multiple regression, so we can better predict the outcomes using more than one predictor

12.1.1. Basic Structure of Simple Regression Equation

is the predictor of interest which can be categorical or continuous.

depends on the outcome type:

1. For continuous outcomes, is the mean of outcome, ;
2. For binary outcomes, is the ln(odds) of binary outcome, ;
3. For time to event outcomes, is ln(hazard rate).

12.2. SIMPLE LINEAR REGRESSION WITH CATEGORICAL PREDICTOR

The categorical predictor can be:

1. Binary predictor with only two values;
2. Categorical predictor with more than two values.

Linear regression is fitted using the lm() function. lm() function has two main arguments, formula takes the form and data takes the dataset that contains the outcome and predictor of interest

12.2.1. Binary Predictor

Can we predict weight of participants using their sex (binary predictor) or their smoking habit (categorical predictor)?

Fitting linear regression model using sex predictor

Step 1: Examine your data, Univariate analysis

```
library(tidyverse)
library(car)
library(knitr)
library(compareGroups)
data("regicor")
str(regicor$sex)
## Factor w/ 2 levels "Male,""Female": 2 2 1 2 2 2 1 2 2 1 ...
## – attr(*, "label")= chr "Sex"
str(regicor$weight)
## num [1:2294] 64 67 70 68 NA 43.5 79.2 45.8 53 62 ...
## – attr(*, "label")= Named chr "Weight (Kg)"
##  ..- attr(*, "names")= chr "weight"
### Plot a histogram

ggplot(data = regicor, aes(x = weight))+geom_histogram()
```

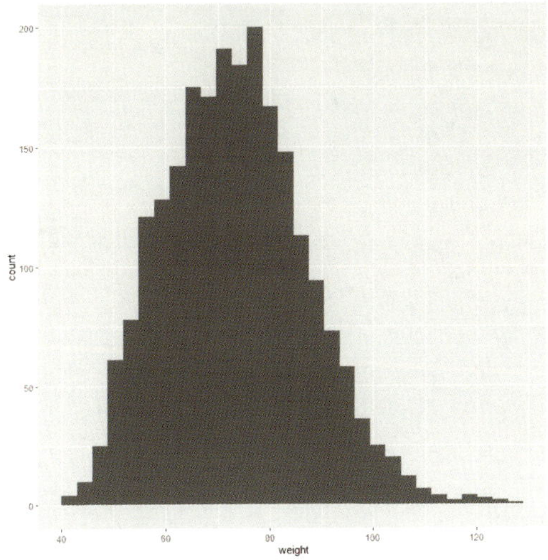

Sex (predictor) is a factor variable while weight (outcome) is a numeric variable. Plotting a histogram of weight has revealed that it is nearly normally distributed.

Step 2: Bivariate analysis between the predictor and outcome.

Linear regression assumes that there are linear relationship between the response and predictors.

```
#### calculating the mean weight for each group

regicor %>% group_by(sex) %>% summarise(mean_wt = mean(weight,
na.rm = TRUE))
## # A tibble: 2 x 2
##  sex  mean_wt
##  <fct> <dbl>
## 1 Male   79.7
## 2 Female  67.6
#### Plot a boxplot and a line between medians

ggplot(data = regicor, aes(x = sex, y = weight, col = sex))+
  geom_boxplot()+ stat_summary(fun = median, geom = "smooth,"
aes(group=1))
```

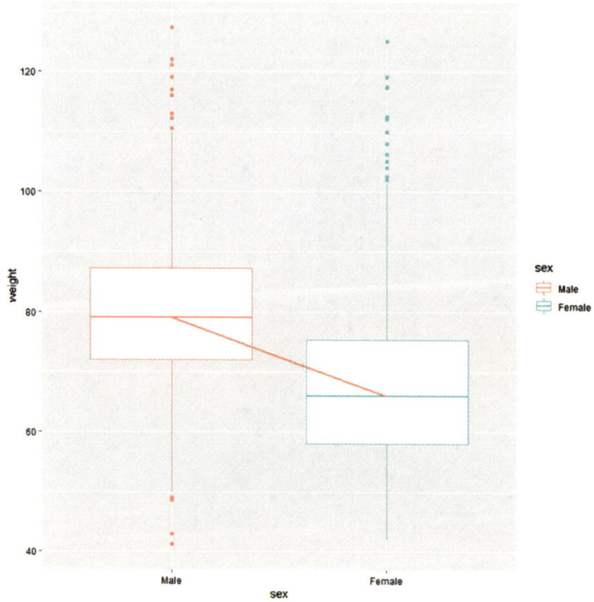

As we move from males to females, the mean (and median) weight decreased from 79.7 kg to 67.6 kg.

Step 3: Fitting the model

```
mod<-lm(formula= weight ~ sex, data = regicor)
summary(mod)
##
## Call:
## lm(formula = weight ~ sex, data = regicor)
##
## Residuals:
##    Min    1Q Median   3Q   Max
## -38.516 -8.580 -1.280  7.734 57.420
##
## Coefficients:
##         Estimate Std. Error t value Pr(>|t|)
## (Intercept) 79.7159   0.3714  214.6  <2e-16 ***
## sexFemale  -12.1356   0.5163  -23.5  <2e-16 ***

## ---
## Signif. codes: 0 '***' 0.001 '**' 0.01 '*' 0.05 .'' 0.1 ' ' 1
##
## Residual standard error: 12.26 on 2257 degrees of freedom
##  (35 observations deleted due to missingness)
## Multiple R-squared: 0.1966, Adjusted R-squared: 0.1963
## F-statistic: 552.4 on 1 and 2257 DF, p-value: < 2.2e-16
```

From the summary of the model, the linear regression equation is estimated as:

We put hats as this regression equation with ,, are estimated from our sample data.

= 1 for Females and =0 for males

This equation estimates two values:

1. Mean weight for males = = 79.7 kg since = 0
2. Mean weight for females = 79.7–12.1 = 67.6 kg since = 1 for females

So in simple regression:

1. or intercept is the mean of continuous outcome for reference category;
2. or slope is the mean difference between the two groups who differ by only one unit of .

Since for binary predictor (as sex), can only takes two values, 0 for males and 1 for females, so –12.1 kg is the mean weight difference between females and males in this population. In other words, the weight of females is lower than the weight of males in this population by 12.1 kg on the average.

12.2.2. Categorical Predictor

Fitting linear regression model using smoker predictor:

Step 1: Examine your data, Univariate analysis

```
str(regicor$smoker)
## Factor w/ 3 levels "Never smoker,"..: 1 1 2 1 NA 2 1 1 3 3 ...
## – attr(*, "label")= Named chr "Smoking status"
## ..- attr(*, "names")= chr "smoker"
```

Smoker (predictor) is a factor variable, while weight (outcome) is a numeric variable. Plotting a histogram of weight has revealed that it is nearly normally distributed.

Step 2: Bivariate analysis between the predictor and outcome

Linear regression assumes that there are linear relationship between the response and predictors.

```
#### calculating the mean weight for each group

dat<- regicor %>% filter(!is.na(smoker))
dat %>% group_by(smoker) %>% summarise(mean_wt = mean(weight,
na.rm = TRUE))
## # A tibble: 3 x 2
## smoker        mean_wt
```

```
## <fct>        <dbl>
## 1 Never smoker       71.1
## 2 Current or former < 1y  74.5
## 3 Former >= 1y       78.2
```

```
#### Plot a boxplot and a line between medians
ggplot(data = dat, aes(x = smoker, y = weight, col = smoker))+
  geom_boxplot()+ stat_summary(fun = median, geom = "smooth," aes(group=1), col = "black")
```

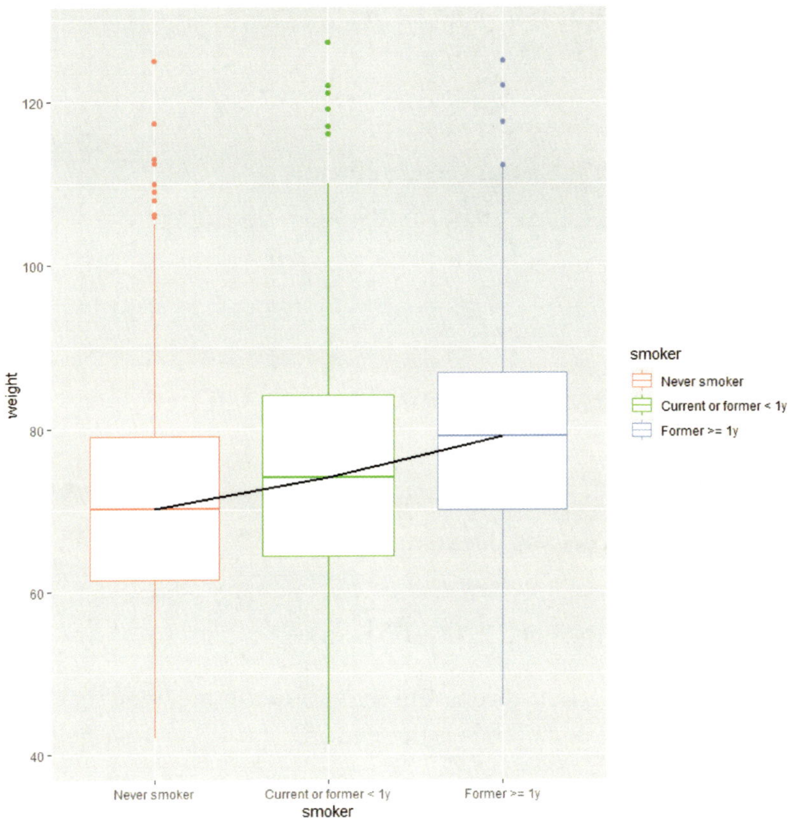

As we move from never smokers to current and then to former smokers, the mean (and median) weight increased from 71.1 kg to 74.5 kg and then to 78.2 kg.

Step 3: Fitting the model

```
mod<-lm(formula= weight ~ smoker, data = dat)
summary(mod)
##
## Call:
## lm(formula = weight ~ smoker, data = dat)
##
## Residuals:
##     Min      1Q  Median      3Q     Max
## -33.311  -9.511  -0.311   8.783  53.873
##
## Coefficients:
##                          Estimate Std. Error t value Pr(>|t|)
## (Intercept)               71.1267     0.3887 182.986  < 2e-16 ***
## smokerCurrent or former < 1y  3.3843  0.6765   5.003  6.1e-07 ***
## smokerFormer >= 1y         7.0905     0.7517   9.433  < 2e-16 ***
## ---
## Signif. codes: 0 '***' 0.001 '**' 0.01 '*' 0.05 '.' 0.1 ' ' 1
##
## Residual standard error: 13.4 on 2206 degrees of freedom
##   (24 observations deleted due to missingness)
## Multiple R-squared: 0.04099,  Adjusted R-squared: 0.04012
## F-statistic: 47.14 on 2 and 2206 DF, p-value: < 2.2e-16
```

- **Interpretation:** From the summary of the model, the linear regression equation is estimated as:

The equation estimated for each smoker level and set never smoker as reference level

= 1 for current smokers and =0 for other smoker levels

= 1 for former smokers and =0 for other smokers levels

This equation estimates 3 mean weights:

1. Mean weight for never smokers, = 71.1 kg since current and former= 0.
2. Mean weight for current smokers, = 71.1+3.4 = 74.5 kg since current = 1 for current smokers and former = 0.
3. Mean weight for former smokers, = 71.1+7.1 = 78.2 kg since former = 1 for former smokers and current = 0.

So in this simple regression,

1. or intercept is the mean of continuous outcome for reference category, never smokers
2. is the mean weight difference between never smokers and current smokers.
3. is the mean weight difference between never smokers and former smokers.

12.3. SIMPLE LINEAR REGRESSION WITH CONTINUOUS PREDICTOR

Can we predict the weight of participants using their physical activity (continuous predictor)?

Step 1: Examine your data, Univariate analysis

```
str(regicor$phyact)
## num [1:2294] 304 160 553 522 NA ...
## – attr(*, "label")= Named chr "Physical activity (Kcal/week)"
## ..- attr(*, "names")= chr "phyact"
str(regicor$weight)
## num [1:2294] 64 67 70 68 NA 43.5 79.2 45.8 53 62 ...
## – attr(*, "label")= Named chr "Weight (Kg)"
## ..- attr(*, "names")= chr "weight"
### Plot a histogram of weight

ggplot(data = regicor, aes(x = weight))+geom_histogram()
```

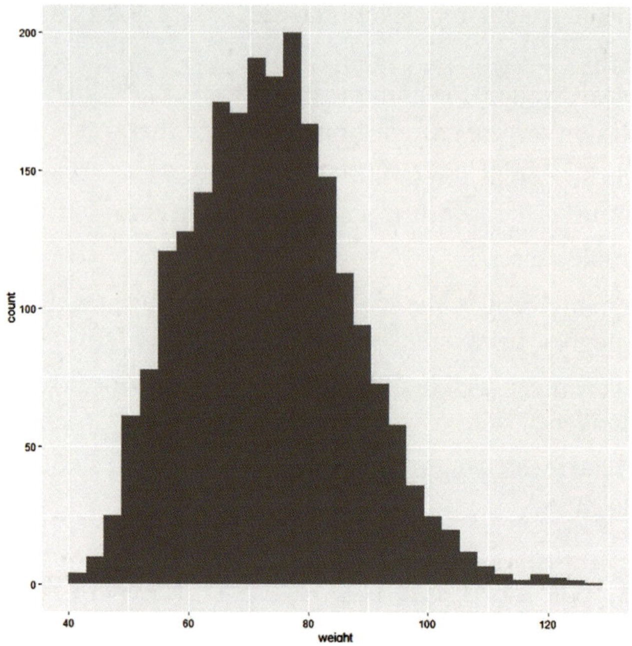

Plot a histogram of physical activity

```
ggplot(data = regicor, aes(x = phyact))+geom_histogram()
```

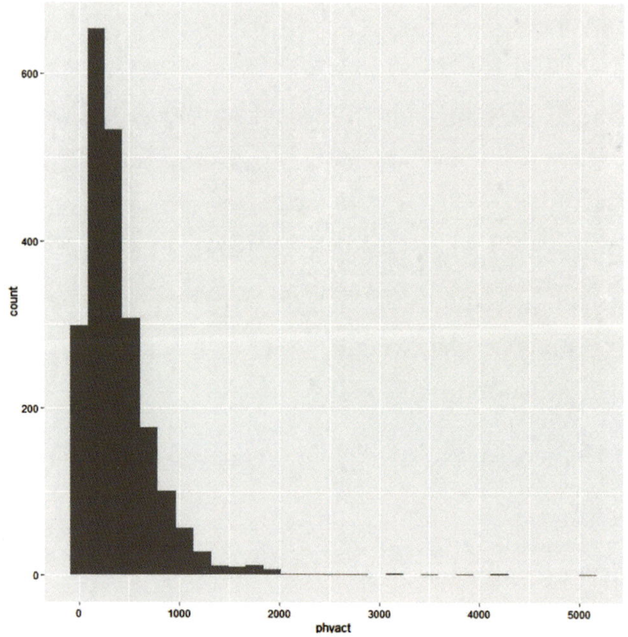

Both variables, weight, and physical activity, are numeric variables. Plotting a histogram of weight has revealed that it is nearly normally distributed, while physical activity is right skewed.

Step 2: Bivariate analysis between the predictor and outcome

Linear regression assumes that there are linear relationship between the response and predictors. We can examine this visually by plotting a scatter plot and numerically by calculating the correlation coefficient. We will use the spearman correlation coefficient as physical activity is a right skewed distribution.

```
library(ggpubr)
dat<- regicor %>% filter(!is.na(weight) & !is.na(phyact))
#### Plot a scatterplot

ggplot(data = dat, aes(x = phyact, y = weight))+ geom_point()
```

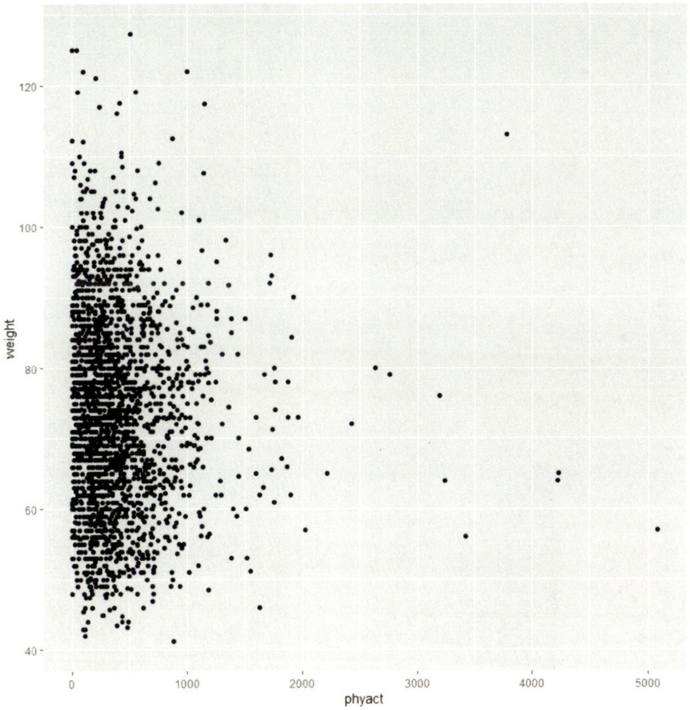


```
### add the linear regression line using geom_smooth and its 95% confidence interval

ggplot(data = dat, aes(x = phyact, y = weight))+
```

```
geom_point() + geom_smooth(method = "lm")
```

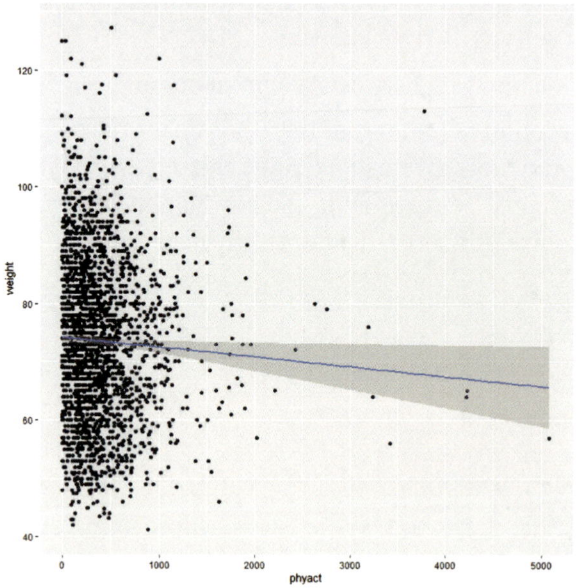


```
### add correlation coefficient by spearman method

ggscatter(data = dat, x = "phyact," y= "weight," add = "reg.line,"
    cor.coeff.args = list(method= "spearman," label.x.npc = "center,"
            label.y.npc = "top"),cor.coef = TRUE)
```

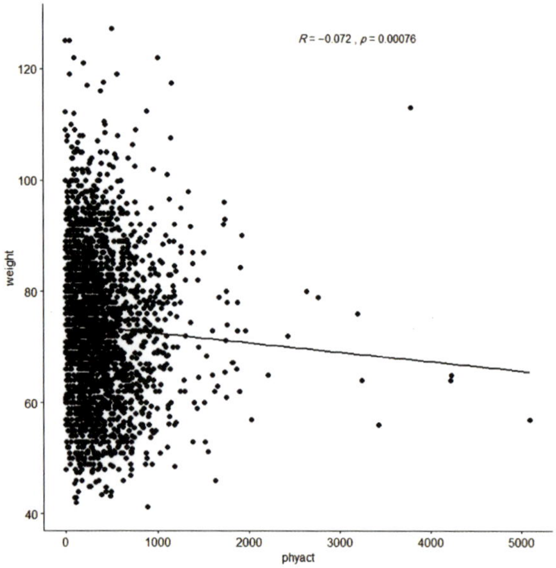

There is a weak negative correlation between physical activity and weight (–0.07) although it is statistically significant (p-value = 0.00076).

Step 3: Fitting the model

```
mod<-lm(formula= weight ~ phyact, data = dat)
summary(mod)
##
## Call:
## lm(formula = weight ~ phyact, data = dat)
##
## Residuals:
##  Min  1Q Median  3Q  Max
## –32.014 –9.933 –0.443  8.909 53.858
##
## Coefficients:
##        Estimate Std. Error t value Pr(>|t|)
## (Intercept) 74.1979165 0.4205901 176.414  <2e–16 ***
## phyact   –0.0016988 0.0007537 –2.254  0.0243 *
## ---
## Signif. codes: 0 '***' 0.001 '**' 0.01 '*' 0.05 .'' 0.1 ' ' 1
##
## Residual standard error: 13.7 on 2185 degrees of freedom
## Multiple R-squared: 0.00232,  Adjusted R-squared: 0.001863
## F-statistic: 5.081 on 1 and 2185 DF, p-value: 0.02429
```

From the summary of the model, the linear regression equation is estimated as

This equation estimates two values:

1. Mean weight for participants with no physical activity, = = 74.2 kg since = 0
2. Slope, = –0.002 , has many meanings:

i. Negative number meaning that linear regression line is of negative slope and as the value of physical activity increases, the weight value decreases.

ii. The average change in participant's weight for one unit change in physical activity.

iii. The mean difference in weight for groups of participants from Girona who differ by 1 physical activity. In other words, each 1 unit increase in physical activity is associated with 0.002 (2 grams) decrease in the body weight.

This estimate is constant across all values of physical activity in our data (0–5083.2) but we cannot extrapolate this relationship to higher values.

- Examples:

1. The mean weight difference between participants with 1 Kcal/week compared to participants with 0 Kcal/week = 0.002 kg or 2 grams.
2. The mean weight difference between participants with 10 Kcal/week compared to participants with 9 Kcal/week = 0.002 kg or 2 grams.
3. The mean weight difference between participants with 5000 Kcal/week compared to participants with 4999 Kcal/week = 0.002 kg or 2 grams.
4. The mean weight difference between participants with 50 Kcal/week compared to participants with 10 Kcal/week = (50–10) = 40*0.002 = 0.08 kg or 80 grams.
5. The mean weight difference between participants with 500 Kcal/week compared to participants with 100 Kcal/week = (500–100) = 400*0.002 = 0.8 kg or 800 grams.
6. The mean weight difference between participants with 5000 Kcal/week compared to participants with 100 Kcal/week = (5000–100) = 4900*0.002 = 9.8 kg.
7. The mean weight difference between participants with 6000 Kcal/week compared to participants with 100 Kcal/week, we cannot estimate this mean weight since in our sample data we have no data points for participants with 6000 Kcal/week.

Another note:

Some continuous predictors do not have a zero value naturally, for example height. In that case, R will also estimate an intercept but this intercept is scientifically meaningless and it is just a placeholder to define the linear equation.

12.4. MULTIPLE REGRESSION

12.4.1. Importance

1. Improves the prediction of an outcome by taking multiple predictors in one equation.
2. Allows for adjustment of potential confounders or predictors.

12.4.2. Basic Structure of Multiple Regression Equation

are the predictors of interest which can be categorical or continuous. In cox regression for time to event data, replace with ln() which is the baseline hazard at time t and it varies over time.

depends on the outcome type:

1. For continuous outcomes, is the mean of outcome, and the model is called multiple linear regression model.
2. For binary outcomes, is the ln(odds) of binary outcome, and the model is called multiple logistic regression model.
3. For time to event outcomes, is ln(hazard rate) and the model is called multiple cox proportional hazard regression model.

Each slope, estimates the difference in for a one unit difference in the corresponding x and adjusted for all other x variables in the model.

12.4.3. Building Multiple Linear Regression Model

For example, we want to build a multiple linear regression model for predicting participants' weight from his sex, smoking status, and physical activity (all previous predictors) in the regicor dataset.

```
mod_mult<- lm(weight ~ sex + smoker + phyact, data = regicor)
summary(mod_mult)
##
## Call:
## lm(formula = weight ~ sex + smoker + phyact, data = regicor)

##
## Residuals:
##    Min     1Q  Median    3Q    Max
## -36.416 -8.639 -0.965  7.826 56.381
```

```
##
## Coefficients:
##                  Estimate Std. Error t value Pr(>|t|)
## (Intercept)       8.149e+01 6.428e-01 126.769 <2e-16 ***
## sexFemale        -1.274e+01 6.111e-01 -20.850 <2e-16 ***
## smokerCurrent or former < 1y -2.047e+00 6.839e-01 -2.992 0.0028 **
## smokerFormer >= 1y    -4.059e-02 7.794e-01 -0.052 0.9585
## phyact          -2.060e-03 6.778e-04 -3.039 0.0024 **
## ---
## Signif. codes: 0 '***' 0.001 '**' 0.01 '*' 0.05 '.' 0.1 ' ' 1
##
## Residual standard error: 12.24 on 2135 degrees of freedom
## (154 observations deleted due to missingness)
## Multiple R-squared: 0.2058, Adjusted R-squared: 0.2043
## F-statistic: 138.3 on 4 and 2135 DF, p-value: < 2.2e-16
confint(mod_mult)
##                  2.5 %    97.5 %
## (Intercept)       80.226806535 8.274796e+01
## sexFemale        -13.939129331 -1.154245e+01
## smokerCurrent or former < 1y -3.387728828 -7.053842e-01
## smokerFormer >= 1y    -1.569025022 1.487850e+00
## phyact          -0.003389182 -7.305678e-04
```

The overall model:

The p-value of the F-statistic for the whole model is < 2.2e–16, which is highly significant. This means that, at least, one of the predictor variables is significantly related to the outcome variable.

The intercept coefficient:

The intercept is the mean weight for participants with reference categories which is 81.5 kg. In this case, it is the mean weight for male and never smoked participants with 0 physical activity. This mean weight can be as low as 80.2 kg and as high as 82.7 kg in the population from which the sample was drawn. This is based on the calculated 95% CI.

Note that hypothesis testing for intercept does not make sense since the average weight of participants who are males, never smoked, and with no physical activity cannot be zero so it is not needed but CIs may be of interest.

The sex coefficient:

The coefficient for female = –12.7 kg which is the mean weight difference between females and males who have the same smoking habit and the same physical activity. In other words, the weight of females is lower than the weight of males in this population by 12.7 kg on the average when males and females have the same smoking habit and physical activity.

Because this is a sample estimate and based on 95% CI, this mean weight difference between females and males, who have the same smoking and physical activity habits, can be as low as 11.5 kg and as high as 13.9 kg.

The p-value for sex was less than 0.05. This means that sex is a significant predictor of weight in this population.

Compared to the unadjusted estimate for simple regression with sex alone (mean difference = –12.1, 95%CI = –13.1 to –11.1), there is a little difference between adjusted (in multiple model) and unadjusted estimate. This means that the association between weight and sex is little confounded by smoking habit or physical activity.

The smoker coefficient:

The coefficient for current smokers = –2.05 kg which is the mean weight difference between current smokers and never smokers for participants having the same sex and the same physical activity. On the other hand, the coefficient for former smokers = –0.04 kg which is the mean weight difference between former smokers and never smokers for participants having the same sex and the same physical activity.

Because this is a sample estimate and based on 95% CI, this mean weight difference between current smokers and never smokers, who have the same sex and physical activity habits, can be as low as 0.7 kg and as high as 3.4 kg. However, the 95% CI for the mean weight difference between former smokers and never smokers contains 0 which means that there is no difference in weights between former smokers and never smokers in this population after adjusting for sex and physical activity.

The p-value for smoker was less than 0.05. This means that smoking habit (especially the current smoking) is a significant predictor of weight in this population.

Compared to the unadjusted estimate for simple regression with smoker alone (mean difference for current smokers = 3.4, 95%CI = 2.06 to 7.4, and mean difference for former smokers = 7.1, 95%CI = 5.6 to 8.6), there is a great difference between adjusted (in multiple model) and unadjusted estimate. This means that the association between weight and smoking is confounded by sex and physical activity.

The physical activity coefficient:

The coefficient for physical activity = –0.002 kg or 2 grams which is the mean weight difference between 2 groups of participants who differ in their physical activity by 1 Kcal/week but have the same sex and smoking habit. In other words, for the same sex and smoking habit, each 1 Kcal/week increase in physical activity is associated with 2 grams decrease in the participant's weight from Girona.

Because this is a sample estimate and based on 95% CI, this mean weight difference between 2 groups of participants who differ in their physical activity by 1 Kcal/week, but have the same sex and smoking habit, can be as low as 0.7 grams and as high as 3 grams.

The p-value for physical activity was less than 0.05. This means that physical activity is a significant predictor of weight in this population.

Compared to the unadjusted estimate for simple regression with physical activity alone (mean difference = –0.002, 95%CI = –0.003 to –0.0002), there is a little difference between adjusted (in multiple model) and unadjusted estimate. This means that the association between weight and physical activity is little confounded by smoking habit or sex. Our model equation can be written as follow: mean(weight) = 81.5 – 12.74**Female –2.05**current smoker – 0.04**former smoker – 0.002**physical activity

12.5. EVALUATING THE REGRESSION MODEL

The algorithm used to estimate the linear equation is called least squares estimation that finds the line that is closest to all points in the sample or that minimizes the residuals. Residuals are defined as the difference between each data point and its predicted point (on the regression line)

By plotting predictions on x-axis and actual values on y-axis, if all points fall on the diagonal line of x = y, it will be a perfect prediction. In addition, there should be an equal number of points on left and right of line and close to it. This means that errors are not systematic or are not correlated with the outcome

When there are regions where all points are above or below line, This is systematic errors where errors correlated with the outcome. This indicates that you do not have all important variables in your model or you need another algorithm that can find more complex relations in the data.

Residual plots plot the residuals (actual outcome-prediction) against the predictions. In a model with no systematic error, the residuals will be evenly distributed between the positive and negative sides and have the same magnitude, above and below. However, when there are systematic errors, there will be clusters of all positive or all negative residuals.

12.5.1. Graphical Evaluation of the Physical Activity Model

1. Plot the model's predictions against the actual weight

```
## fitting the model

mod<-lm(formula= weight ~ phyact, data = dat)
summary(mod)
##
## Call:
## lm(formula = weight ~ phyact, data = dat)
##
## Residuals:
##   Min   1Q Median   3Q   Max
## –32.014 –9.933 –0.443  8.909 53.858
##
## Coefficients:
##       Estimate Std. Error t value Pr(>|t|)
## (Intercept) 74.1979165 0.4205901 176.414  <2e–16 ***
## phyact   –0.0016988 0.0007537 –2.254  0.0243 *
## ---
## Signif. codes: 0 '***' 0.001 '**' 0.01 '*' 0.05 .'' 0.1 ' ' 1
##
## Residual standard error: 13.7 on 2185 degrees of freedom
## Multiple R-squared: 0.00232,  Adjusted R-squared: 0.001863
## F-statistic: 5.081 on 1 and 2185 DF, p-value: 0.02429
```

```
# Make predictions from the model

dat$predictions <- predict(mod)
# plot predictions (on x-axis) versus the actual weight on y-axis

ggplot(dat, aes(x = predictions, y = weight)) +
 geom_point() +
 geom_abline()
```

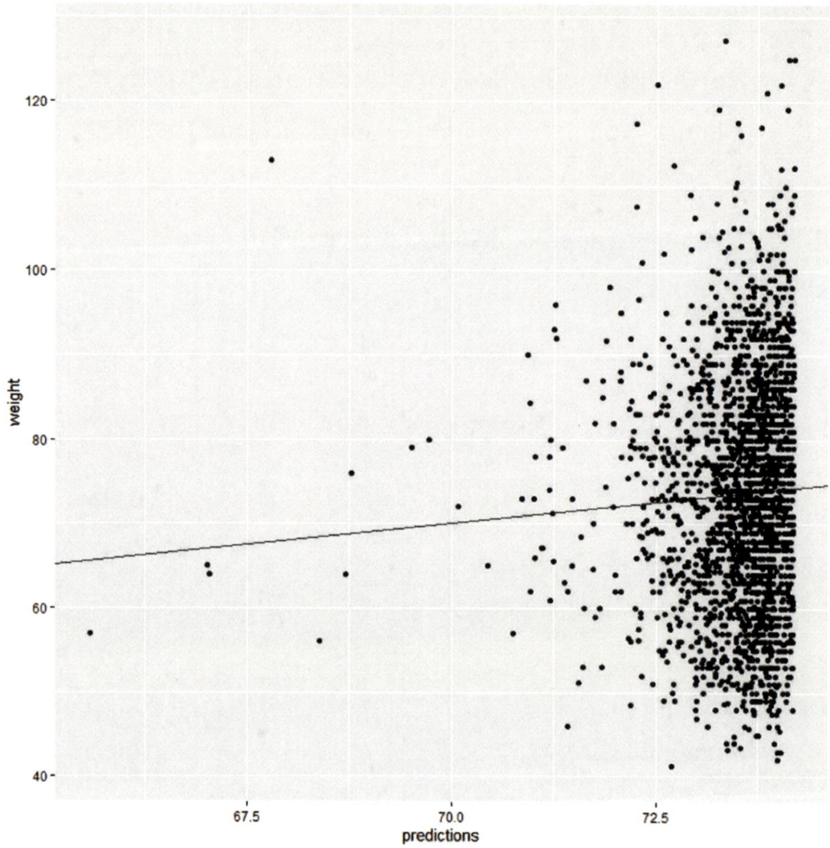

2. Calculate the residuals
3. Plot predictions against residuals.

```
# Calculate residuals

dat$residuals <- dat$weight – dat$predictions
summary(dat$residuals)
```

```
## Min. 1st Qu. Median  Mean 3rd Qu.  Max.
## –32.014 –9.933 –0.443  0.000  8.909 53.858
# plot predictions (on x-axis) versus the residuals

ggplot(dat, aes(x = predictions, y = residuals)) +
 geom_pointrange(aes(ymin = 0, ymax = residuals)) +
 geom_hline(yintercept = 0, linetype = 3) +
 ggtitle("residuals vs. linear model prediction")
```

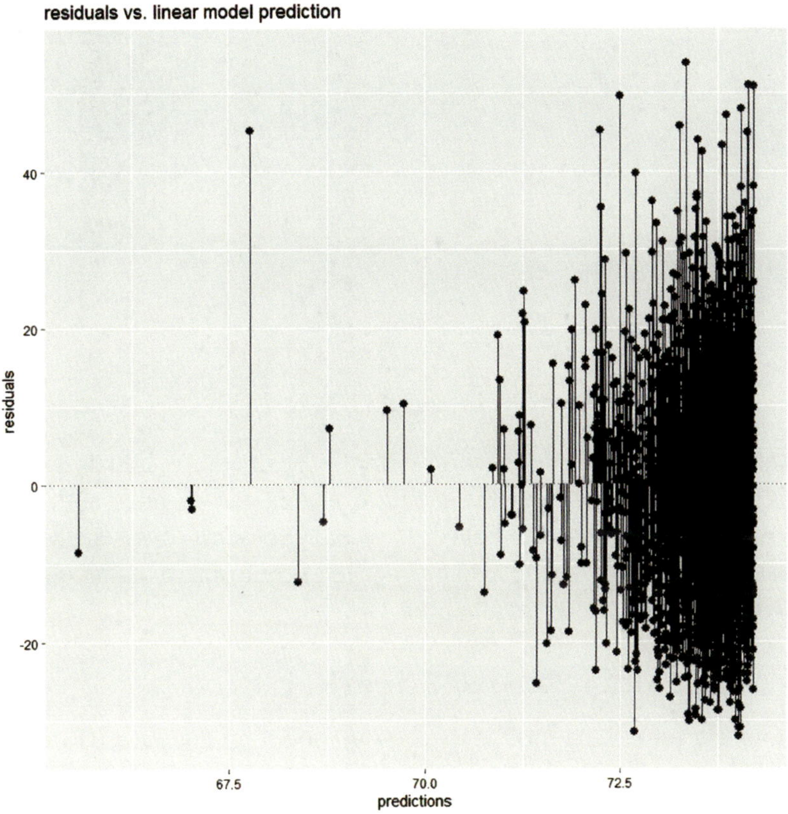


```
# you can also plot the residual plot using plot() function

plot(mod, which = 1)
```

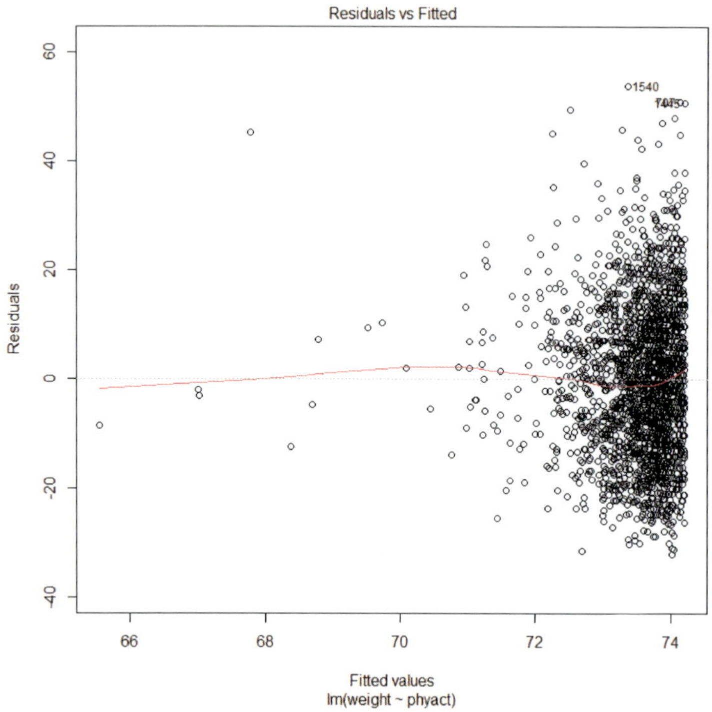

As we can see, there is no systematic error, the residuals are nearly evenly distributed between the positive and negative sides and have the same magnitude, above and below. However, some residuals are very large (> 50). The residuals should also be checked for normality using qqplot. See below for multiple model evaluation

12.5.2. Root Mean Squared Error (RMSE)

It is a metric for evaluating the regression model prediction performance

= , is the error or residuals vector

RMSE is the square root of the mean squared error of the model on the data. RMSE is the typical prediction error of the model on your data. The lower RMSE, the better is our model.

```
# For convenience put the residuals in the variable res

res <- dat$residuals

# Calculate RMSE, assign it to the variable rmse and print it
```

```
(rmse <- sqrt(mean(res^2)))
## [1] 13.69703
# Calculate the standard deviation of participants weights and print it

(sd_wt <- sd(dat$weight))
## [1] 13.71609
```

The RMSE = 13.7 then the typical prediction error will be 13.7 kg. Comparing the RMSE to sd of the outcome (weight) which equals 13.7 also, so RMSE is similar to sd (standard deviation is the typical difference between a specific weight and the average weight) so our model is not better than the average to estimate the weights. In other words, when RMSE is much smaller than the outcome's standard deviation, this suggests that our model predicts well.

12.5.3. Confidence Intervals (CI) for Regression Estimates of the Physical Activity Model

The values of = 74.2 and = –0.002 printed from the summary(mod) were estimates based on a single sample of 2187 participants. If we have a different sample, the estimates for and will be different

Therefore, all regression coefficients or estimates, and have an associated standard error (SE) (printed from the summary(mod) also) that can be used to get the CIs that contain the true population parameters for and and get p-values for hypothesis testing (printed also from summary(mod))

Hypothesis testing for

Ho:

Ha:

Hypothesis testing for

Ho:

Ha:

```
## summary(model) to get useful results

summary(mod)
##

## Call:
## lm(formula = weight ~ phyact, data = dat)
```

```
##
## Residuals:
##     Min      1Q  Median      3Q     Max
## -32.014  -9.933  -0.443   8.909  53.858
##
## Coefficients:
##               Estimate Std. Error t value Pr(>|t|)
## (Intercept) 74.1979165  0.4205901 176.414   <2e-16 ***
## phyact      -0.0016988  0.0007537  -2.254   0.0243 *
## ---
## Signif. codes: 0 '***' 0.001 '**' 0.01 '*' 0.05 '.' 0.1 ' ' 1
##
## Residual standard error: 13.7 on 2185 degrees of freedom
## Multiple R-squared: 0.00232,  Adjusted R-squared: 0.001863
## F-statistic: 5.081 on 1 and 2185 DF, p-value: 0.02429
##estimating the 95% confidence intervals

confint(mod)
##                    2.5 %        97.5 %
## (Intercept) 73.373118244 75.0227148278
## phyact      -0.003176752 -0.0002208126
```

Using simple linear regression to estimate the association between physical activity and weight based on a sample of 2187 participants from Girona, the estimate for was 74.2 kg which is the average weight of participants having no physical activity. This estimate was significantly different from zero and it can be as low as 73.4 kg and as high as 75.02 kg in the population from which the sample was drawn. This is based on the calculated 95% CI.

Note that hypothesis testing for does not make sense since the average weight of participants with no physical activity cannot be zero so it is not needed but CIs may be of interest.

The estimate for was –0.002 kg or 2 grams which is the average weight difference between two groups of participants who differ by 1 physical activity. A statistically significant negative association between physical activity and weight was found. This estimate was significantly different from zero and it can be as low as 0.2 grams and as high as 3 grams in the population from which the sample was drawn. This is based on the calculated 95% CI.

12.5.4. Measuring the Strength of Linear Association by R-Squared

R-squared is a measure of how the model fits or explains the data. It has values between 0–1. Values near 1 mean that model fits the data well (or data points are closer to regression line). Values near 0 mean that the model is not better predicting than using the average value

is total sum of squares (data variance)

is residual sum of squares (model variance)

R-squared is variance explained by the model. If is small compared to then R-squared is large and the model fits the data well.

For lm models, you can read the r-squared from the summary of the model or call summary(lm)$r.squared will show R-squared. For models that minimize the squared error (as linear regression), R-squared is the square of the correlation between the outcome and prediction.

```
# Get R-squared from summary. Print it

(rsq_mod<- summary(mod)$r.squared)
## [1] 0.00231986
# Get the correlation between the prediction and true outcome: rho and print it

(rho <- cor(dat$phyact, dat$weight))
## [1] –0.04816492
# Square rho: rho2 and print it

(rho2 <- rho^2)
## [1] 0.00231986
# compare r-squared with other models
```

```
mod_sex <-lm(weight ~ sex, data = regicor)
(rsq_mod_sex <- summary(mod_sex)$r.squared)
## [1] 0.196631
mod_smoker<-lm(weight ~ smoker, data = regicor)
(rsq_mod_smoker <- summary(mod_smoker)$r.squared)
## [1] 0.04099042
```

The R-squared for linear model with physical activity = 0.002 or 0.2 %. This means that physical activity explains only 0.2% of the variation in participants' weight. This also means that major remaining 99.8% of variation in weight is not explained by physical activity.

The R-squared for linear model with sex was 19.7% and with smoker was 4.1%. This means that the linear model with sex as a predictor explains greater variation in participants' weight than using their smoking status or physical activity.

The unexplained variation, in each model, can be explained by other variables or combining all these variables together. This is the case of Multiple regression which allow us to predict the participant's weight using several predictors in one analysis.

12.5.5. Evaluating Multiple Linear Regression Model

1. Plot the model's predictions against the actual weight

```
dat<- regicor %>% filter(!is.na(weight) & !is.na(sex) & !is.na(smoker) &
!is.na(phyact))
mod_mult<- lm(weight ~ sex + smoker + phyact, data = dat)
# Make predictions from the model

dat$predictions <- predict(mod_mult)
# plot predictions (on x-axis) versus the actual weight on y-axis

ggplot(dat, aes(x = predictions, y = weight)) +
 geom_point() +
 geom_abline()
```

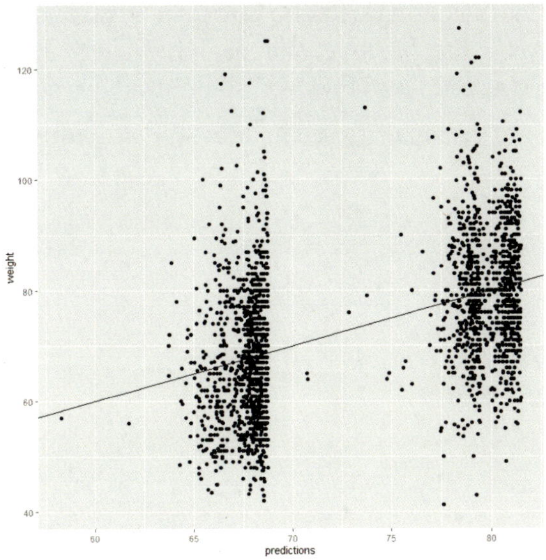

Although the points are nearly equally distributed above the line, there are a strange 3 clusters in our data.

2. Plot predictions against residuals.

```
# you can also plot the residual plot using plot() function

plot(mod_mult, 1)
```

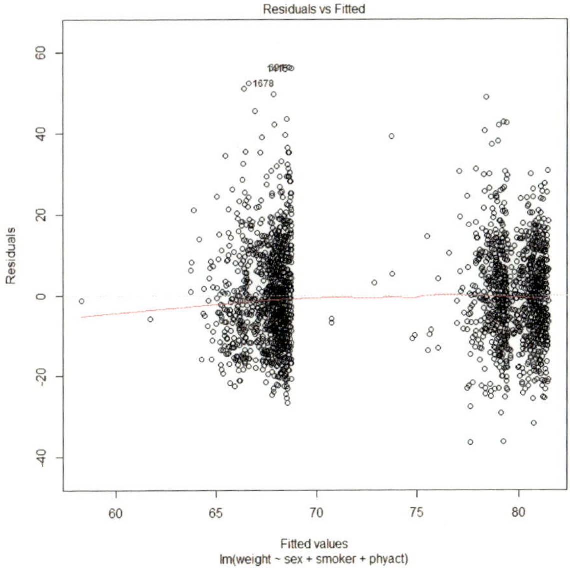

As we can see, there is no systematic error, the residuals are nearly evenly distributed between the positive and negative sides and have the same magnitude, above and below. However, some residuals are very large (> 50)

3. qqplot of residuals to see its normal distribution

```
plot(mod_mult, 2)
```

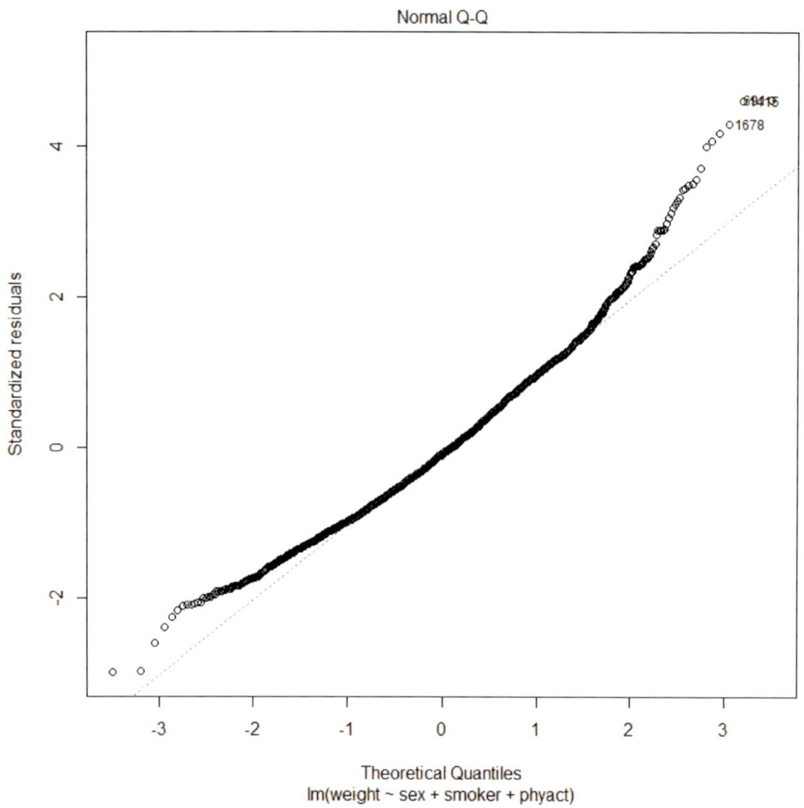

The residuals are nearly normally distributed with some deviation at the extremities:

4. Root mean squared error (RMSE) is the square root of the mean squared error of the model on the data. RMSE is the typical prediction error of the model on your data. The lower RMSE, the better is our model.

```
# Calculate RMSE, assign it to the variable rmse and print it
```

```
(rmse <- sqrt(mean(mod_mult$residuals^2)))
## [1] 12.23032
# Calculate the standard deviation of participants weights and print it

(sd_wt <- sd(dat$weight))
## [1] 13.72667
```

The RMSE = 12.2 then the typical prediction error will be 12.2 kg. Comparing the RMSE to sd of the outcome (weight) which equals 13.7, so RMSE is slightly smaller than standard deviation. Therefore, our model is not better than the average to estimate the weight.

5. R-squared:

```
# Get R-squared from summary. Print it

(rsq_mod<- summary(mod_mult)$r.squared)
## [1] 0.2057664
# compare r-squared with other models

mod_phyact <-lm(weight ~ phyact, data = regicor)
(rsq_mod_phyact <- summary(mod_phyact)$r.squared)
## [1] 0.00231986
mod_sex <-lm(weight ~ sex, data = regicor)
(rsq_mod_sex <- summary(mod_sex)$r.squared)
## [1] 0.196631
mod_smoker<-lm(weight ~ smoker, data = regicor)
(rsq_mod_smoker <- summary(mod_smoker)$r.squared)
## [1] 0.04099042
```

The R-squared for linear model with 3 predictors = 0.21 or 21 %. This means that this model explains 21 % of the variation in participants' weight. This is compared to the R-squared of 0.2% for physical activity, R-squared of 19.7% for sex, and R-squared of 4.1% for smoker.

This means that the linear model with 3 predictors explains greater variation in participants' weight than using only one of these predictors, although it is nearly similar to the sex model.

CHAPTER 13

SIMPLE AND MULTIPLE LOGISTIC REGRESSION

CONTENTS

13.1. SIMPLE LOGISTIC REGRESSION WITH CATEGORICAL PREDICTOR

Simple logistic regression is a method for relating a binary outcome to a single predictor by a linear equation.

In logistic regression, the equation will be:

ln(odds of binary outcome)= =

p is the probability of our binary outcome to occur.

13.1.1. Binary Predictor

Define a simple logistic regression for estimating the relation between cardiovascular events (cv) and sex in the regicor dataset.

The logistic regression is defined by glm() function with argument family = binomial()

```
# load packages

library(tidyverse)
library(compareGroups)
data("regicor")
# data structure

str(regicor)
## 'data.frame': 2294 obs. of 25 variables:
## $ id   : num 2.26e+03 1.88e+03 3.00e+09 3.00e+09 3.00e+09 ...
##  ..- attr(*, "label")= Named chr "Individual id"
##  .. ..- attr(*, "names")= chr "id"
## $ year  : Factor w/ 3 levels "1995,""2000,"..: 3 3 2 2 2 2 2 1 3 1 ...
##  ..- attr(*, "label")= Named chr "Recruitment year"
##  .. ..- attr(*, "names")= chr "year"
## $ age   : int 70 56 37 69 70 40 66 53 43 70 ...
##  ..- attr(*, "label")= Named chr "Age"
##  .. ..- attr(*, "names")= chr "age"
## $ sex   : Factor w/ 2 levels "Male,""Female": 2 2 1 2 2 2 1 2 2 1 ...
##  ..- attr(*, "label")= chr "Sex"
```

```
## $ smoker : Factor w/ 3 levels "Never smoker,"..: 1 1 2 1 NA 2 1 1 3 3 ...
## ..- attr(*, "label")= Named chr "Smoking status"
## .. ..- attr(*, "names")= chr "smoker"
## $ sbp   : int 138 139 132 168 NA 108 120 132 95 142 ...
## ..- attr(*, "label")= Named chr "Systolic blood pressure"
## .. ..- attr(*, "names")= chr "sbp"
## $ dbp   : int 75 89 82 97 NA 70 72 78 65 78 ...
## ..- attr(*, "label")= Named chr "Diastolic blood pressure"
## .. ..- attr(*, "names")= chr "dbp"
## $ histhtn : Factor w/ 2 levels "Yes,""No": 2 2 2 2 2 2 1 2 2 2 ...
## ..- attr(*, "label")= Named chr "History of hypertension"
## .. ..- attr(*, "names")= chr "histbp"
## $ txhtn  : Factor w/ 2 levels "No,""Yes": 1 1 1 1 1 1 2 1 1 1 ...
## ..- attr(*, "label")= chr "Hypertension treatment"
## $ chol  : num 294 220 245 168 NA NA 298 254 194 188 ...
## ..- attr(*, "label")= Named chr "Total cholesterol"
## .. ..- attr(*, "names")= chr "chol"
## $ hdl   : num 57 50 59.8 53.2 NA ...
## ..- attr(*, "label")= Named chr "HDL cholesterol"
## .. ..- attr(*, "names")= chr "hdl"
## $ triglyc : num 93 160 89 116 NA 94 71 NA 68 137 ...
## ..- attr(*, "label")= Named chr "Triglycerides"
## .. ..- attr(*, "names")= chr "triglyc"
## $ ldl   : num 218.4 138 167.4 91.6 NA ...
## ..- attr(*, "label")= Named chr "LDL cholesterol"
## .. ..- attr(*, "names")= chr "ldl"
## $ histchol: Factor w/ 2 levels "Yes,""No": 2 2 2 2 NA 2 1 2 2 2 ...
## ..- attr(*, "label")= chr "History of hyperchol."
## $ txchol : Factor w/ 2 levels "No,""Yes": 1 1 1 1 NA 1 1 1 1 1 ...
## ..- attr(*, "label")= Named chr "Cholesterol treatment"
## .. ..- attr(*, "names")= chr "txchol"
```

```
## $ height : num 160 163 170 147 NA ...
## ..- attr(*, "label")= Named chr "Height (cm)"
## .. ..- attr(*, "names")= chr "height"
## $ weight : num 64 67 70 68 NA 43.5 79.2 45.8 53 62 ...
## ..- attr(*, "label")= Named chr "Weight (Kg)"
## .. ..- attr(*, "names")= chr "weight"
## $ bmi   : num 25 25.2 24.2 31.5 NA ...
## ..- attr(*, "label")= Named chr "Body mass index"
## .. ..- attr(*, "names")= chr "bmi"
## $ phyact : num 304 160 553 522 NA ...
## ..- attr(*, "label")= Named chr "Physical activity (Kcal/week)"
## .. ..- attr(*, "names")= chr "phyact"
## $ pcs   : num 54.5 58.2 43.4 54.3 NA ...
## ..- attr(*, "label")= Named chr "Physical component"
## .. ..- attr(*, "names")= chr "pcs"
## $ mcs   : num 58.9 48 62.6 57.9 NA ...
## ..- attr(*, "label")= chr "Mental component"
## $ cv   : Factor w/ 2 levels "No,""Yes": 1 1 1 1 NA 1 1 1 1 1 ...
## ..- attr(*, "label")= chr "Cardiovascular event"
## $ tocv  : num 1025 2757 1906 1055 NA ...
## ..- attr(*, "label")= chr "Days to cardiovascular event or end of follow-up"
## $ death  : Factor w/ 2 levels "No,""Yes": 2 1 1 1 NA 1 2 1 1 1 ...
## ..- attr(*, "label")= chr "Overall death"
## $ todeath : num 1299.2 39.3 858.4 1833.1 NA ...
## ..- attr(*, "label")= chr "Days to overall death or end of follow-up"
# logistic model

# remove NA values

dat<- regicor %>% filter(!is.na(cv) & !is.na(age))
mod_sex<- glm(cv ~ sex, data = dat, family = binomial())
```

```
summary(mod_sex)

##
## Call:
## glm(formula = cv ~ sex, family = binomial(), data = dat)

##
## Deviance Residuals:
##   Min    1Q  Median    3Q   Max
## -0.3005 -0.3005 -0.2895 -0.2895  2.5272
##
## Coefficients:
##       Estimate Std. Error z value Pr(>|z|)
## (Intercept) -3.07511  0.15081 -20.391  <2e-16 ***
## sexFemale  -0.07633  0.21310 -0.358   0.72
## ---
## Signif. codes: 0 '***' 0.001 '**' 0.01 '*' 0.05 '.' 0.1 ' ' 1
##
## (Dispersion parameter for binomial family taken to be 1)
##
##   Null deviance: 761.00 on 2162 degrees of freedom
## Residual deviance: 760.88 on 2161 degrees of freedom
## AIC: 764.88
##
## Number of Fisher Scoring iterations: 6
# to get the odds and odds ratio

exp(coef(mod_sex))
## (Intercept)  sexFemale
## 0.04618474 0.92651163
```

From the summary of the model, the logistic regression equation is estimated as:

ln(odds of cv)=

p is the probability of cv event. We put hats as these regression coefficients with , are estimated from our sample data. = 1 for Females and =0 for males.

This equation estimates two values:

1. ln(odds of cv for males)= =–3.08 since = 0. By exponentiation, we get 0.046 which is the odds of cv for males in this population.
2. ln(odds of cv for females) = =–3.08–0.08 =–3.16 since = 1 for females. By exponentiation, we get 0.042 which is the odds of cv for females in this population. In this population, females have lower odds of cv events than males.

So in simple logistic regression,

1. or intercept is the ln(odds of desired outcome for reference category).
2. or slope is the ln(odds ratio of females to males) or ln(odds ratio) between the two groups who differ by only one unit of .

Since for binary predictor (as sex), can only takes two values, 0 for males and 1 for females, so –0.08 is the ln(odds ratio of females to males) in this population. By exponentiation, we get 0.93 so females have lower odds of cv events than males by (1–0.93 = 0.07) 7%. In other words, the odds of cv events for females is 0.93 times the odds of cv events for males.

The column marked “Pr(>|z|)” gives a p-value larger than 0.05. So, we can conclude that sex type is not a significant predictor of cv events in this population.

13.1.2. Categorical Predictor

Define a simple logistic regression for estimating the relation between cardiovascular events (cv) and smoker status in the regicor dataset.

```
# logistic model

# remove NA values

dat2<- regicor %>% filter(!is.na(cv) & !is.na(smoker))
mod_smok<- glm(cv ~ smoker, data = dat2, family = binomial())
summary(mod_smok)
##
```

```
## Call:
## glm(formula = cv ~ smoker, family = binomial(), data = dat2)
##
## Deviance Residuals:
##   Min    1Q  Median    3Q   Max
## -0.4215 -0.2573 -0.2573 -0.2573  2.8204
##
## Coefficients:
##              Estimate Std. Error z value Pr(>|z|)
## (Intercept)        -3.3912   0.1671 -20.289 < 2e-16 ***
## smokerCurrent or former < 1y  1.0148   0.2263  4.485 7.27e-06 ***
## smokerFormer >= 1y      -0.5672   0.3941 -1.439   0.15
## ---
## Signif. codes: 0 '***' 0.001 '**' 0.01 '*' 0.05 .'' 0.1 ' ' 1
##
## (Dispersion parameter for binomial family taken to be 1)
##
##   Null deviance: 756.87 on 2115 degrees of freedom
## Residual deviance: 727.29 on 2113 degrees of freedom
## AIC: 733.29
##
## Number of Fisher Scoring iterations: 6
# to get the odds and odds ratio

exp(coef(mod_smok))
##       (Intercept) smokerCurrent or former < 1y
##        0.03366697         2.75894669
##     smokerFormer >= 1y
##        0.56711605
```

From the summary of the model, the logistic regression equation is estimated as:

ln(odds of cv)=

p is the probability of cv event. We put hats as these regression coefficients with „ are estimated from our sample data. = 1 for current smokers and =0 for other two categories. = 1 for former smokers and =0 for other two categories.

This equation estimates three values:

1. ln(odds of cv for never smokers)= = –3.39. By exponentiation, we get 0.03 which is the odds of cv for never smokers in this population.
2. ln(odds of cv for current smokers) = = –3.39+1.01 = –2.38. By exponentiation, we get 0.09 which is the odds of cv for current smokers in this population. In this population, current smokers have higher odds of cv events than never smokers.
3. ln(odds of cv for former smokers) = = –3.39–0.57 = –3.96. By exponentiation, we get 0.019 which is the odds of cv for former smokers in this population. In this population, former smokers have lower odds of cv events than never smokers.

So in this simple logistic regression,

1. or intercept is the ln(odds of cv events for reference category or never smokers).
2. is the ln(odds ratio of current to never smokers)= 1.015. By exponentiation, we get 2.76 so current smokers have higher odds of cv events than never smokers by (2.76–1 = 1.76) 176%. In other words, the odds of cv events for current smokers is 2.76 times the odds of cv events for never smokers. The column marked "Pr(>|z|)" gives a p-value smaller than 0.05. So, we conclude that current smokers have significantly greater odds of cv events than never smokers in this population.
3. is the ln(odds ratio of former to never smokers)= –0.57. By exponentiation, we get 0.57 so former smokers have lower odds of cv events than never smokers by (1–0.57 =) 43%. In other words, the odds of cv events for former smokers is 0.57 (about half) times the odds of cv events for never smokers. The column marked "Pr(>|z|)" gives a p-value larger than 0.05. So, we conclude that former smokers have equivalent odds of cv events to that of never smokers in this population.

13.2. SIMPLE LOGISTIC REGRESSION WITH CONTINUOUS PREDICTOR

Examine the association between cardiovascular (cv) events the participant's age:

```
# logistic model

# remove NA

dat3<- regicor %>% filter(!is.na(age) & !is.na(cv))
mod_age<- glm(cv ~ age, data = dat3, family = binomial())
summary(mod_age)
##
## Call:
## glm(formula = cv ~ age, family = binomial(), data = dat3)
##
## Deviance Residuals:
##   Min    1Q  Median    3Q   Max
## –0.3618 –0.3229 –0.2880 –0.2567  2.6958
##
## Coefficients:
##       Estimate Std. Error z value Pr(>|z|)
## (Intercept) –4.426704  0.572048 –7.738 1.01e–14 ***
## age     0.023421  0.009802  2.389  0.0169 *
## ---
## Signif. codes: 0 '***' 0.001 '**' 0.01 '*' 0.05 .'' 0.1 ' ' 1
##
## (Dispersion parameter for binomial family taken to be 1)
##
##   Null deviance: 761.00 on 2162 degrees of freedom
```

```
## Residual deviance: 755.19 on 2161 degrees of freedom
## AIC: 759.19
##
## Number of Fisher Scoring iterations: 6
# to get the odds and odds ratio

exp(coef(mod_age))
## (Intercept)     age
## 0.01195383 1.02369788
```

From the summary of the model, the logistic regression equation is estimated as

ln(odds of cv)=

This equation estimates two values:

1. or intercept is the ln(odds of cv events for participants with age = 0). As our data contains no observations with age = 0, so this intercept is scientifically meaningless and it is just a placeholder to define the linear equation; and
2. is the ln(odds ratio of two persons who differs by 1 year in age)= 0.0234. It is a positive number meaning that as age increases, the odds of cv events also increase.

By exponentiation, we get 1.02 which is the odds ratio (OR) for groups of participants from Girona who differ by 1 year of age. In other words, each 1 year increase in age is associated with (1.02–1 = 0.2) 2% increase in the odds of cv events. This estimate is constant across all values of age in our data (35–74) but we cannot extrapolate this relationship to higher or lower values. Also, we assume that there is a linear relation between ln(odds) and age which we will explore in the next section.

- Examples:

1. The OR for participants with 36 years old compared to participants with 35 years old = 1.02.
2. The OR for participants with 41 years old compared to participants with 41 years old = 1.02.
3. The OR for participants with 74 years old compared to participants with 73 years old = 1.02.

4. The OR for participants with 74 years old compared to participants with 64 years old = (74–64)*0.02 = 0.2 which is ln(OR). By exponentiation, we get OR = 1.22.
5. The OR for participants with 70 years old compared to participants with 50 years old = (70–50)*0.02 = 0.4 which is ln(OR). By exponentiation, we get OR = 1.5. So the group with 70 years old have 50% more odds of cv events that the group with 50 years old.
6. The OR for participants with 80 years old compared to participants with 50 years old, we cannot estimate this OR since in our sample data we have no data points for participants with 80 years old.

The column marked "Pr(>|z|)" gives a p-value smaller than 0.05. So, we conclude that age is a significant predictor of cv events in this population.

13.3. MULTIPLE LOGISTIC REGRESSION

For example, we want to build a multiple logistic regression model for predicting cv events from the participant's sex, smoking status, and age (all previous predictors) in the regicor dataset.

```
# remove NA

dat4<- regicor %>% filter(!is.na(age) & !is.na(cv) & !is.na(smoker) &
!is.na(sex))
mod_mult<- glm(cv ~ sex + smoker + age, data = dat4, family = binomial())
summary(mod_mult)
##
## Call:
## glm(formula = cv ~ sex + smoker + age, family = binomial(), data = dat4)
##
## Deviance Residuals:
##   Min    1Q Median    3Q   Max
## –0.6956 –0.3328 –0.2576 –0.1924  3.0827
##
## Coefficients:
##              Estimate Std. Error z value Pr(>|z|)
```

```
## (Intercept)        -5.97316  0.72453 -8.244 < 2e-16 ***
## sexFemale           0.32095  0.25792  1.244 0.213373
## smokerCurrent or former < 1y 1.41115  0.27607  5.112 3.2e-07 ***
## smokerFormer >= 1y      -0.32191  0.42862 -0.751 0.452625
## age             0.03980  0.01041  3.823 0.000132 ***
## ---
## Signif. codes: 0 '***' 0.001 '**' 0.01 '*' 0.05 '.' 0.1 ' ' 1
##
## (Dispersion parameter for binomial family taken to be 1)
##
##   Null deviance: 756.87 on 2115 degrees of freedom

## Residual deviance: 711.86 on 2111 degrees of freedom
## AIC: 721.86
##
## Number of Fisher Scoring iterations: 6
# to get OR

exp(coef(mod_mult))
##        (Intercept)          sexFemale
##        0.002546175         1.378432383
## smokerCurrent or former < 1y      smokerFormer >= 1y
##        4.100664428         0.724762897
##           age
##        1.040606279
```

- The intercept coefficient:

  The intercept = –5.97 which is the ln(odds) of cv events for participants with reference categories. In this case, it is the ln(odds) of cv events for male and never smoked participants with age = 0. As we do not have age = 0 in this data, so this value is scientifically meaningless and just a placeholder.

- The sex coefficient:

  The coefficient for female = 0.32 which is the ln(OR) of cv events for females and males who have the same smoking habit and the same age. By exponentiation. We get 1.38, so females have higher odds of cv events than males in this population by 38% when males and females have the same smoking habit and age.

The p-value for sex was larger than 0.05. This means that sex is not a significant predictor of cv events in this population.

Compared to the unadjusted estimate for simple regression with sex alone (ln(OR) = –0.076), there is a great difference in magnitude and direction between adjusted (in multiple model) and unadjusted estimate. This means that the association between cv events and sex is confounded by smoking habit and age.

– The smoker coefficient:

The coefficient for current smokers = 1.4 which is ln(OR) between current smokers and never smokers for participants having the same sex and the same age. On the other hand, the coefficient for former smokers = –0.32 which is ln(OR) between former smokers and never smokers for participants having the same sex and the same age.

By exponentiation, we get 4.1 OR for current smokers and 0.72 OR for former smokers. Therefore, current smokers have higher odds of cv events than never smokers in this population by 310% when current and never smokers have the same sex and age. Similarly, former smokers have lower odds of cv events than never smokers in this population by 28% when former and never smokers have the same sex and age.

The p-value for smoker was less than 0.05 for current smokers and larger than 0.05 for the former smokers. This means that current smokers have significantly greater odds of cv events than never smokers in this population, while former smokers have equivalent odds of cv events than never smokers in this population when they have the same sex and age.

Compared to the unadjusted estimate for simple regression with smoker alone (ln(OR) for current smokers = 1.01 and ln(OR) for former smokers = –0.57), there is some difference in magnitude between adjusted (in multiple model) and unadjusted estimate. This means that the association between cv events and smoking is confounded by sex and age in this population.

- The age coefficient:

  The coefficient for age = 0.04 which is the ln(OR) between two groups of participants who differ in their age by 1 year but have the same sex and smoking habit. By exponentiation, we get OR = 1.04, so, for the same sex and smoking habit, each 1 year increase in age is associated with 4% increase in the odds of cv events.

  The p-value for age was less than 0.05. This means that age is a significant predictor of cv events in this population.

  Compared to the unadjusted estimate for simple regression with age alone (ln(OR) = 0.23), there is some difference between adjusted (in multiple model) and unadjusted estimate. This means that the association between cv events and age is somewhat confounded by smoking habit and sex.

  Our model equation can be written as follow: ln(odds of cv) = –5.97 + 0.32 Female + 1.4 current smoker – 0.32 former smoker + 0.04 age

13.4. EVALUATION OF THE REGRESSION MODEL

13.4.1. The Model's Accuracy

We will use the predict() to estimate the probability of cv event for each observation. Then, according to the average cv events in our data, determine if each predicted probability corresponds to a cv event or not.

For sex model:

```
# Estimate the cv probability

dat$cv_sex_prob <- predict(mod_sex, type = "response")
head(dat$cv_sex_prob)
## [1] 0.04103479 0.04103479 0.04414587 0.04103479 0.04103479
0.04414587
# Find the average cv probability

mean(dat$cv=="Yes")
## [1] 0.04253352
# Predict a cv if predicted probability of cv is greater than average (0.0425)
```

```
dat$cv_sex_pred <- ifelse(dat$cv_sex_prob > 0.0425, "Yes,""No")
# Calculate the model's accuracy

mean(dat$cv == dat$cv_sex_pred)

## [1] 0.5182617
```

The model's accuracy when using sex alone as single predictor is 0.52 or 52%.

For smoker model:

```
# Estimate the cv probability

dat2$cv_smok_prob <- predict(mod_smok, type = "response")
head(dat2$cv_smok_prob)
## [1] 0.03257042 0.03257042 0.08499096 0.03257042 0.08499096 0.03257042
# Find the average cv probability

mean(dat2$cv=="Yes")
## [1] 0.04347826
# Predict a cv if predicted probability of cv is greater than average (0.0435)

dat2$cv_smok_pred <- ifelse(dat2$cv_smok_prob > 0.0435, "Yes,""No")
# Calculate the model's accuracy

mean(dat2$cv == dat2$cv_smok_pred)
## [1] 0.739603
```

The model's accuracy when using smoker alone as single predictor is 0.74 or 74%, so it is a better model than the sex model.

For age model:

```
# Estimate the cv probability

dat3$cv_age_prob <- predict(mod_age, type = "response")
head(dat3$cv_age_prob)
## [1] 0.05801967 0.04248857 0.02764946 0.05675278 0.02960253 0.05310646
```

```
# Find the average cv probability

mean(dat3$cv==”Yes”)
## [1] 0.04253352
# Predict a cv if predicted probability of cv is greater than average (0.0425)

dat3$cv_age_pred <- ifelse(dat3$cv_age_prob > 0.0425, “Yes,””No”)
# Calculate the model’s accuracy

mean(dat3$cv == dat3$cv_age_pred)
## [1] 0.5520111
```

The model’s accuracy when using age alone as single predictor is 0.55 or 55%, so it is a less accurate model than the smoker model.

For multiple model:

```
# Estimate the cv probability

dat4$cv_mult_prob <- predict(mod_mult, type = “response”)
head(dat4$cv_mult_prob)
## [1] 0.05386162 0.03157762 0.04355144 0.05186886 0.06605519 0.03402207
# Find the average cv probability

mean(dat4$cv==”Yes”)
## [1] 0.04347826
# Predict a cv if predicted probability of cv is greater than average (0.0435)

dat4$cv_mult_pred <- ifelse(dat4$cv_mult_prob > 0.0435, “Yes,””No”)
# Calculate the model’s accuracy

mean(dat4$cv == dat4$cv_mult_pred)
## [1] 0.6375236
```

The model’s accuracy when using all 3 predictors is 0.64 or 64%, so it is a less accurate model than the smoker model.

13.4.2. ROC Curve (Receiver Operator Characteristic Curve)

Rare events as cv events create challenges for classification models. When one outcome is rare, predicting the opposite can result in very high accuracy, so we sacrifice the overall accuracy to better target the outcome of interest.

ROC curve better understands the model ability to distinguish the outcome of interest versus the opposite outcome. ROC curve shows the % of positive outcomes (y-axis) against the % of opposite outcomes (x-axis)

If the classifier, or logistic model, is not good, the ROC curve will run on the diagonal. If the model classifies correctly, it will have a bell shape as it identify several positive examples for each negative examples. To quantify this performance, AUC is used, AUC=0.5 for diagonal or random model and the best AUC is 1.

ROC help to visualize the tradeoffs between the true positives and false positives for the outcome of interest

We will use the pROC package to plot the ROC curve and calculate AUC for each model

```
library(pROC)
# sex model

# Create a ROC curve

ROC <- roc(dat$cv, dat$cv_sex_prob)
# Plot the ROC curve

plot(ROC, col = "blue")
```

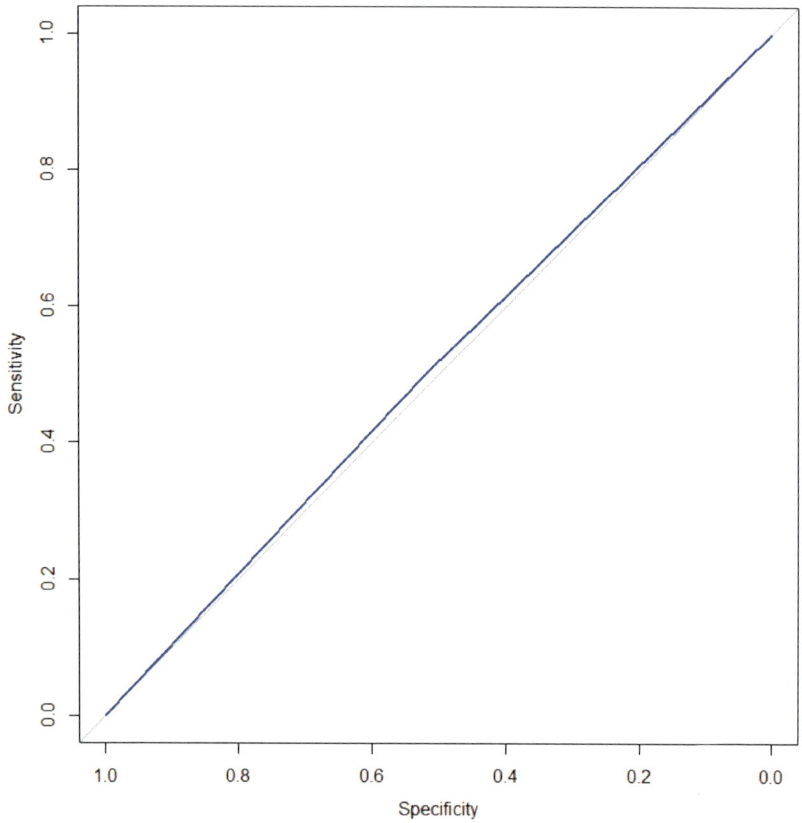


```
# Calculate the area under the curve (AUC)

auc(ROC)
## Area under the curve: 0.5095
# smoker model

# Create a ROC curve

ROC <- roc(dat2$cv, dat2$cv_smok_prob)
# Plot the ROC curve

plot(ROC, col = "blue")
```

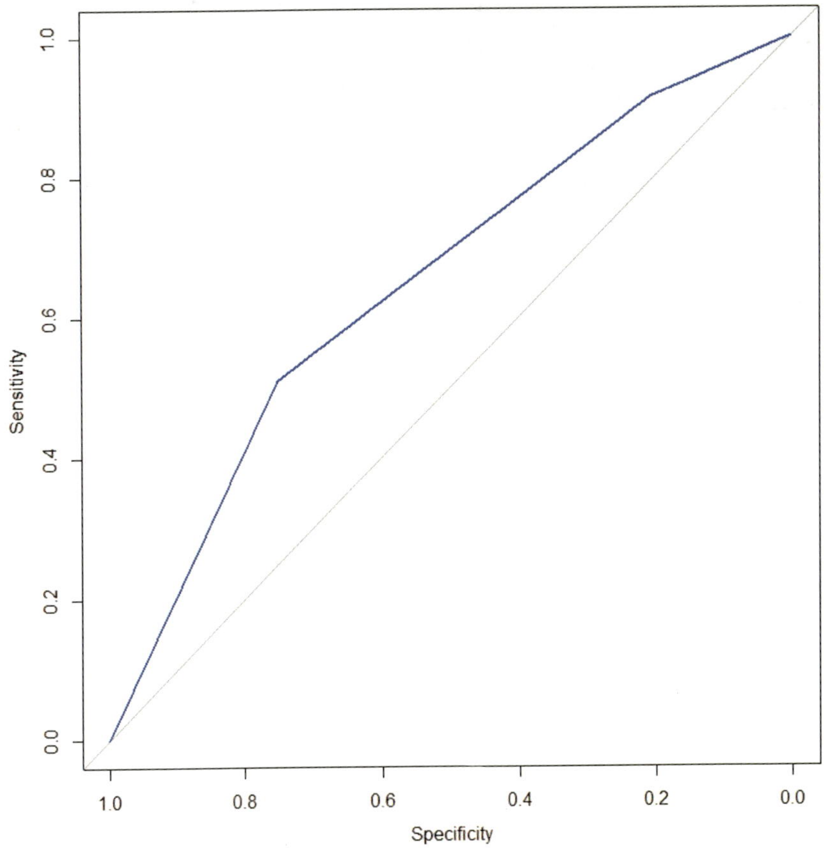


```
# Calculate the area under the curve (AUC)

auc(ROC)
## Area under the curve: 0.6485

# age model

# Create a ROC curve

ROC <- roc(dat3$cv, dat3$cv_age_prob)
# Plot the ROC curve

plot(ROC, col = "blue")
```

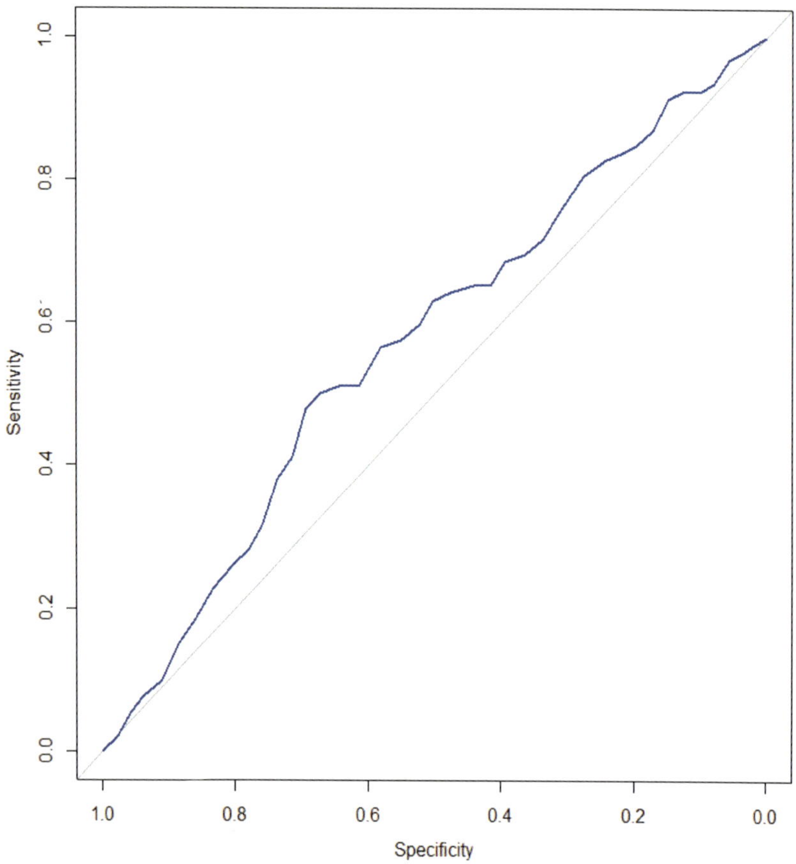


```
# Calculate the area under the curve (AUC)

auc(ROC)
## Area under the curve: 0.5739
# Multiple model

# Create a ROC curve

ROC <- roc(dat4$cv, dat4$cv_mult_prob)
# Plot the ROC curve

plot(ROC, col = "blue")
```

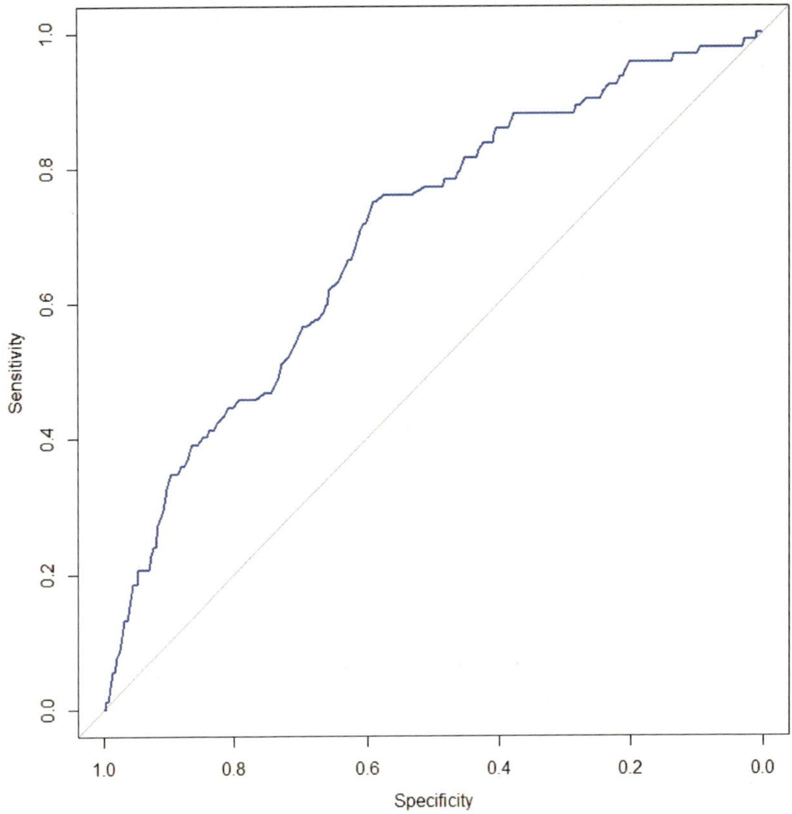

 # Calculate the area under the curve (AUC)

auc(ROC)

Area under the curve: 0.6969

AUC for sex model = 0.51, for the smoker model = 0.65, for the age model = 0.57, and for the multiple model = 0.70. So the best model is the multiple model with three predictors.

13.4.3. Linearity Assumption

For continuous predictors, logistic regression assumes a linear relationship between the continuous predictor variables and the ln(odds) of the outcome. This can be done by visually inspecting the scatter plot between the predictor and the ln(odds).

We will examine this relation for the age model and multiple model and use the predict() function that calculates ln(odds) by default.

```
# age model

dat3 <- dat3 %>% mutate(ln_odds = predict(mod_age))
# plot the curve with loess smoothing

ggplot(dat3, aes(x = age, y = ln_odds))+
 geom_point() + geom_smooth(method = "loess")
```

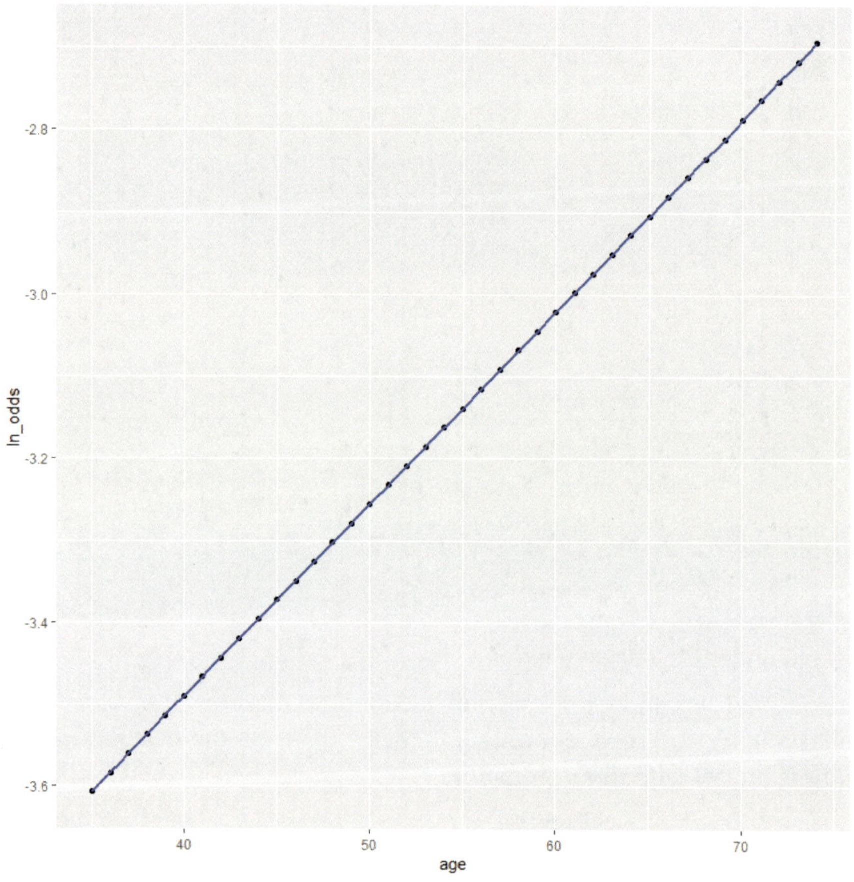


```
# multiple model

dat4 <- dat4 %>% mutate(ln_odds = predict(mod_mult))
# plot the curve with loess smoothing
```

```
ggplot(dat4, aes(x = age, y = ln_odds))+
 geom_point() + geom_smooth(method = "loess")
```

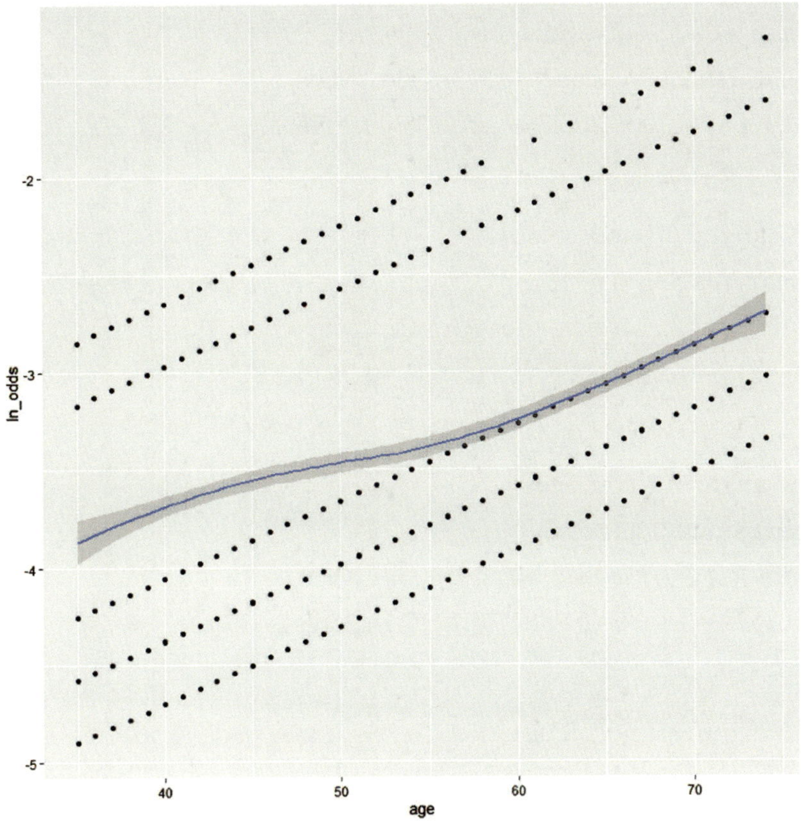

We see that age in both models is linearly related to ln(odds of cv) so this assumption holds.

13.4.4. Influential or Outlier Values

Influential values are extreme individual data points that can alter the quality of the logistic regression model.

Data points with an absolute standardized residuals above three represent possible outliers and deserve more attention.

The augment() function from the broom package will be used to calculate these standardized residuals for all models.

```
library(broom)
```

```
# sex model

dat5<- augment(mod_sex)
summary(dat5$.std.resid)
##  Min. 1st Qu. Median  Mean 3rd Qu.  Max.
## –0.3006 –0.3006 –0.2896 –0.1755 –0.2896 2.5283
# smoker model

dat5<- augment(mod_smok)
summary(dat5$.std.resid)
##  Min. 1st Qu. Median  Mean 3rd Qu.  Max.
## –0.4219 –0.2575 –0.2575 –0.1673 –0.2575 2.8237
# age model

dat5<- augment(mod_age)
summary(dat5$.std.resid)
##  Min. 1st Qu. Median  Mean 3rd Qu.  Max.
## –0.3622 –0.3230 –0.2881 –0.1737 –0.2568 2.6979
#multiple model

dat5<- augment(mod_mult)
summary(dat5$.std.resid)
##  Min. 1st Qu. Median  Mean 3rd Qu.  Max.
## –0.7024 –0.3331 –0.2582 –0.1627 –0.1925 3.0850
```

Nearly, all of our models do not have standardized residuals greater than 3.

CHAPTER 14

SIMPLE AND MULTIPLE COX REGRESSION

CONTENTS

14.1. INTRODUCTION

Kaplan-Meier curves and logrank tests are examples of univariate analysis. They describe the survival according to one factor (predictor) under investigation, but ignore the impact of any others. In addition, Kaplan-Meier curves and logrank tests are useful only when the predictor variable is categorical (e.g.: treatment A vs. treatment B; males vs. females). They don't work easily for quantitative predictors such as weight or age.

An alternative method is the Cox proportional hazards (PH) regression analysis, which works for both quantitative predictor variables and for categorical variables. Furthermore, the Cox regression model extends survival analysis methods to assess simultaneously the effect of several risk factors on survival time.

The equation is similar to that of logistic regression:

ln(hazard of y = 1, at time = t) = ln(

where; t represents the survival time

are different predictors

which is the baseline hazard at time t. It corresponds to the value of the hazard if all the x_i are equal to zero

The coefficients or slopes () measure the difference in ln(hazard) or ln(hazard ratio) per unit difference of the corresponding x at a given time (t) in the follow up period. This difference in ln(hazard) or ln(hazard ratio) is constant regardless of the time of comparison and so called proportional hazard. This is unlike which varies with time.

Hazard is the instantaneous risk of having our binary event at a given time in the follow up period. In other words, hazard is the time specific risk of having the event of interest.

The equation estimates ln(hazard) of our binary outcome at any given time of the follow-up period.

14.2. COX REGRESSION WITH CATEGORICAL PREDICTOR

14.2.1. Binary Predictor

From the veteran dataset of the survival package, test the association between survival and treatment type (binary predictor).

```
library(tidyverse)
## -- Attaching packages ---------------------------------------------------------
tidyverse 1.3.0 --
## v ggplot2 3.3.0   v purrr  0.3.4
## v tibble 3.0.1    v dplyr  0.8.5
## v tidyr  1.1.0    v stringr 1.4.0
## v readr  1.3.1    v forcats 0.5.0
##    --    Conflicts    -----------------------------------------------------------
tidyverse_conflicts() --
## x dplyr::filter() masks stats::filter()
## x dplyr::lag()  masks stats::lag()
library(survival)

library(survminer)
## Loading required package: ggpubr
data("veteran")
#Surv function to build the survival object

km<-Surv(veteran$time,veteran$status)
##To plot the survival curves for the two treatments

trt_fit<-survfit(km ~trt, data = veteran)
ggsurvplot(trt_fit,pval = TRUE,pval.method = TRUE)
```

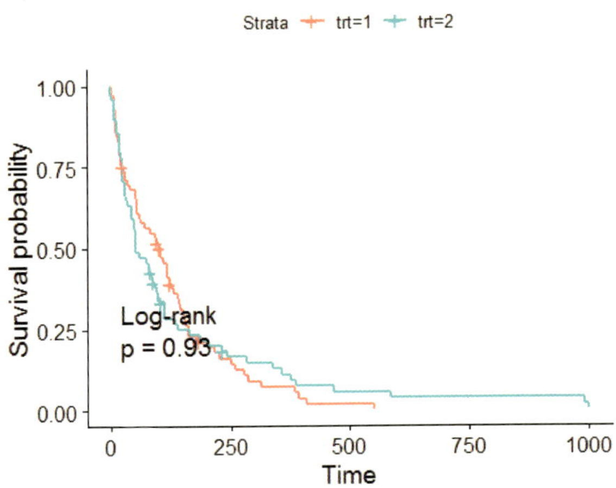

```
##To run the log-rank test

survdiff(formula = km ~ trt, data = veteran)
## Call:
## survdiff(formula = km ~ trt, data = veteran)
##
##       N Observed Expected (O-E)^2/E (O-E)^2/V
## trt=1 69     64     64.5   0.00388   0.00823
## trt=2 68     64     63.5   0.00394   0.00823
##
## Chisq= 0 on 1 degrees of freedom, p= 0.9
# Cox regression

veteran$trt<- factor(veteran$trt)
trt.cox <- coxph(Surv(time,status) ~ trt, data = veteran)
summary(trt.cox)
## Call:
## coxph(formula = Surv(time, status) ~ trt, data = veteran)
##
##   n= 137, number of events= 128
##
##         coef exp(coef) se(coef)     z Pr(>|z|)
## trt2 0.01774   1.01790  0.18066 0.098    0.922
##
##      exp(coef) exp(-coef) lower .95 upper .95
## trt2     1.018     0.9824    0.7144      1.45

##
## Concordance= 0.525 (se = 0.026 )
## Likelihood ratio test= 0.01 on 1 df,   p=0.9
## Wald test            = 0.01 on 1 df,   p=0.9
## Score (logrank) test = 0.01 on 1 df,   p=0.9
```

 # forest plot

```
ggforest(trt.cox)
```

Warning in .get_data(model, data = data): The `data` argument is not provided.

Data will be extracted from model fit.

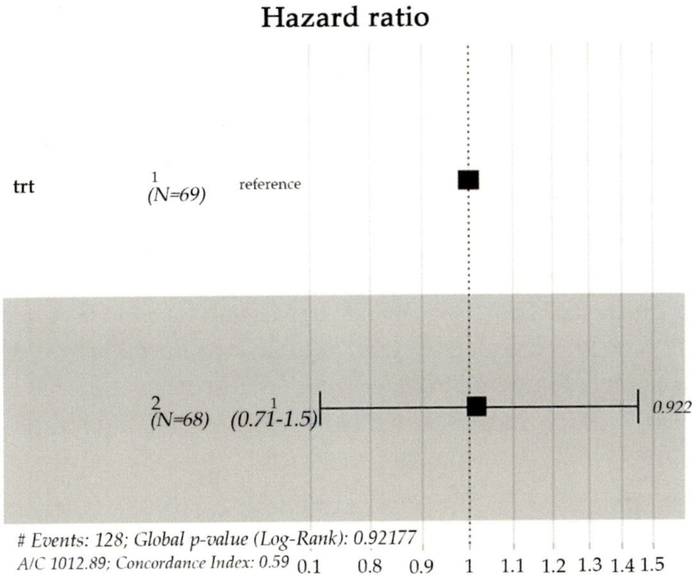

 # to see predicted survival curve for each treatment

```
ggadjustedcurves(trt.cox, data = veteran, variable = "trt")
```

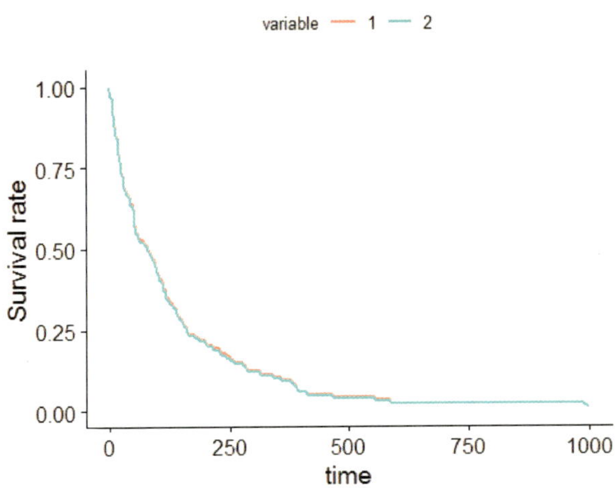

There is no difference between the two treatments in survival or mortality rates according to log-rank test.

The Cox regression equation will be:

ln(hazard of mortality at time = t) = ln(hazard of mortality for trt = 1 at time = t) +

ln(hazard of mortality at time = t) = ln(hazard of mortality for trt = 1 at time = t)) + 0.018*trt2

0.018 = ln(hazard of trt=2/hazard of trt = 1) and Hazard ratio = exp(0.018)= 1.018

The column marked "Pr(>|z|)" gives a p-value larger than 0.05. So, we can conclude that treatment type is not a significant predictor of mortality in this population.

The sign of the regression coefficients (coef) = 0.018. A positive sign means that the hazard (risk of death) is higher, and thus the prognosis worse, for subjects with higher values of that variable. The variable trt is encoded as a numeric vector, 1=standard treatment and 2= test treatment. The coefficient for treatment = 0.018 indicates that test treatment has higher risk of death (lower survival rates) than standard treatment.

The exponentiated coefficients (exp(coef) = exp(0.018) = 1.018), also known as hazard ratio, give the effect size of covariates. Treatment 2 has an elevated risk of mortality than treatment 1 by (1.018–1 = 0.018) 1.8% at any time of the follow up period. So, treatment 2 is associated with bad prognosis. The 95% CI for the hazard ratio = 0.71–1.5 that contains 1 so treatment 2 is not different from treatment 1 in mortality risk.

Finally, the output gives p-values for three alternative tests for overall significance of the model: The likelihood-ratio test, Wald test, and score logrank statistics. These tests evaluate the null hypothesis that all of the betas (β) are 0. In the above example, the null hypothesis is not rejected. The Likelihood ratio test has better behavior for small sample sizes, so it is generally preferred.

All these results are evident from the forest plot.

14.2.2. Categorical Predictor

Is there a significant difference in survival curves according to cancer cell type in veteran dataset?

```
cancer_fit<-survfit(km ~celltype, data = veteran)
ggsurvplot(cancer_fit,pval = TRUE,pval.method = TRUE)
```

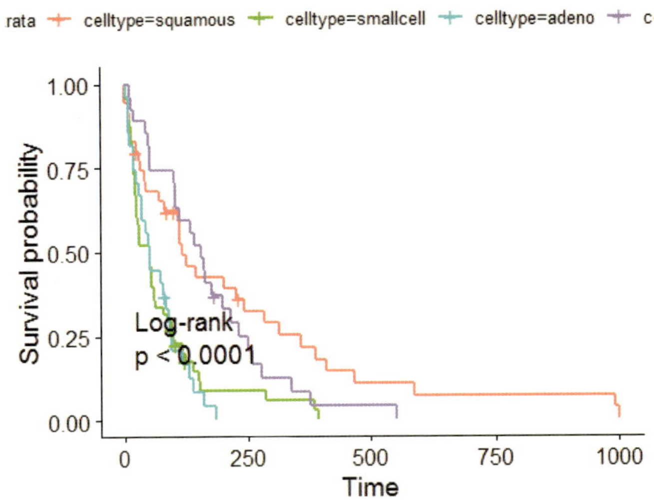


```
##Run the log-rank test
survdiff(formula = km ~ celltype, data = veteran)
## Call:
## survdiff(formula = km ~ celltype, data = veteran)
##
##                     N Observed Expected (O-E)^2/E (O-E)^2/V
## celltype=squamous 35    31    47.7    5.82    10.53
## celltype=smallcell 48    45    30.1    7.37    10.20
## celltype=adeno   27    26    15.7    6.77    8.19
## celltype=large   27    26    34.5    2.12    3.02
##
## Chisq= 25.4 on 3 degrees of freedom, p= 1e-05
##Step 4: Pairwise comparison
pairwise_survdiff(Surv(time, status) ~ celltype, data = veteran)
##
## Pairwise comparisons using Log-Rank test
```

```
##
## data: veteran and celltype
##
##      squamous smallcell adeno
## smallcell 0.00134 -     -
## adeno   0.00134 0.75565 -
## large   0.43731 0.00331  0.00016
##
## P value adjustment method: BH
```

```
# Cox regression

cancer.cox <- coxph(km ~ celltype, data = veteran)
summary(cancer.cox)
## Call:
## coxph(formula = km ~ celltype, data = veteran)
##
##  n= 137, number of events= 128
##
##          coef exp(coef) se(coef)   z Pr(>|z|)
## celltypesmallcell 1.0013  2.7217  0.2535 3.950 7.83e-05 ***
## celltypeadeno   1.1477  3.1510  0.2929 3.919 8.90e-05 ***
## celltypelarge   0.2301  1.2588  0.2773 0.830  0.407
## ---
## Signif. codes: 0 '***' 0.001 '**' 0.01 '*' 0.05 .'' 0.1 ' ' 1
##
##         exp(coef) exp(-coef) lower .95 upper .95
## celltypesmallcell   2.722   0.3674   1.656   4.473
## celltypeadeno     3.151   0.3174   1.775   5.594
## celltypelarge     1.259   0.7944   0.731   2.168
##
## Concordance= 0.608 (se = 0.028 )
## Likelihood ratio test= 24.85 on 3 df,  p=2e-05
```

Wald test = 24.09 on 3 df, p=2e–05

Score (logrank) test = 25.51 on 3 df, p=1e–05

 # forest plot

ggforest(cancer.cox)

Warning in .get_data(model, data = data): The `data` argument is not provided.

Data will be extracted from model fit.

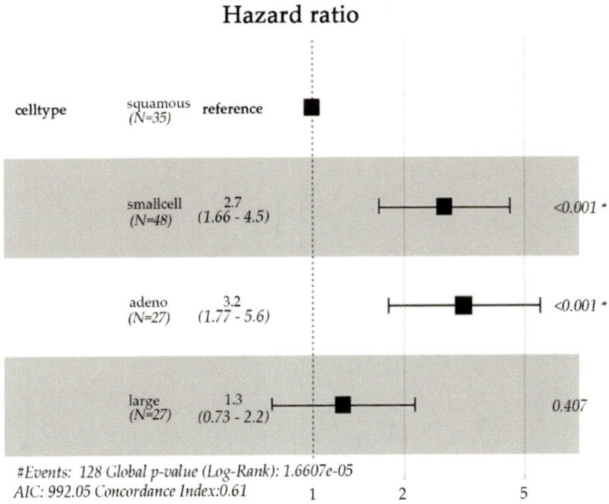

 # to see predicted survival curve for each cell type

ggadjustedcurves(cancer.cox, data = veteran, variable = "celltype")

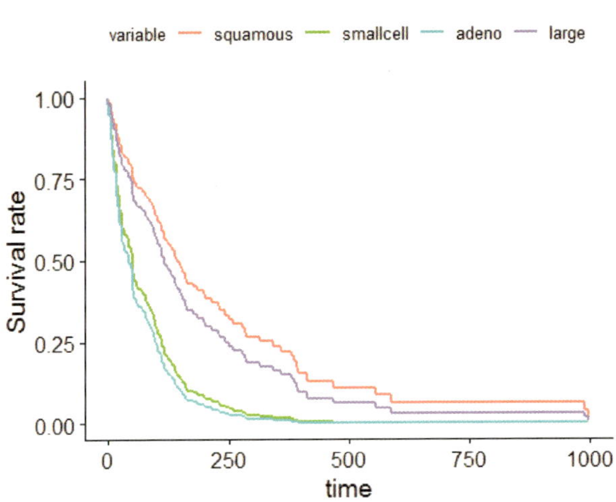

Log_rank test and pairwise comparison has shown that:

1. Large cell and squamous cell cancers are equivalent in their survival rates. Also, Small cell and Adeno cell cancers are equivalent in their survival rates.

2. Large cell cancer has significantly higher survival rate than small cell and adeno cell cancers.

3. Adeno cell cancer has significantly lower survival rate (more deaths) than squamous cell cancer.

4. Small cell cancer has significantly lower survival rate (more deaths) than squamous cell cancer.

Cox regression:

The Cox regression equation will be:

ln(hazard of mortality at time = t) = ln(hazard of mortality for squamous cancer at time = t) + + +

ln(hazard of mortality at time = t) = ln(hazard of mortality for squamous cancer at time = t) + 1.0013* small + 1.1477 * adeno + 0.23 * large

ln(hazard of small/hazard of squamous) = 1.0013 and Hazard ratio = exp(1.0013)= 2.722

ln(hazard of adeno/hazard of squamous) = 1.1477 and Hazard ratio = exp(1.1477)= 3.151

ln(hazard of large/hazard of squamous) = 0.2301 and Hazard ratio = exp(0.2301)= 1.259

The column marked "Pr(>|z|)" gives a p-value smaller than 0.05 for small and adeno cancer cell types. So, we can conclude that small and adeno cancers are significantly larger than squamous cancer in mortality rate and celltype is a significant predictor of mortality in this population.

The sign of all regression coefficients (coef) are positive meaning that the hazard (risk of death) is higher, and thus the prognosis is worse, for subjects with small, adeno, or large than subjects with squamous cancers.

The exponentiated coefficients for small = 2.72. Smallcell cancer has an elevated risk of mortality than squamous cancer by (2.72–1 = 1.72) 172% at any time of the follow up period. The 95% CI = 1.66 – 4.47. This means that the elevated mortality risk of small cell cancer compared to squamous can be between 66% and 347%.

The exponentiated coefficients for adeno = 3.15. Adeno cancer has an elevated risk of mortality than squamous cancer by (3.15 – 1 = 2.15) 215% at any time of the follow up period. The 95% CI = 1.78 – 5.59. This means that the elevated mortality risk of adeno cancer compared to squamous can be between 78% and 459%.

The exponentiated coefficients for large = 1.26. Large cancer has an elevated risk of mortality than squamous cancer by (1.26 – 1 = 0.26) 26% at any time of the follow up period. The 95% CI = 0.73 – 2.17 that contains 1 and means that large cancer is not significantly from squamous cancer in the mortality risk.

In conclusion, the worst prognostic cancer is the Adeno cancer.

The three tests for overall significance of the model: The likelihood-ratio test, Wald test, and score logrank statistics. These tests evaluate the null hypothesis that all of the betas (β) are 0. In the above example, the null hypothesis is rejected meaning that at least one beta is significantly different from zero.

This is also evident from the forest plot.

14.3. COX REGRESSION WITH CONTINUOUS PREDICTOR

Is there a difference in survival curves according to diagnosis time (months from diagnosis to randomization)? We are expecting that increasing time from diagnosis to randomization (treatment) will decrease the survival rate but does data provide evidence for that?

```
##Define the boundary limits for quantiles, we put 0 to include 1 values of diagtime

summary(veteran$diagtime)
## Min. 1st Qu. Median Mean 3rd Qu. Max.
## 1.000 3.000 5.000 8.774 11.000 87.000
veteran$quant<-cut(veteran$diagtime, breaks = c(0,3,5,11,87), labels=c("Q1,""Q2,""Q3,""Q4"))
table(veteran$quant)
##
```

```
## Q1 Q2 Q3 Q4
## 42 33 30 32
##To plot the survival curves for each quantile

quantile_fit<-survfit(km ~quant, data = veteran)
ggsurvplot(quantile_fit,pval = TRUE,pval.method = TRUE)
```

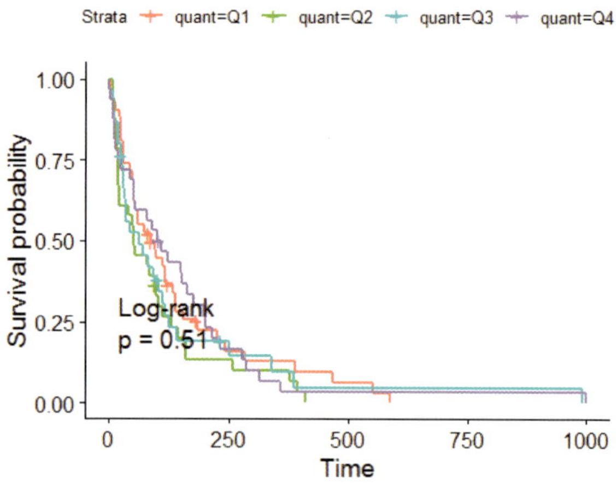


```
# to see median survival

quantile_fit
## Call: survfit(formula = km ~ quant, data = veteran)
##
##          n events median 0.95LCL 0.95UCL
## quant=Q1 42   38   87   52   139
## quant=Q2 33   32   52   21   111
## quant=Q3 30   27   63   33   118
## quant=Q4 32   31  105   53   200
##To run the log-rank test

survdiff(formula = km ~ quant, data = veteran)
## Call:
## survdiff(formula = km ~ quant, data = veteran)
##
```

```
##          N Observed Expected (O-E)^2/E (O-E)^2/V
## quant=Q1 42      38     42.1    0.3968     0.605
## quant=Q2 33      32     25.9    1.4239     1.825
## quant=Q3 30      27     25.5    0.0828     0.105
## quant=Q4 32      31     34.4    0.3443     0.493
##
## Chisq= 2.3 on 3 degrees of freedom, p= 0.5
# Cox regression for categorical diagnosis time

diag_cat_cox <- coxph(km ~ quant, data = veteran)
summary(diag_cat_cox)
## Call:
## coxph(formula = km ~ quant, data = veteran)
##
##   n= 137, number of events= 128
##
##              coef exp(coef) se(coef)      z Pr(>|z|)
## quantQ2  0.318864  1.375564 0.241310  1.321    0.186
## quantQ3  0.155789  1.168579 0.253096  0.616    0.538
## quantQ4 -0.009293  0.990750 0.246120 -0.038    0.970
##
##         exp(coef) exp(-coef) lower .95 upper .95
## quantQ2    1.3756     0.7270    0.8572     2.207
## quantQ3    1.1686     0.8557    0.7116     1.919
## quantQ4    0.9908     1.0093    0.6116     1.605
##
## Concordance= 0.545  (se = 0.03 )
## Likelihood ratio test= 2.24  on 3 df,   p=0.5
## Wald test            = 2.31  on 3 df,   p=0.5
## Score (logrank) test = 2.32  on 3 df,   p=0.5
ggforest(diag_cat_cox)
```

Warning in .get_data(model, data = data): The `data` argument is not provided.

Data will be extracted from model fit.

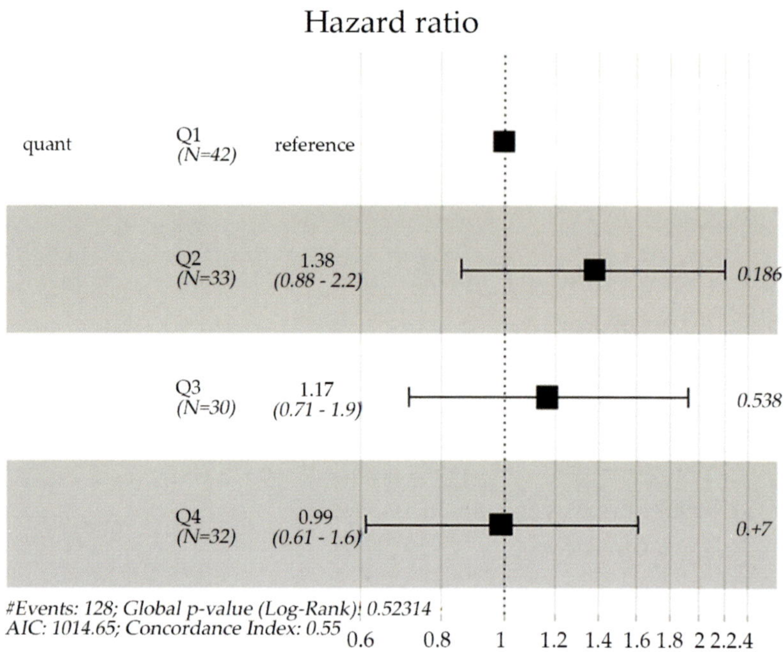


```
# Cox regression for continuous diagnosis time

diag.cox <- coxph(km ~ diagtime, data = veteran)
summary(diag.cox)
## Call:
## coxph(formula = km ~ diagtime, data = veteran)
##
##  n= 137, number of events= 128
##
##          coef exp(coef) se(coef)     z Pr(>|z|)
## diagtime 0.009100 1.009142 0.008978 1.014  0.311
##
##        exp(coef) exp(-coef) lower .95 upper .95
## diagtime    1.009     0.9909    0.9915     1.027
```

```
##
## Concordance= 0.509 (se = 0.03 )
## Likelihood ratio test= 0.91 on 1 df,  p=0.3
## Wald test       = 1.03 on 1 df,  p=0.3
## Score (logrank) test = 1.02 on 1 df,  p=0.3
# forest plot

ggforest(diag.cox)
## Warning in .get_data(model, data = data): The `data` argument is not provided.
## Data will be extracted from model fit.
```

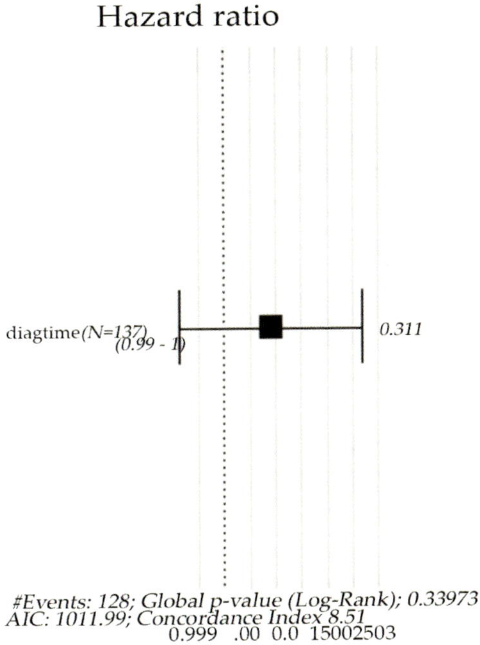

Running log-rank and Cox regression tests has showed that there is no difference between the diagnosis time quantiles and survival rates.

Assuming the relation between ln(hazard of mortality) and diagnosis time is linear, the Cox regression equation will be:

ln(hazard of mortality at time = t) = ln(hazard of mortality for diagnosis time = 0 at time = t) +

ln(hazard of mortality at time = t) = ln(hazard of mortality for diagnosis time = 0 at time = t) + 0.009* diagtime ln(hazard of diagtime = 2/hazard of diagtime = 1) = 0.009 and Hazard ratio = 1.009

ln(hazard of mortality for diagnosis time = 0 at time = t) is a placeholder as we do not have diagtime = 0 in our data and this placeholder varies with time.

The column marked "Pr(>|z|)" gives a p-value larger than 0.05. So, we conclude that diagnosis time is not a significant predictor of mortality in this population.

The sign of the regression coefficients (coef) is positive meaning that the hazard (risk of death) is higher, and thus the prognosis is worse, for subjects with longer diagnosis time than subjects with shorter diagnosis time.

The exponentiated coefficients = 1.009. Each one month increase in diagnosis time is associated with (1.009 – 1 = 0.009) 0.9% increase in the hazard of mortality at any time of the follow up period. For a person with 80 months diagnosis time compared to a person with 1 month diagnosis time, difference = 80 – 1 = 79*0.009 = 0.711. By exponentiation, we get a hazard ratio = 2.04. This means that the person with 80 month diagnosis has (2.04–1 = 1.04) 104% more mortality risk than the person with 1 month diagnosis at any time of the follow up period.

The three tests for overall significance of the model, The likelihood-ratio test, Wald test, and score logrank statistics, which evaluate the null hypothesis that all of the betas (β) are 0. In the above example, the null hypothesis is not rejected meaning that diagnosis time coefficient is not different from (equal to) zero. This is also evident from the forest plot.

However, when we assessed linearity by categorizing the diagnosis time to 4 quartiles, we found that Q4 has highest survival rate followed by Q1, Q3, then Q2 so the relation between ln(hazard) and diagnosis time is not linear and it is better to model the diagnosis time as a categorical predictor.

Is there a difference in survival curves according to patients' Karnofsky performance score?

```
##Define the boundary limits for quantiles, we put 9 to include 10 values of karno score

summary(veteran$karno)
## Min. 1st Qu. Median Mean 3rd Qu. Max.
## 10.00 40.00 60.00 58.57 75.00 99.00
veteran$quant_karno <-cut(veteran$karno, breaks = c(9,40,60,75,99),
```

```
labels=c("Q1,""Q2,""Q3,""Q4"))
table(veteran$quant_karno)
##
## Q1 Q2 Q3 Q4
## 38 41 25 33
##To plot the survival curves for each quantile

karno_quantile_fit<-survfit(km ~quant_karno, data = veteran)
ggsurvplot(karno_quantile_fit,pval = TRUE,pval.method = TRUE, data =
veteran)
```

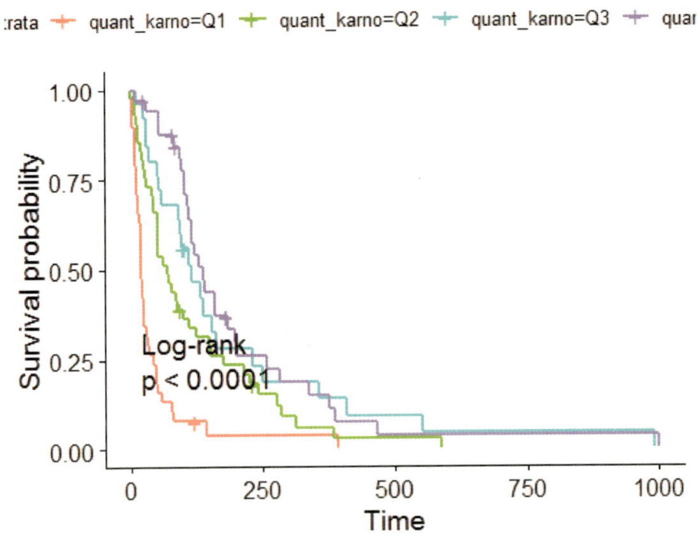


```
# to see median survival

karno_quantile_fit
## Call: survfit(formula = km ~ quant_karno, data = veteran)
##
##          n events median 0.95LCL 0.95UCL
## quant_karno=Q1 38   37   20   16   33
## quant_karno=Q2 41   39   72   51   126
## quant_karno=Q3 25   23  118   92   231
## quant_karno=Q4 33   29  139  112   201
```

```
##To run the log-rank test

survdiff(formula = km ~ quant_karno, data = veteran)
## Call:
## survdiff(formula = km ~ quant_karno, data = veteran)
##
##        N Observed Expected (O-E)^2/E (O-E)^2/V
## quant_karno=Q1 38    37   14.1  36.9402  44.4950
## quant_karno=Q2 41    39   37.5  0.0603  0.0872
## quant_karno=Q3 25    23   31.4  2.2257  3.0260
## quant_karno=Q4 33    29   45.0  5.6931  9.1115

##
## Chisq= 48.5 on 3 degrees of freedom, p= 2e–10
# Cox regression for categorical karno

karno_cat_cox <- coxph(km ~ quant_karno, data = veteran)
summary(karno_cat_cox)
## Call:
## coxph(formula = km ~ quant_karno, data = veteran)
##
##  n= 137, number of events= 128
##
##        coef exp(coef) se(coef)   z Pr(>|z|)
## quant_karnoQ2 –1.0316  0.3564  0.2379 –4.336 1.45e–05 ***
## quant_karnoQ3 –1.4124  0.2436  0.2770 –5.098 3.43e–07 ***
## quant_karnoQ4 –1.5540  0.2114  0.2617 –5.939 2.86e–09 ***
## ---
## Signif. codes: 0 '***' 0.001 '**' 0.01 '*' 0.05 .'' 0.1 ' ' 1
##
##       exp(coef) exp(-coef) lower .95 upper .95
## quant_karnoQ2  0.3564   2.806  0.2236  0.5682
## quant_karnoQ3  0.2436   4.106  0.1415  0.4192
```

```
## quant_karnoQ4  0.2114   4.730  0.1266  0.3530
##
## Concordance= 0.697 (se = 0.022 )
## Likelihood ratio test= 38.1 on 3 df,  p=3e–08
## Wald test      = 43.12 on 3 df,  p=2e–09
## Score (logrank) test = 48.72 on 3 df,  p=1e–10
ggforest(karno_cat_cox)
## Warning in .get_data(model, data = data): The `data` argument is not provided.
## Data will be extracted from model fit.
```

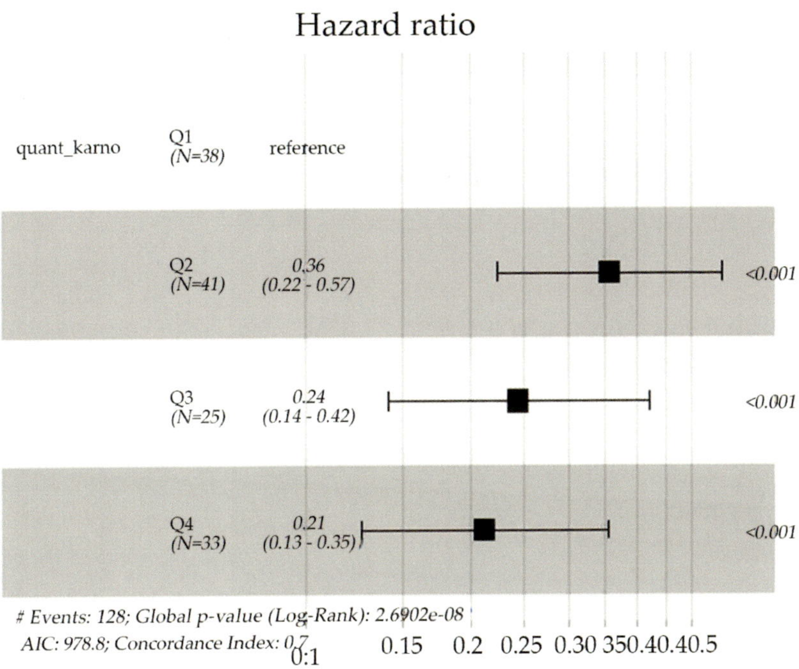

 # Cox regression for continuous karno

```
karno_cox <- coxph(km ~ karno, data = veteran)
summary(karno_cox)
## Call:
## coxph(formula = km ~ karno, data = veteran)
##
```

```
## n= 137, number of events= 128
##
##      coef exp(coef) se(coef)   z Pr(>|z|)
## karno –0.033424 0.967129 0.005075 –6.586 4.51e–11 ***
## ---
## Signif. codes: 0 '***' 0.001 '**' 0.01 '*' 0.05 '.' 0.1 ' ' 1
##
##    exp(coef) exp(-coef) lower .95 upper .95
## karno  0.9671    1.034  0.9576  0.9768
##
## Concordance= 0.709 (se = 0.023 )
## Likelihood ratio test= 42.03 on 1 df,  p=9e–11
## Wald test       = 43.38 on 1 df,  p=5e–11
## Score (logrank) test = 45.32 on 1 df,  p=2e–11
# forest plot

ggforest(karno_cox)
## Warning in .get_data(model, data = data): The `data` argument is not provided.
## Data will be extracted from model fit.
```

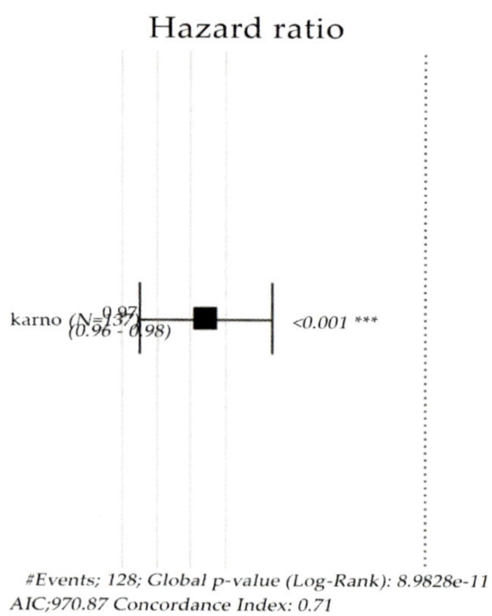

Running log-rank and Cox regression tests has showed that there is significant difference among the karno quantiles in survival rates.

Assuming the relation between ln(hazard of mortality) and Karnofsky score is linear, the Cox regression equation will be:

ln(hazard of mortality at time = t) = ln(hazard of mortality for karno = 0 at time = t) +

ln(hazard of mortality at time = t) = ln(hazard of mortality for karno = 0 at time = t) – 0.03* karno

ln(hazard of karno = 11/hazard of karno = 10) = –0.03 and Hazard ratio = 0.97

ln(hazard of mortality for karno = 0 at time = t) is a placeholder as we do not have karno = 0 in our data and this placeholder varies with time.

The column "Pr(>|z|)" gives a p-value less than 0.05. So, we conclude that Karnofsky score is a significant predictor of mortality in this population.

The sign of the regression coefficients (coef) is negative meaning that the hazard (risk of death) is lower, and thus the prognosis is better, for subjects with higher scores than subjects with lower scores.

The exponentiated coefficients = 0.97. Each one unit increase in the score is associated with (1 – 0.97 = 0.03) 3% decrease in the hazard of mortality at any time of the follow up period. This relation is constant over our entire score range (10–99). The 95% CI = 0.96–0.98 so this decrease in risk can be between 2–4%. For example, a person with 80 score compared to 40-score person, the difference = 40*–0.03 = –1.2. By exponentiation, we get 0.3 so the person with 80 score has 70% lower hazard or mortality than the person with 40 score.

The three tests for overall significance of the model, The likelihood-ratio test, Wald test, and score logrank statistics, which evaluate the null hypothesis that all of the betas (β) are 0. In the above example, the null hypothesis is rejected meaning that score coefficient is smaller than zero. This is also evident from the forest plot.

In addition, when we assessed linearity by categorizing the score to 4 quartiles, we found that Q4 has highest survival rate (median = 139) followed by Q3 (median = 118), Q2 (median = 72), then Q1 (median = 20) so the relation between ln(hazard) and score is linear and it is safe to model the score as a continuous predictor.

14.4. MULTIPLE COX REGRESSION

We checked using simple Cox regression and found that celltype (categorical predictor) and Karnofsky performance score (continuous predictor) are significant predictors of mortality in this population.

We want to describe how these two predictors (celltype and karno) jointly impact the survival in the veteran data

```
cox_mult <- coxph(Surv(time, status) ~ celltype + karno, data = veteran)
summary(cox_mult)
## Call:
## coxph(formula = Surv(time, status) ~ celltype + karno, data = veteran)
##
##  n= 137, number of events= 128
##
##                        coef exp(coef) se(coef)      z Pr(>|z|)
## celltypesmallcell 0.715334 2.044870 0.252686 2.831 0.00464 **
## celltypeadeno     1.157733 3.182711 0.292937 3.952 7.74e–05 ***
## celltypelarge     0.325645 1.384923 0.276680 1.177 0.23921
## karno            –0.031057 0.969421 0.005177 –5.999 1.99e–09 ***
## ---
## Signif. codes: 0 '***' 0.001 '**' 0.01 '*' 0.05 '.' 0.1 ' ' 1

##
##                   exp(coef) exp(-coef) lower .95 upper .95
## celltypesmallcell    2.0449     0.4890    1.2462    3.3555
## celltypeadeno        3.1827     0.3142    1.7925    5.6513
## celltypelarge        1.3849     0.7221    0.8052    2.3820
## karno                0.9694     1.0315    0.9596    0.9793
##
## Concordance= 0.734 (se = 0.023 )
## Likelihood ratio test= 59.37 on 4 df,  p=4e–12
## Wald test            = 61.26 on 4 df,  p=2e–12
## Score (logrank) test = 63.94 on 4 df,  p=4e–13
```

```
ggforest(cox_mult, data = veteran)
```

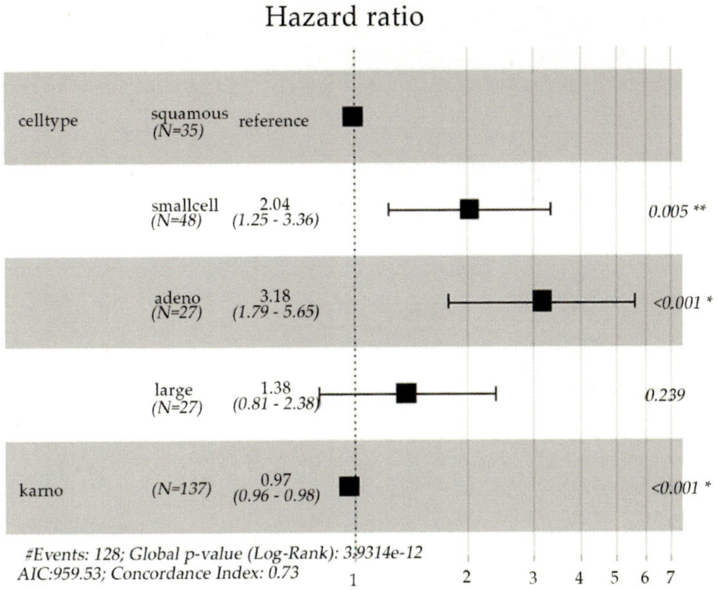


```
# to see predicted survival curves for each celltype

ggadjustedcurves(cox_mult, data = veteran, variable = "celltype")
```

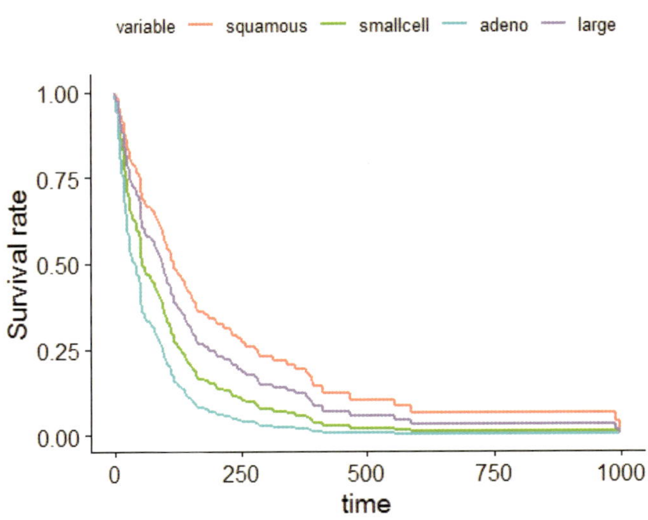

 # to see predicted survival curves for each karno score

ggadjustedcurves(cox_mult, data = veteran, variable = "karno")

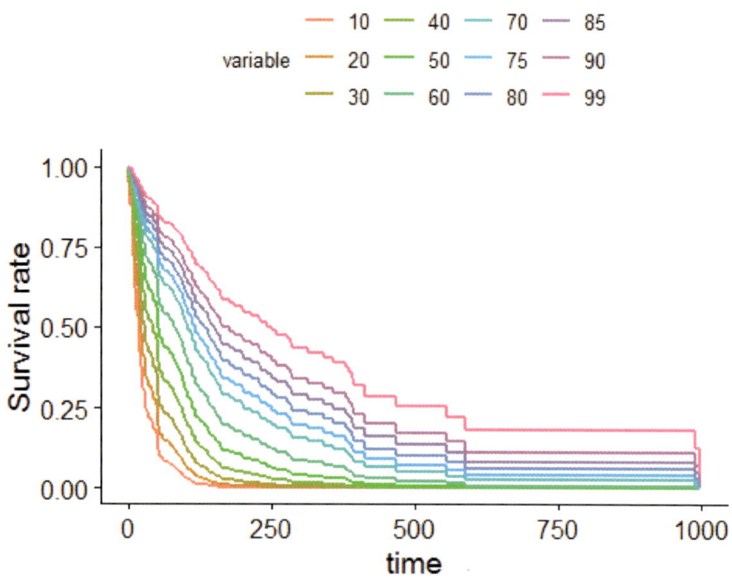

The whole model:

The p-value for all three overall tests (likelihood, Wald, and score) are significant, indicating that the model is significant. These tests evaluate the null hypothesis that all of the betas (β) are 0. In the above example, the null hypothesis is rejected which means that, at least, one of the predictor variables is significantly associated with the outcome variable. This is also evident from the forest plot.

The Cox regression equation will be:

ln(hazard of mortality at time = t) = ln(hazard of mortality for karno = 0 and squamous cancer at time = t) + 0.7 * small + 1.15 * adeno + 0.33 * larger – 0.03* karno

ln(hazard of mortality for karno = 0 and squamous cancer at time = t) is the ln(hazard) for a person with 0 score and squamous cancer. This hazard varies with time. Both predictors (celltype and karno scores) remained significant in the multiple model.

The celltype coefficients:

The column "Pr(>|z|)" gives p-values less than 0.05 for the small and adeno cancers. So, we conclude that celltype is a significant predictor of mortality in this population.

The sign of all regression coefficients (coef) are positive meaning that the hazard (risk of death) is higher, and thus the prognosis is worse, for subjects with small, adeno, or large cancers than subjects with squamous cancers when the karno scores are the same.

The exponentiated coefficients (Hazard ratio, HR) for small = 2.04. A person with smallcell cancer has an elevated risk of mortality than a person with squamous cancer by (2.04–1 = 1.04) 104% at any time of the follow up period when they have the same karno scores. The 95% CI = 1.25 – 3.36. This means that the elevated mortality risk of small cell cancer compared to squamous can be between 25% and 236% for the same karno scores in this population.

HR for adeno = 3.18. A person with adeno cancer has an elevated risk of mortality than a person with squamous cancer by (3.18 –1 = 2.18) 218% at any time of the follow up period when they have the same karno scores. The 95% CI = 1.79 – 5.65. This means that the elevated mortality risk of adeno cancer compared to squamous can be between 79% and 465% for the same karno scores in this population.

HR for large = 1.38. A person with large cancer has an elevated risk of mortality than a person with squamous cancer by (1.38 – 1 = 0.38) 38% at any time of the follow up period when they have the same karno scores. The 95% CI = 0.81 – 2.38 that contains 1 and means that large cancer is not significantly different from squamous cancer in the mortality risk. In conclusion, the worst prognostic cancer is the Adeno cancer in this population.

Compared to the unadjusted estimate for simple regression with celltype alone (HR for small = 2.72, HR for adeno = 3.15, and HR for large = 1.26), there is a difference between adjusted (in multiple model) and unadjusted estimate. This means that the association between mortality hazard and cancer celltype is confounded by karno scores to some extent.

The Karno coefficient:

The column "Pr(>|z|)" gives p-values less than 0.05;so, we conclude that karno score is a significant predictor of mortality in this population.

The sign of the regression coefficient is negative meaning that the hazard (risk of death) is lower, and thus the prognosis is better, for subjects with high scores than subjects with low scores when they have the same celltype cancer.

The HR = 0.97. Each one unit increase in the score is associated with (1 – 0.97 = 0.03) 3% decrease in the hazard of mortality at any time of the follow up period and for the same cancer type. This relation is constant over our entire score range (10–99). The 95% CI = 0.96–0.98 so this decrease in risk can be between 2–4%. For example, a person with 80 score compared to 40-score person and both have the same cancer type, the difference = 40*–0.03 = –1.2. By exponentiation, we get 0.3 so the person with 80 score has 70% lower hazard or mortality than the person with 40 score.

Compared to the unadjusted estimate for simple regression with karno score alone (HR = 0.97), there is nearly no difference between adjusted (in multiple model) and unadjusted estimate. This means that the association between mortality hazard and karno scores is not confounded by cancer cell type in this population.

14.5. EVALUATION OF THE COX MODEL

The Cox PH model makes several assumptions; so, it is important to assess whether a fitted Cox regression model adequately describes the data. The three types of diagnostics for the Cox model:

- Testing the PH assumption.
- Examining influential observations (or outliers).
- Detecting nonlinearity in relationship between the ln(hazard) and the predictors.

The residuals for the Cox model are used to check these assumptions and include:

- Schoenfeld residuals to check the PH assumption
- Martingale residual to assess nonlinearity
- Deviance residual (symmetric transformation of the Martingale residuals), to examine influential observations

14.5.1. Testing the Proportional Hazards (PH) Assumption

The PH assumption can be checked using statistical tests and graphical diagnostics based on the scaled Schoenfeld residuals. The Schoenfeld residuals are independent of time. A plot that shows a non-random pattern against time is evidence of violation of the PH assumption.

The function cox.zph() provides a solution to test the PH assumption for each predictor included in a multiple Cox regression model. For each predictor, the function cox.zph() correlates the corresponding set of scaled Schoenfeld residuals with time, to test for independence between residuals and time. Additionally, it performs a global test for the model as a whole.

The proportional hazard assumption is supported by a non-significant relationship between residuals and time.

```
cox_mult <- coxph(Surv(time, status) ~ celltype + karno, data = veteran)
summary(cox_mult)
## Call:
## coxph(formula = Surv(time, status) ~ celltype + karno, data = veteran)
##
## n= 137, number of events= 128
##
##             coef exp(coef) se(coef)   z Pr(>|z|)
## celltypesmallcell 0.715334 2.044870 0.252686 2.831 0.00464 **
## celltypeadeno   1.157733 3.182711 0.292937 3.952 7.74e–05 ***
## celltypelarge  0.325645 1.384923 0.276680 1.177 0.23921
## karno      –0.031057 0.969421 0.005177 –5.999 1.99e–09 ***
## ---
## Signif. codes: 0 '***' 0.001 '**' 0.01 '*' 0.05 .'' 0.1 ' ' 1
##
##          exp(coef) exp(-coef) lower .95 upper .95
## celltypesmallcell  2.0449   0.4890  1.2462  3.3555
## celltypeadeno    3.1827   0.3142  1.7925  5.6513
## celltypelarge    1.3849   0.7221  0.8052  2.3820
## karno        0.9694   1.0315  0.9596  0.9793
##
## Concordance= 0.734 (se = 0.023 )
## Likelihood ratio test= 59.37 on 4 df,  p=4e–12
## Wald test      = 61.26 on 4 df,  p=2e–12
## Score (logrank) test = 63.94 on 4 df,  p=4e–13
```

```
# test PH

test_ph <- cox.zph(cox_mult)
test_ph
##        chisq df    p
## celltype 14.1 3 0.00282
## karno    14.4 1 0.00015
## GLOBAL  23.6 4 9.4e–05
# diagnostic plot

ggcoxzph(test_ph)
```

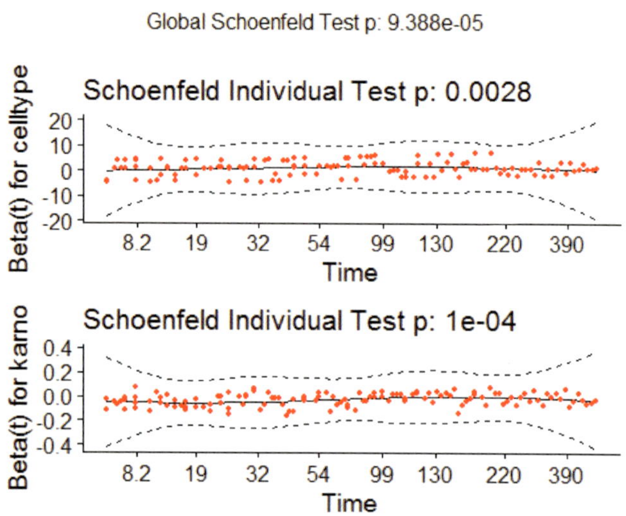

From the output above, the test is statistically significant for each predictor, and the global test is statistically significant. Therefore, the PH is violated in this model and cannot be used.

A common solution is to add a predictor*time interaction term and we increase the iterations to 100 to attempt for convergence.

```
cox_mult <- coxph(Surv(time, status) ~ celltype + celltype:time +

        karno + karno:time, data = veteran,

      control = coxph.control(iter.max = 100))
```

Warning in fitter(X, Y, istrat, offset, init, control, weights = weights, :
Loglik converged before variable 1,2,3,4,5,6,7 ; coefficient may be infinite.

summary(cox_mult)

Call:

coxph(formula = Surv(time, status) ~ celltype + celltype:time +

karno + karno:time, data = veteran, control = coxph.control(iter.max = 100))

##

n= 137, number of events= 128

##

coef exp(coef) se(coef) z Pr(>|z|)

celltypesmallcell –0.008248 0.991786 6.969598 –0.001 0.999056

celltypeadeno –0.100028 0.904812 8.001455 –0.013 0.990026

celltypelarge –0.032250 0.968264 7.284184 –0.004 0.996467

karno –0.016562 0.983574 0.050028 –0.331 0.740600

celltypesquamous:time –0.866466 0.420435 0.232629 –3.725 0.000196 ***

celltypesmallcell:time –0.860144 0.423101 0.227276 –3.785 0.000154 ***

celltypeadeno:time –0.859800 0.423247 0.206203 –4.170 3.05e–05 ***

celltypelarge:time NA NA 0.000000 NA NA

time:karno NA NA 0.000000 NA NA

Signif. codes: 0 '***' 0.001 '**' 0.01 '*' 0.05 .'' 0.1 ' ' 1

##

exp(coef) exp(-coef) lower .95 upper .95

celltypesmallcell 0.9918 1.008 1.158e–06 8.491e+05

celltypeadeno 0.9048 1.105 1.399e–07 5.853e+06

celltypelarge 0.9683 1.033 6.105e–07 1.536e+06

karno 0.9836 1.017 8.917e–01 1.085e+00

celltypesquamous:time 0.4204 2.378 2.665e–01 6.633e–01

```
## celltypesmallcell:time  0.4231   2.364 2.710e-01 6.605e-01
## celltypeadeno:time    0.4232   2.363 2.825e-01 6.340e-01
## celltypelarge:time       NA      NA     NA      NA
## time:karno               NA      NA     NA      NA
##
## Concordance= 1 (se = 0 )
## Likelihood ratio test= 883.9 on 7 df,  p=<2e-16
## Wald test         = 25484 on 7 df,  p=<2e-16
## Score (logrank) test = 179.3 on 7 df,  p=<2e-16
# test PH

test_ph <- cox.zph(cox_mult)
test_ph
##         chisq df  p
## celltype   0.342 3 0.95
## karno      0.196 1 0.66
## celltype:time 0.128 3 0.99
## GLOBAL     2.478 7 0.93
# diagnostic plot

ggcoxzph(test_ph)
```

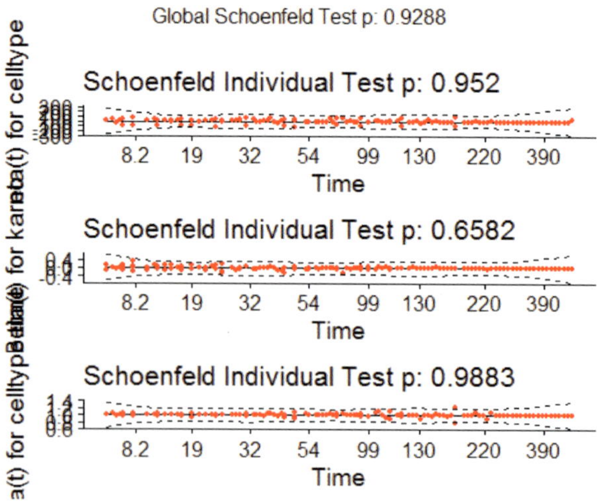

From the output above, the test is not statistically significant for each predictor, and the global test is also not statistically significant. Therefore, the PH is not violated in this model and can be used. However, we see that celltype and karno predictors lose their significance as predictors of mortality hazard in this data. Only the interaction terms between time and celltype are the significant predictors.

14.5.2. Examining Influential Observations (or Outliers)

To test outliers, we can visualize the deviance residuals. The function ggcoxdiagnostics() provides a convenient solution for checking outliers by specifying the argument type = "deviance" and ox.scale = "observation.id" to plot the observation.id on x-axis. The deviance residual is a normalized transform of the martingale residual. These residuals should be roughly symmetrically distributed about zero with a standard deviation of 1.

Positive values correspond to individuals that "died too soon" compared to expected survival times. Negative values correspond to individual that "lived too long." Very large or small values are outliers, which are poorly predicted by the model.

```
ggcoxdiagnostics(cox_mult, type = "deviance," ox.scale = "observation.
id")
## `geom_smooth()` using formula 'y ~ x'
## Warning: Removed 2 rows containing non-finite values (stat_smooth).
## Warning: Removed 2 rows containing missing values (geom_point).
```

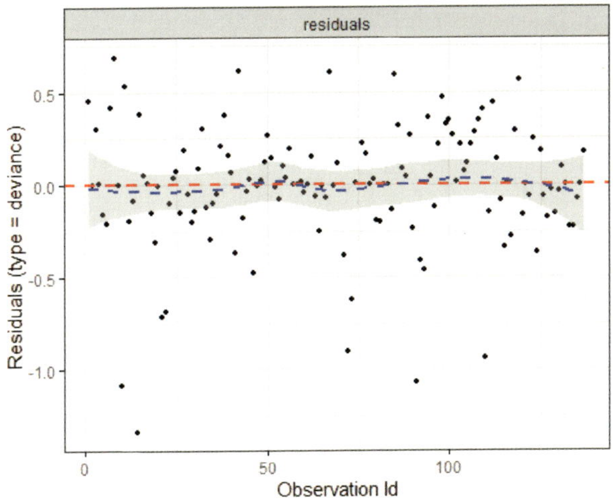

The residual pattern looks symmetric around 0.

14.5.3. Detecting Nonlinearity in Relationship Between the ln(hazard) and the Predictors

It is assumed that continuous predictors have a linear relation with ln(hazard) and this should be checked. One common method is splitting the continuous predictor to 4 quantiles and compare the survival curve for each quantile. Another method is plotting the Martingale residuals against continuous predictors to detect nonlinearity. Nonlinearity is not an issue for categorical variables, so we only examine plots of martingale residuals against a continuous variable. Martingale residuals can take any value in the range (-INF, +1) A value of martingale residuals near 1 represents individuals that "died too soon" and large negative values correspond to individuals that "lived too long."

We will use the function ggcoxfunctional() that displays graphs of continuous predictors against martingale residuals of null cox PH model. This might help to properly choose the functional form of continuous variable in the Cox model. Fitted lines with lowess function should be linear to satisfy the Cox PH model assumptions.

For example, to assess the functional form of age

```
ggcoxfunctional(Surv(time, status) ~ age + log(age) + sqrt(age), data = veteran)
```

Warning: arguments formula is deprecated; will be removed in the next version;

please use fit instead.

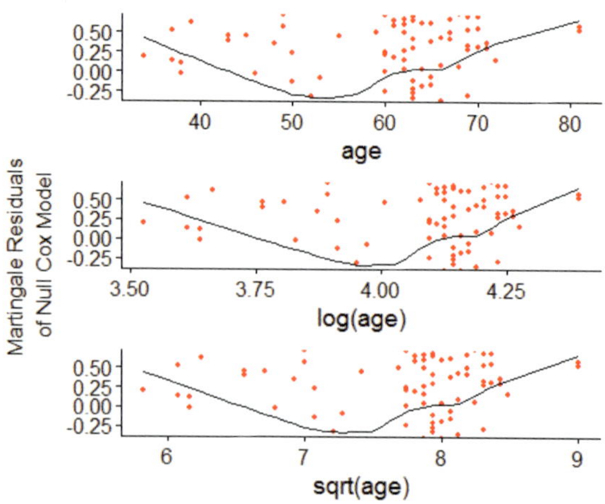

It appears that nonlinearity is present here so it is better to use age as quantiles and not as a continuous predictor.

For example, to assess the functional form of karno scores

ggcoxfunctional(**Surv**(time, status) ~ karno + **log**(karno) + **sqrt**(karno), data = veteran)

Warning: arguments formula is deprecated; will be removed in the next version;

please use fit instead.

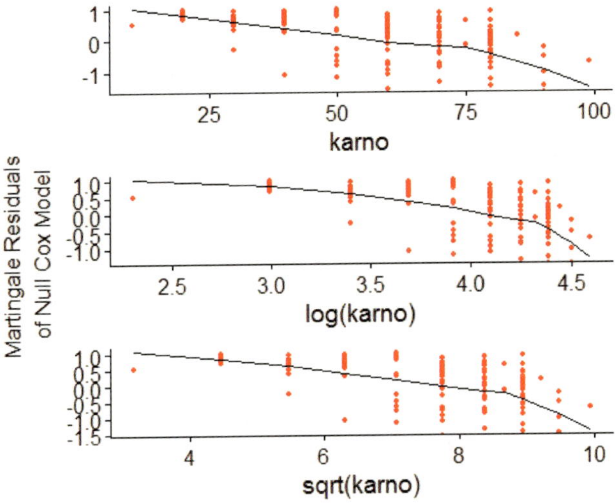

It appears that linearity between karno scores and martingale residuals is present here so karno scores can be used as a continuous predictor in this model.

BIBLIOGRAPHY

1. Alboukadel, K., (2020). *ggpubr: 'ggplot2' Based Publication Ready Plots*. R package version 0.3.0. https://CRAN.R-project.org/package=ggpubr (accessed on 20 March 2021).
2. Alboukadel, K., Marcin, K., & Przemyslaw, B., (2019). *Survminer: Drawing Survival Curves Using 'ggplot2.'* R package version 0.4.6. https://CRAN.R-project.org/package=survminer (accessed on 20 March 2021).
3. Barret, S., Jason, C., Di, C., Francois, B., Moritz, M., Edwin, T., Amos, E., & Joseph, L., (2020). *GGally: Extension to 'ggplot2.'* R package version 1.5.0. https://CRAN.R-project.org/package=GGally (accessed on 20 March 2021).
4. Brian, G. P., & Peter, C., (2020). *Performance Analytics: Econometric Tools for Performance and Risk Analysis*. R package version 2.0.4. https://CRAN.R-project.org/package=PerformanceAnalytics (accessed on 20 March 2021).
5. David, R., & Alex, H., (2020). *Broom: Convert Statistical Analysis Objects into Tidy Tibbles*. R package version 0.5.6. https://CRAN.R-project.org/package=broom (accessed on 20 March 2021).
6. Fay, M. P., (2010). Two-sided exact tests and matching confidence intervals for discrete data. *R Journal, 2*(1), 53–58.
7. Hadley, W., & Dana, S., (2019). *Scales: Scale Functions for Visualization*. R package version 1.1.0. https://CRAN.R-project.org/package=scales (accessed on 20 March 2021).
8. Hadley, W., Romain, F., Lionel, H., & Kirill, M., (2019). *dplyr: A Grammar of Data Manipulation*. R package version 0.8.3. https://CRAN.R-project.org/package=dplyr (accessed on 20 March 2021).

9. Isaac, S., Hector, S., & Joan, V., (2014). *Building Bivariate Tables: The Compare Groups Package for R. Journal of Statistical Software, 57*(12), 1–16. http://www.jstatsoft.org/v57/i12/ (accessed on 20 March 2021).
10. John, F., & Sanford, W., (2019). *An {R} Companion to Applied Regression* (3rd edn.). Thousand Oaks CA: Sage. https://socialsciences.mcmaster.ca/jfox/Books/Companion/ (accessed on 20 March 2021).
11. Mark, S., Telmo, N., Cord, H., Jonathon, M., Javier, S., Ron, T., Jeno, R., Robison-Cox, J., et al., (2020). *epiR: Tools for the Analysis of Epidemiological Data.* R package version 1.0–14. https://CRAN.R-project.org/package=epiR (accessed on 20 March 2021).
12. McDonald, J. H., (2014). *Handbook of Biological Statistics* (3rd edn.) Sparky House Publishing, Baltimore, Maryland.
13. Myles, H., & Douglas A. W., (1973). *Nonparametric Statistical Methods* (pp. 115–120). New York: John Wiley and Sons.
14. R Core Team, (2019). *R: A Language and Environment for Statistical Computing*. R Foundation for Statistical Computing, Vienna, Austria. https://www.R-project.org/ (accessed on 20 March 2021).
15. Rosner, B., (2006). *Fundamentals of Biostatistics* (6th edn.). Thomson Brooks/Cole.
16. Salvatore, M., (2020). *R Companion: Functions to Support Extension Education Program Evaluation.* R package version 2.3.25. https://CRAN.R-project.org/package=rcompanion (accessed on 20 March 2021).
17. Therneau, T., (2020). *Survival: A Package for Survival Analysis in R_*. R package version 3.1–12, https://CRAN.R-project.org/package=survival (accessed on 20 March 2021).
18. Wickham, et al., (2019). Welcome to the tidy verse. *Journal of Open Source Software, 4*(43), 1686. https://doi.org/10.21105/joss.01686.
19. Xavier, R., Natacha, T., Alexandre, H., Natalia, T., Frédérique, L., Jean-Charles, S., & Markus, M., (2011). pROC: An open-source package for R and S+ to analyze and compare ROC curves. *BMC Bioinformatics, 12*, 77. doi: 10.1186/1471-2105-12-77. http://www.biomedcentral.com/1471-2105/12/77/ (accessed on 20 March 2021).

Index

IPW0122688